Treatment of Waters With Elevated Organic Content

Awwa Research Foundation

About the Awwa Research Foundation

The Awwa Research Foundation (AwwaRF) is a member-supported, international, nonprofit organization that sponsors research to enable water utilities, public health agencies, and other professionals to provide safe and affordable drinking water to consumers.

The Foundation's mission is to advance the science of water to improve the quality of life. To achieve this mission, the Foundation sponsors studies on all aspects of drinking water, including supply and resources, treatment, monitoring and analysis, distribution, management, and health effects. Funding for research is provided primarily by subscription payments from approximately 1,000 utilities, consulting firms, and manufacturers in North America and abroad. Additional funding comes from collaborative partnerships with other national and international organizations, allowing for resources to be leveraged, expertise to be shared, and broad-based knowledge to be developed and disseminated. Government funding serves as a third source of research dollars.

From its headquarters in Denver, Colorado, the Foundation's staff directs and supports the efforts of more than 800 volunteers who serve on the board of trustees and various committees. These volunteers represent many facets of the water industry, and contribute their expertise to select and monitor research studies that benefit the entire drinking water community.

The results of research are disseminated through a number of channels, including reports, the Web site, conferences, and periodicals.

For subscribers, the Foundation serves as a cooperative program in which water suppliers unite to pool their resources. By applying Foundation research findings, these water suppliers can save substantial costs and stay on the leading edge of drinking water science and technology. Since its inception, AwwaRF has supplied the water community with more than $300 million in applied research.

More information about the Foundation and how to become a subscriber is available on the Web at **www.awwarf.org.**

Treatment of Waters With Elevated Organic Content

Prepared by:
Simon A. Parsons, **Bruce Jefferson**, **Peter Jarvis**, and **Emma Sharp**
School of Water Sciences
Cranfield University, Cranfield, Bedfordshire, MK43 OAL, United Kingdom
and
David Dixon, **Brian Bolto**, and **Peter Scales**
PFPC, Department of Chemical & Bimolecular Engineering
University of Melbourne, Melbourne 3010, Australia

Sponsored by:
Awwa Research Foundation
6666 West Quincy Avenue, Denver, CO 80235-3098

Yorkshire Water Services

Severn Trent Water

and

Thames Water Utilities

Published by:

Awwa Research Foundation

American Water Works Association

IWA Publishing

DISCLAIMER

This study was jointly funded by the Awwa Research Foundation (AwwaRF), Yorkshire Water Services, Severn Trent Water, and Thames Water Utilities. AwwaRF and the co-funding utilities assume no responsibility for the content of the research study reported in this publication or for the opinions or statements of fact expressed in this report. The mention of trade names for commercial products does not represent or imply the approval or endorsement of AwwaRF or the co-funding utilities. This report is presented solely for informational purposes.

CONTENTS

LIST OF TABLES

LIST OF FIGURES

FOREWORD

The Awwa Research Foundation is a nonprofit corporation that is dedicated to the implementation of a research effort to help utilities respond to regulatory requirements and traditional high-priority concerns of the industry. The research agenda is developed through a process of consultation with subscribers and drinking water working professionals. Under the umbrella of a Strategic Research Plan, the Research Advisory Council prioritizes the suggested projects based upon current and future needs, applicability, and past work; the recommendations are forwarded to the Board of Trustees for final selection. The foundation also sponsors research projects through the unsolicited proposal process; the Collaborative Research, Research Applications, and Tailored Collaboration programs; and various joint research efforts with organizations such as the U.S. Environmental Protection Agency, the U.S. Bureau of Reclamation, and the Association of California Water Agencies.

This publication is a result of one of these sponsored studies, and it is hoped that its findings will be applied in communities throughout the world. The following report serves not only as a means of communicating the results of the water industry's centralized research program, but also as a tool to enlist the further support of the nonmember utilities and individuals.

Projects are managed closely from their inception to the final report by the foundation's staff and large cadre of volunteers who willingly contribute their time and expertise. The foundation serves a planning and management function and awards contracts to other institutions such as water utilities, universities, and engineering firms. The funding for this research effort comes primarily from the Subscription Program, through which water utilities subscribe to the research program and make an annual payment proportionate to the volume of water they deliver and consultants and manufacturers subscribe based on their annual billings. The program offers a cost-effective and fair method for funding research in the public interest.

A broad spectrum of water supply issues is addressed by the foundation's research agenda: resources, treatment and operations, distribution and storage, water quality and analysis, toxicology, economics, and management. The ultimate purpose of the coordinated effort is to assist water suppliers to provide the highest possible quality of water economically and reliably. The true benefits are realized when the results are implemented at the utility level. The foundation's trustees are pleased to offer this publication as a contribution toward that end.

Walter J. Bishop
Chair, Board of Trustees
Awwa Research Foundation

Robert C. Renner, P.E.
Executive Director
Awwa Research Foundation

ACKNOWLEDGMENTS

The authors wish to thank Awwa Research Foundation (AwwaRF) for funding this project and My-Linh Nguyen and Kenan Ozekin for their excellent support as project managers. The authors would like to thank the following people and organizations for their cooperation, participation and support of this project.

- Derek Wilson and Jenny Banks, Yorkshire Water
- Kevin Gertig and Judy Billica, Fort Collins Utilities
- Ed Irvine, Scottish Water
- Brian Drage, Severn Trent Water
- Michael Chipps, Thames Water
- Peter Hillis and Andrew Campbell, United Utilities
- Prof John Gregory, University College London
- Hung Nguyen and Steve Carroll of Orica Watercare for supplying MIEX® samples.

In addition we would like to pass on our thanks to the PAC members Prof Jean-Philippe Croué and Dr Harold Walker for helpful comments and advice through the project. The authors would also like to thank Dr Emma Goslan, Dr David Fearing, Richard Hill and Rita Henderson of Cranfield University for their contributions to the final report.

EXECUTIVE SUMMARY

BACKGROUND

A number of water utilities in the UK and the US are faced with problems associated with treating waters with elevated dissolved organic carbon (DOC) and color levels. These periods of increased organic loading are often linked to periods of heavy rainfall or snow melt, and lead to a number of operational problems as the waters are difficult to treat. Whilst the treatment of raw waters under normal operating conditions is well established, it is known that under conditions of high organic loading the performance of treatment plants can rapidly deteriorate with respect to DOC and color removal with the formation of weak and easy to break flocs. There is a need to investigate the impact of elevated organic content waters on conventional and alternative treatment processes, to understand why weak flocs are formed and to assess the impact of floc strength on downstream clarification processes.

RESEARCH OBJECTIVES

The overall aim of the project was to identify treatment strategies for dealing with elevated organic carbon levels in terms of organic load reduction and floc characteristics at elevated loads, and specifically to identify:

- The conditions under which conventional and alternative processes are suitable for use in the treatment of waters with a higher organic load.
- The effect of organic content on floc size, structure, strength, growth and break up.
- The effects of changes in floc structure on removal in physicochemical processes.

APPROACH

To meet the project objectives, the approach taken in this 24 month long project was to combine both practical and mechanistic investigations, firstly to provide real practical solutions for sponsoring water utilities and, secondly, to allow a better understanding of the role that organic matter plays in floc formation.

MONITORING

A number of bulk water and fractionation methods have been used to monitor seasonal changes in NOM. UV_{254} and DOC/TOC measurement are the most suitable online measurements and give information on the load of organics entering a treatment works. UV_{254} in particular has been reported to correlate well with THMs in distribution. HPSEC and fractionation allowed a more detailed investigation into the character of NOM entering the works and are more suitable for linking character with treatability.

TREATMENT

A review of the literature identified inorganic coagulants, cationic polymers and ion exchange resins as the most effective in treating water rich in organic matter. A comparion of the processes identified inorganic coagulants and MIEX as the two most suitable options and further work focused on these two options. Coagulation trials focused on the impact of changes in NOM charge density and defining a target zeta potential range required for optimum organic and turbidity removal. The tests were repeated using three pilot plants of varying scale to investigate the effect of scale on the identified operational envelope. The MIEX® ion exchange process was evaluated as a pre-treatment to coagulation and appears to offer a very robust option for the treatment of organic rich waters. It was evaluated over a number of water sources and seasons which appear to have little effect in the efficacy of the process, giving increased organic removal, lower THMs and an overall reduction in coagulant use.

DIAGNOSTICS

If treated under conventional operating conditions, an increase in organic loading on a coagulation/flocculation process leads to the formation of weaker flocs. A review was undertaken on techniques for measuring the size, strength and structure of flocs. From this a toolkit of experimental techniques have been developed and used to assess size, strength, settling rate and regrowth potential of flocs and looked at how operational decisions such as coagulant choice and impact on the properties measured. The chapter also includes a more mechanistic section which attempts to link raw water characteristics, through fractionation, to the physical properties of flocs.

An insight into how hydrophobic and hydrophilic organics interact with iron surfaces was gained from the use of rheology and Atomic Force Microscope (AFM) trials. The dynamic rheological tests indicate that the hydrophilic fractions tend to orientate outwards as opposed to the hydrophobic fractions which orientate along the surface of the substrate. This is corroborated by the AFM measurements which reveal a stepped release curve during the hydrophilic force measurements characteristic of molecules orientated perpendicular to the surface of the substrate. Furthermore the AFM data demonstrates that the hydrophilic fraction generates the largest adhesive bonding force and that orientation of the molecules would seem to be more important than bonding strength. Comparison between all the different diagnostic tests does not reveal a clear picture of the significance of each fraction on the resultant floc properties. In particular, the hydrophobic fractions produce the extremes in each of the floc structural properties observed (size, compaction and settling rate). Such a view reflects the importance of the interactions between the different fractions during coagulation and suggests that small concentration of one fraction within the overall raw water may be responsible for the observed behaviour.

FUTURE RESEARCH

This project identified 7 future research studies that will benefit drinking water utilities and continue much of the monitoring and treatment tool development undertaken during this project. Key future research activities include reproducing part of the project on a difficult to treat water and the development of treatment options for hydrophilic organics.

RECOMMENDATIONS FOR DRINKING WATER UTILITIES

The recommendations for utilities section of the report is a practical approach to dealing with elevated organic carbon waters. The project has extended previous characterisation and treatability studies and provides a diagnostic approach to monitoring changes in organics, optimising coagulation and when and how to use alternative processes. Operator checklists are provided for controlling organic residual levels and weak flocs are proposed.

CHAPTER 1
INTRODUCTION AND OBJECTIVES

The phrase "natural organic matter" (NOM) is used to describe the complex mixture of organic material, such as humic acids, hydrophilic acids, proteins, lipids, amino acids and hydrocarbons, present in all drinking water sources. The variety of the components in NOM changes from water to water depending on season, climatic effects and drought conditions, and this leads to variation in the reactivity with chemical disinfectants such as chlorine, ozone and chlorine dioxide. Many of the disinfection by-products (DBPs) formed by the reaction of NOM with disinfectants, usually chlorine, are reported to be toxic, carcinogenic and mutagenic to humans if ingested over an extended period of time (Singer, 1999). The first DBPs to be identified were trihalomethanes (THMs) and are usually the most abundant in drinking water. The removal of NOM and hence reduction in DBPs is, therefore, a major goal in the treatment of any water source.

Even though many of the studies into the effects of DBPs on health have proven inconclusive, the United States Environmental Protection Agency (USEPA) recently proposed a two stage Disinfectants/Disinfection by-products (D/DBP) rule. During stage I the consent level for the total of four THMs (trichloro-, bromodichloro-, dibromochloro-, and tribromo-methane) was reduced from 100 μg/l to 80 μg/l and the consent on the total of five haloacetic acids (monochloro-, dichloro-, trichloro-, monobromo-, and dibromo-acetic acids) was set at 60 μg/l. The standard for THMs in the UK moves from a three-month average of 100 μg/l to a limit of 100 μg/l by 2005. There is no regulation for haloacetic acids currently in EU or UK legislation.

The main focus of attention in the prevention of these DBPs is the removal of the NOM before disinfection or oxidation is used. The most commonly used processes for the reduction in TOC or DOC are coagulation, carbon adsorption, biofiltration and membrane filtration. Whilst the treatment of raw waters under normal operating conditions is well established, it is known that under conditions of high organic loading the performance of treatment plants can rapidly deteriorate with respect to DOC, color and THM precursor removal. This problem is typically observed with low alkalinity (<30 mg/l) waters that have a high humic or fulvic acid content (SUVA>4).

The project proposal arose from discussion with water utilities in the UK and the US, all of whom have faced problems treating waters with elevated DOC and color levels. These periods of increased loading were linked to periods of heavy rainfall or snow melt and led to a number of operational problems as the waters are difficult to treat. An additional issue in the UK is the degradation of the upland moorland catchment where there have been substantial increases in both color and corresponding DOC year on year (Figure 1.1).

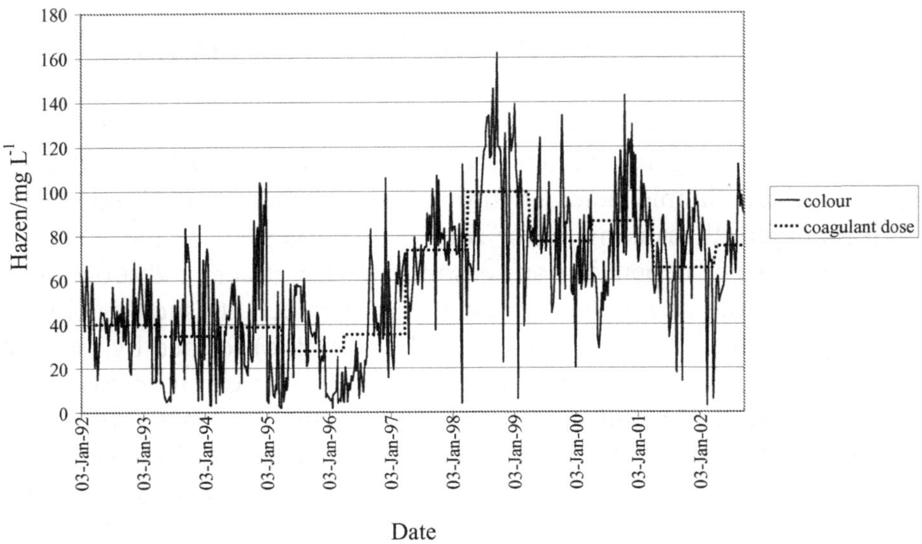

Figure 1.1 Seasonal and annual changes in raw water color at an upland moorland water treatment works.

If treated under conventional operating conditions, the additional organic loading leads to the formation of weaker flocs. These flocs are not removed efficiently by physico-chemical processes such as sedimentation and flotation, and carry over on to downstream processes leading to problems such as increased disinfection by-product formation. This can be seen on many moorland water treatment works (WTWs) where although chemical dosing was optimized by jar testing for both NOM and turbidity removal, when applied on the WTWs, problems occurred with floc separation. Flash mixing and flocculation appeared to perform adequately, producing a floc of good appearance, but this floc then proved to be very fragile and was not adequately removed in the subsequent dissolved air flotation (DAF) and filtration processes. The turbidity from the DAF treated water rose and the iron coagulant residuals were very finely divided (colloidal) and proved very difficult to filter. This resulted in short filter runs and early breakthrough of iron, turbidity and DBPs in the final water (Figure 1.2).

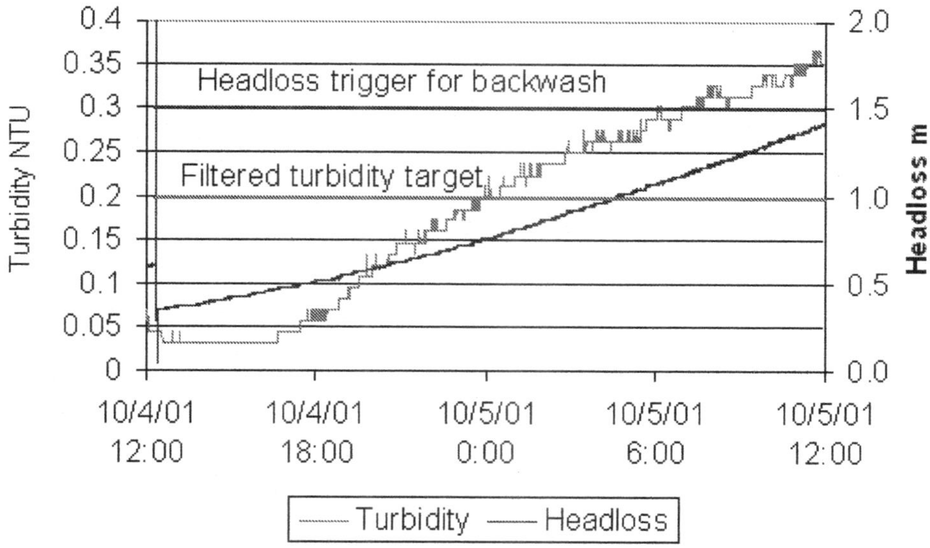

Figure 1.2 Impact of elevated organics on rapid gravity filter operation.

There was clearly a need to develop remediation strategies both for reducing the total organic load but also to allow an investigation of the effect that the organics have on the size, density and strength of flocs. These variables play a major role in the removal of contaminants from water in downstream clarification processes, and this project aimed to investigate the effect they have on floc formation and process performance.

The approach taken in this 24 month long project was to combine both practical and mechanistic investigations, firstly to provide real practical solutions for sponsoring water utilities and secondly to allow a better understanding of the role organic matter plays in floc formation.

The overall aim of the project was to identify treatment strategies for dealing with elevated organic carbon levels in terms of organic load reduction and floc characteristics especially at elevated loads and then specifically to identify:

- The conditions under which conventional and alternative processes are suitable for use in the treatment of waters with a high organic load.
- The effect of organic content on floc size, structure, strength, growth and break up.
- The effect of changes in floc structure on removal in physicochemical processes.

CHAPTER 2
REVIEW OF NATURAL ORGANIC MATTER REMOVAL PROCESS OPTIONS

NATURAL ORGANIC MATTER

Potable waters containing large amounts of low molecular weight (MW) humic substances can be difficult to treat by coagulation processes (Edzwald, 1993). The smaller molecules and the uncharged species are less effectively removed (Thébault *et al.,* 1981; Semmens and Ayers, 1985; Lefebvre and Legube, 1993). For waters that are low in their content of high MW organic compounds, the impurities can be taken out by adsorbents (Johannsen *et al.,* 1993) or, after oxidative degradation, by biological methods (Stumm, 1992). There is scope for a vastly improved approach for such waters.

In general terms, impurities in source waters can be present as suspended material, like clay, silica, microbial cells or algae, as dissolved and colloidal natural organic matter, and as dissolved salts. Drinking water sources typically contain 2-10 mg/L of NOM, of which only 10-30% has been identified (Boggs *et al.,* 1985). There are many reasons why it poses a problem for the water treatment industry, among them the formation of potentially carcinogenic chlorinated hydrocarbons such as trihalomethanes (THMs) when disinfection is carried out with chlorine. There are also bacterial regrowth problems and aesthetic issues like color, taste and odor (Duguet and Mallevialle, 1997). NOM fouls membranes and adsorbents, interferes with the oxidation of dissolved iron and manganese to insoluble, easily removed forms, and encourages corrosion (Huang and Yeh, 1993).

Raw water processing normally involves physicochemical procedures, based on coagulation and flocculation of suspended solids and colloids, and the adsorption of soluble material on solid substrates such as metal hydroxide flocs. In this context, the focus of this report will be on the removal of soluble NOM by adsorption onto ion exchange resins.

Natural contaminants in water supplies

Impurities may be dissolved compounds as well as insoluble particles, and may be of organic or inorganic origin (Bolto, 1995). Some of the more commonly found natural components containing organic material are, in decreasing order of size, zooplankton, phytoplankton, bacteria, viruses, clay-humic acid complexes, humic acids, proteins, polysaccharides, fulvic acids, and very small molecules such as fatty acids, carbohydrates, amino acids and hydrocarbons. The organic impurities are formed by the biological degradation of organic life substances (Thurman, 1985), and include highly colored compounds. Inorganic salts of natural origin are also present to some degree. Ion exchange will adsorb predominantly dissolved and colloidal ionic matter, but will not adsorb insoluble material to any extent.

Dissolved compounds, defined as those that will pass through a membrane having pores of 0.45 μm size, when measured as DOC, have levels in the range 0.1-115 mg/l, with 5.75 mg/l being reported as a global average for streams (Boggs *et al.,* 1985). DOC poses a problem for the water treatment industry for a number of reasons. DOC can be made up of amino acids, fatty acids, phenols, sterols, sugars, hydrocarbons, urea, porphyrins or polymers (Thurman, 1985). The polymers include polypeptides, lipids, polysaccharides and humic substances. The variation of DOC can be from 2.2 mg/L for an oligotrophic lake, 5 mg/L for a river, 10 mg/L for a eutrophic lake, 15 mg/L for a marsh and ≥30 mg/L for bog water. DOC varies from one river to another, and variations along the length of a river can range from 0.30-30.0 mg/L. There are also seasonal variations with summer levels up to four times higher than in winter. In Arizona, spring snowmelt and late summer monsoon rains give the highest DOC concentrations (Nguyen *et al.,* 2002). The make up of the organic compounds for an average river water is 40% fulvic acids, 10% humic acids, 30% hydrophilic acids, 10% carbohydrates, 6% carboxylic acids and 4% amino acids. Although humic substances, encompassing fulvic and humic acids, constitute about 50% of the total DOC of a typical surface water, the proportion can be as high as 80% (Lawrence, 1989). Humic substances are unique and troublesome materials in that they have quite variable properties, in terms of acidity (pK_a 3-5), MW (several hundred to more than ten thousand) and molecular structure (mostly phenolic and carboxylic acid functionalities, but also alcohol, quinone, ether, ester and ketone groups). They behave as negatively charged colloids or anionic polyelectrolytes at natural pH levels, and have surface-active properties, but can interact via their hydrophobic aromatic and aliphatic regions with non-polar pollutants such as pesticides and polychlorinated biphenyls. Humic substances are often present as stable complexes with metal ions.

These variable properties influence reactivity, which changes spatially and temporally. If the smaller charged organic molecules are first removed by ion exchange, from raw water, the performance of a subsequent alum clarification stage is greatly enhanced: larger flocs are formed that settle more rapidly, organic residuals in the product water are lower, and only 25% of the original alum dose is required in a conventional clarification process (Bursill *et al.,* 1985; Singer and Bilyk, 2002).

Soluble inorganic matter comprising dissolved salts that contribute to salinity are acceptable up to certain prescribed limits. By-products of agriculture in run-off water that have consequences for eutrophication of natural waters include ammonia, nitrate and phosphate. Other soluble inorganic matter can be removed by physicochemical treatment, such as phosphate removal by precipitation or ion exchange. There are heavy metals in cationic and anionic forms that can be natural in origin, but are more frequently a result of mining and manufacturing. Cationic heavy metals can be precipitated as the hydroxide, sulfide or carbonate; anionic species such as arsenate, selenate and their reduced forms adsorb on metal oxides.

Previous reports of this type on the removal of NOM from drinking water supplies have focused on the use of ion exchange (Bolto, 2002), cationic polyelectrolytes (Bolto, 2003a) and other adsorbents such as clays, metal oxides and activated carbon (Bolto, 2003b). In this report an attempt is made to identify which NOM compounds are mainly responsible for the production of disinfection by-products (DBPs) when chlorine is employed as the disinfectant. Compliance with DBP guidelines is one of the major objectives for water treatment plant operators and therefore removal of DBP precursors is a critical target for the various treatment options.

NOM can be fractionated into various classes of compounds depending on molecular size or on chemical type. It is then possible to determine which compounds are responsible for most of the potentially harmful disinfection by-products formed on chlorination, and the process that is best for their removal. The answer can depend on the source of the water and the time of year. The broad types of compounds present in waters from a number of locations around the world have been established to some extent. The use of membranes to fractionate NOM on the basis of molecular size has received considerable attention (Goslan *et al.*, 2004; Symons *et al.,* 1992; Kainulainen *et al.,* 1994). Another approach is to carry out the fractionation by chemical adsorption. Chemical fractionation is carried out using established adsorption procedures (Leenheer, 1981; Croué *et al.,* 1994; Bolto *et al.,* 1998, 1999, 2002a). Three stages are used: loading onto the non-ionic resins Amberlite DAX-8 and XAD-4, followed by desorption with 0.1 M NaOH; conversion of the desorbed species to the free acid form with the cation exchange resin Amberlite IR-120; subsequent loading of the species that do not adsorb on DAX-8 or XAD-4 onto the anion exchange resin Amberlite IRA-958.

REMOVAL OF NATURAL ORGANIC MATTER WITH COAGULANTS

The treatment of surface waters is traditionally focused on the removal of turbidity; however, water treatment facilities have recently started to optimize their works for the removal of NOM (Chow *et al.,* 2000). As NOM is almost always anionic at the pH of natural water, it has a strong affinity to cationic additives such as metal coagulants and cationic polyelectrolytes, consequently coagulation is regarded as a vital unit process when treating a water source for the removal of organic matter (Volk *et al.,* 2000; Lind, 1995). The definition of coagulation as a process varies but three definitions include:

- A process for combining small particles into larger aggregates' (Amirtharajah and O'Melia, 1990);
- Coagulation involves the addition of chemicals into a water to produce chemical species that act to destabilize contaminants and improve their removal' (Dennett *et al.,* 1995); and
- Coagulation is a process for combining colloidal materials and small particles into larger aggregates and for adsorbing dissolved organic matter on to these aggregates, thereby facilitating their removal in subsequent sedimentation/flotation and filtration stages' (Jiang and Graham, 1998).

Whilst there are several mechanisms proposed for coagulation with metal salts, it is likely that a combination of two (i) charge neutralization by complexation with the metal species and (ii) adsorption onto hydroxide species are observed during the coagulation of NOM. (Gregor *et al.,* 1997; Bell-Ajy *et al.,* 2000; Cheng *et al.,* 1995; Dennet *et al.,* 1995). Iron and aluminum salts are primarily used as they promote the formation of strong aggregates that are resistant to shear, which is important in effective coagulation (Gregory, 1989).

Detailed descriptions of the mechanisms involved are reviewed elsewhere (Dennet *et al.,* 1995; Amirtharajah and O'Melia, 1990), and it is generally accepted that the pH of the system determines which of the two mechanisms is dominant (Cheng *et al.,* 1995; Dennet *et al.,*1995; Krasner and Amy, 1995). The type of coagulant employed at a water treatment works varies greatly, but generally metal salts, such as iron (Dennett *et al.,* 1995; Wahlroos, 1991; Sinsabaugh *et al.,* 1986) and aluminum (Lind, 1995; Billica and Gertig, 2000; Gregor *et al.,* 1997; Marhaba and Pipada, 2000; Chow *et al.,* 1999) are favored. Recently the use of cationic polyelectrolytes with high charge densities have been reported as being effective for the reduction of NOM (Kam and Gregory, 2001; Bolto *et al.,* 1999).

Performance of any coagulation process is dependant on:

- Choice of coagulant chemical;
- pH;

- coagulant dose; and
- mixing time.

The choice of whether to use iron or aluminum salts as coagulants is dependent on the water type and its effectiveness in treating the individual water. Here published data from 72 waters using either iron or aluminum based coagulants was analyzed to identify any trends linking character to performance. There was no overall trend linking coagulant type to DOC removal efficiency although the results did show that on average the use of iron tends to produce higher removal efficiencies than would be expected if aluminum were used, Figure 2.1.

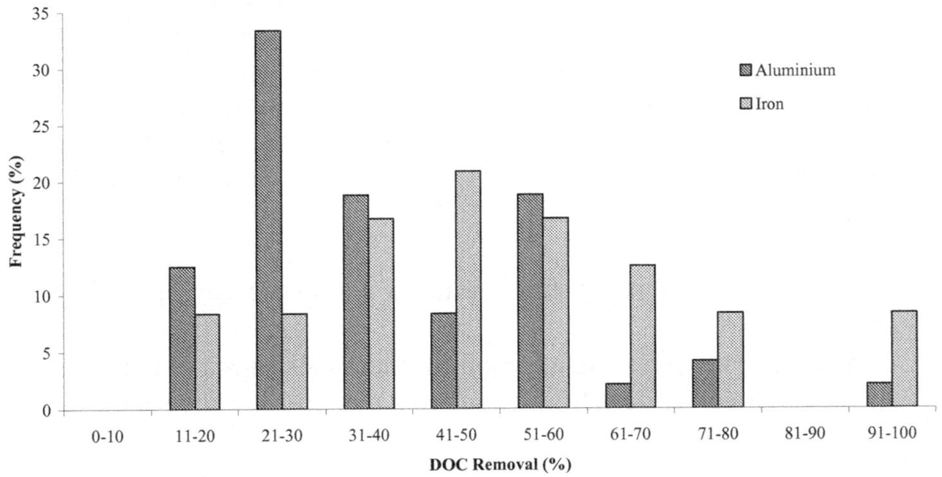

Figure 2.1 Comparison of percentage DOC removal by aluminum and iron based coagulants (Croué *et al.*, 1993; Owen *et al.*, 1993; Dennett *et al.*, 1995; Edzwald, 1993; Crozes *et al.*, 1995; Volk *et al.*, 2000)

Crozes *et al.* (1995) suggested several theories as to why iron may be a more effective coagulant than aluminum:

- ferric chloride presents approximately two times more active positive charges per dry weight of coagulant than aluminum sulfate (alum);
- ferric chloride solution is more acidic than alum solution and the alkalinity consumed by ferric chloride during coagulation is twice as much as for alum. Therefore, for a similar coagulant dose, the pH will be lower with ferric chloride (Equation 2.1).

$$\text{Amount of Coagulant}(\text{mg L}^{-1}) \times \frac{\text{Alkalinity "as } CaCO_3\text{" }(\text{g mol}^{-1})}{\text{MW of Coagulant}(\text{g mol}^{-1})} = \text{Alkalinity Required}(\text{mg L}^{-1}) \quad (2.1)$$

For example 9.26 mg L^{-1} and 4.5 mg L^{-1} of alkalinity will be consumed for a 10 mg L^{-1} dose of ferric chloride and alum respectively; and finally, it has been reported that the specific surface area of aluminum and ferric hydroxide flocs are similar; however, due to there being a higher concentration of active metal in iron (III) chloride solution and the higher MW of iron, iron produces approximately 2.8 times more metal hydroxide than for a similar dose of aluminum. This results in the surface area available for adsorption being considerably higher than that if aluminum was used.

A number of individual studies have been carried out where iron and aluminum based coagulants are compared (Lind, 1995; Volk *et al.,* 2000). Lind (1995) compared 36 water treatment works using a variety of aluminum and iron based coagulants. This demonstrated that aluminum coagulants provided considerably greater DOC removal than iron coagulants. However, Volk *et al.* (2000) showed that for some water sources iron is a more effective coagulant than aluminum. No clear conclusions could be gained as to what conditions made each more suitable.

Jiang and Graham (1998) studied the effect of partially pre-hydrolyzing aluminum and iron coagulants on the formation of polymeric species. This had the advantage of the treatment plant being able to operate over a wider range of conditions of pH and water temperature. This is useful in parts of the world where there is a significant variation in the temperature from season to season. The results of this study showed that for a natural water source DOC, a 28% increase in removal of the humic fraction (DOC of 2.8 mg L^{-1}) was observed when polyferric sulfate was compared to ferric sulfate although little improvement was observed for the non-humic fraction when either the conventional or the hydrolyzed coagulant was used.

Gregory and Dupont (2001) also compared conventional and pre-hydrolyzed coagulants, in this study the rate of growth of the floc and the floc size formed from alum and three commercial polyaluminium chloride (PACl) products were compared. The residual turbidity obtained with the three PACl products was lower in all cases than with alum, with reductions from ~7 to 2 NTU observed. Dynamic monitoring of the coagulants showed that the PACl products were also faster to form flocs and produced a floc size almost double that of alum in terms of flocculation index.

A key parameter in optimizing coagulant performance is pH. Amirtharajah and O'Melia (1990) reported that the conditions for optimum removal of DOC using aluminum or iron based coagulants were acidic (pH 4.5–5.5 using iron and 5–6 for aluminum). The explanation offered was that, at lower pH, the humic substances become more protonated and the coagulating species becomes more positively charged, which in turn makes adsorption more favorable (Crozes *et al.,* 1995). Chadik and Amy (1983) reported that lowering the pH from 8.5 to 6.0 for iron coagulation of Biscayne Aquifer water led to a reduction in the amount of metal coagulant required and an increase in DOC and THM removal compared to coagulation at ambient conditions.

Whilst there is no overall relationship between coagulant type and removal efficiency, a comparison of DOC vs. dose for 57 different water sources shows a relationship between initial DOC and coagulant dose. As expected, the greater the initial DOC of the raw water the greater the coagulant dose required for optimum removal (Bell-Ajy *et al.*, 2000; Singer and Bilyk, 2002; Volk *et al.*, 2000; Croue *et al.*, 1993b; Owen *et al.*, 1993; Crozes *et al.*, 1995) (Figure 2.2).

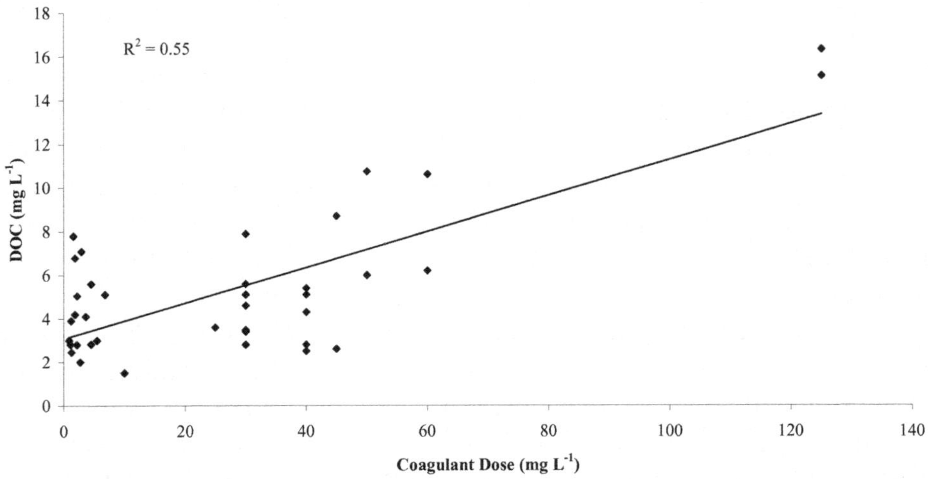

Figure 2.2 Effect of DOC vs. coagulant dose for a range of 57 waters (Bell-Ajy *et al.*, 2000; Singer and Bilyk, 2002; Volk *et al.*, 2000; Croue *et al.*, 1993b; Owen *et al.*, 1993; Crozes *et al.*, 1995)

Where removal of DOC is of prime concern, the United States Environmental Protection Agency (USEPA) has acknowledged enhanced coagulation or granular activated carbon as the best available technology for the reduction of disinfection by-products (Volk *et al.,* 2000; Cheng *et al.,* 2000). Enhanced coagulation is described as the treatment strategy with the single objective of removing TOC, rather than color and turbidity, in order that DBP consents in the US can be achieved (Edzwald and Tobiason, 1999; Volk *et al.,* 2000; Crozes *et al.,* 1995; Krasner and Amy, 1995). The requirements used by the US EPA to decide whether enhanced coagulation is to be employed at a water treatment works are as follows:

Step 1 If the source water has a TOC concentration of 2 mg L^{-1} or less the water treatment works does not have to practice enhanced coagulation. For influent TOC levels greater than 2 mg L^{-1}, the TOC removal required decreases with increasing alkalinity. In the event of the removal criteria not being technically feasible, an alternative TOC performance criteria must be determined (step 2).

Step 2 TOC removal is determined by using jar tests and/or pilot studies under differing pH conditions depending on the alkalinity of the water. The step 2 TOC removal alternative is established by determining the coagulant dosage required when an additional dose (10 mg L^{-1} for alum or 9.1 mg L^{-1} for ferric chloride) results in a TOC removal of less than 0.3 mg L^{-1} or the settled water has a SUVA of less than 2 (Edzwald and Tobiason, 1999; Volk *et al.,* 2000).

A few problems have been highlighted with the use of enhanced coagulation (Edzwald and Tobiason, 1999):

- Step 2 does not allow for coagulants such as PACl, which may be an effective coagulant at neutral pHs;

- TOC measurements are required when determining enhanced coagulation and it is DOC not TOC that is shown to be of major interest in the removal of NOM, as filtration will remove any particulate matter in a source water;
- Overdosing of coagulant may lead to higher purchasing costs and greater sludge production;
- There may also be a problem with poor downstream solid-liquid separation and residual aluminum when coagulation with aluminum based coagulants at low pH is employed;
- It assumes water with SUVA<2 will not generate THM problems; and
- Under enhanced coagulation conditions optimum turbidity removal may not be achieved (Crozes *et al.,* 1995).

An alternative to enhanced coagulation is multiple stage coagulation (Chow *et al.,* 1999; Wahlroos, 1991; Carlson and Gregory, 2000; Billica and Gertig, 2000). This can be further split into sequential and independently optimized multiple stage coagulation. Chow *et al.* (1999) investigated the removal of DOC from three raw waters using alum as the coagulant. A series of jar tests controlled at pH 6 were performed and dosed repeatedly with the same alum dose five times. The results showed DOC removal for a water with an initial TOC of 9 mg L^{-1} increased from 50% to 60% after the second dose. Subsequent alum doses showed no additional removal. The data did show an increased removal of UV absorbing compounds after each dose, although this was only significant for the high TOC, high SUVA water tested and was correlated with an increased removal of high MW organics.

Alternatively, a number of researchers have looked at two-stage coagulation where each stage is optimized independently. Wahlroos (1991) used two-stage coagulation with iron coagulants to improve organics removal (measured as chemical oxygen demand-COD) from 50 to 90% using an initial dose at pH 4.8-5 followed by a secondary dose at pH 8.0. Carlson and Gregory (2000) reported similar experiments on snow melt water using sequential coagulation. They reported that the first stage was conducted at pH 4.8–5.1, intended for the removal of humic substances, and the second stage was at pH 8.0–8.5, intended for particulate removal. They identified that current treatment efficiency was reduced at elevated TOC levels, and this was thought to be due to NOM restabilizing the alum flocs. Two-stage coagulation was shown to improve the treatability of the source water especially when it contained high levels of humic material. Billica and Gertig (2000) following on from the work by Carlson and Gregory looked at the impact of this sequential coagulation on filter performance. They reported that when using single stage coagulation with alum during periods of elevated TOC, breakthrough of particles would be observed after 5 hours and would rise from 5 counts ml^{-1} to 40 counts ml^{-1} after 9 hours. With two-stage coagulation no particle breakthrough was observed.

Effect of bulk water character on removal

The character of organic material in terms of hydrophobicity, hydrophilicity and MW greatly affects its removal with metal coagulants (Chow *et al.,* 2000). Whilst few studies have been undertaken on fractionated organic material a number have reported the SUVA value. Generally as SUVA increase so does the SUVA removal (Figure 2.3). Water sources with a SUVA value between 3 and 4 can be regarded as a mixture of hydrophobic and hydrophilic material. The removal of DOC in this range is generally difficult to predict, as is indicated by the scatter in removal efficiencies (Figure 2.3A). If we consider those waters as strictly hydrophilic or hydrophobic, the difference in removal efficiencies can clearly be observed (Figure 2.4), which shows that when a water is hydrophobic in nature (SUVA>4) 50% removal of DOC will normally be achieved. If the water is hydrophilic in nature (SUVA<2.5) much lower removal efficiencies will be achieved. This can be related back to Figure 2.2 where the water sources that only achieve 11-20% DOC removal all have SUVA values less than 3 and the sources that realize greater than 70% removal tend to have SUVA values of 4.5 $m^{-1}.L\ mg^{-1}$ or above.

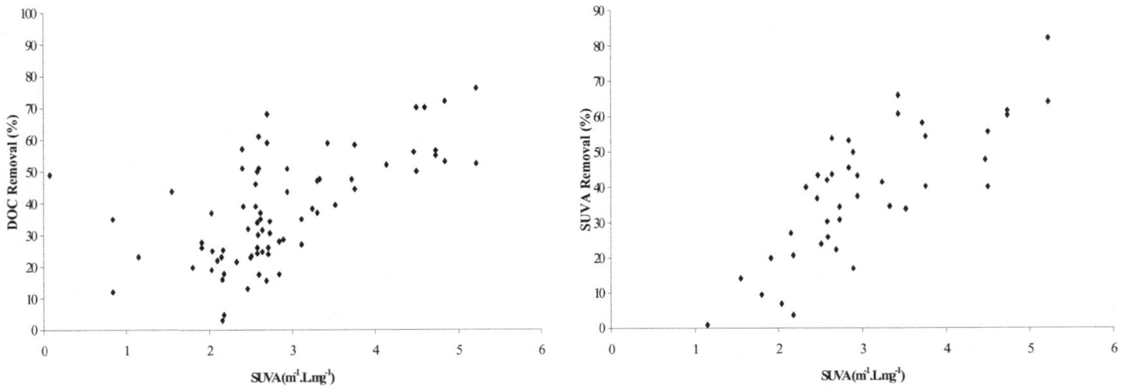

Figure 2.3 Relationship between SUVA and (A) DOC removal and (B) SUVA removal for a range of raw water sources (Bell-Ajy *et al.*, 2000; Singer and Bilyk, 2002; Volk *et al.*, 2000; Croue *et al.*, 1993b; Owen *et al.*, 1993; Crozes *et al.*, 1995; Chow *et al.*, 1999; Edzwald, 1993)

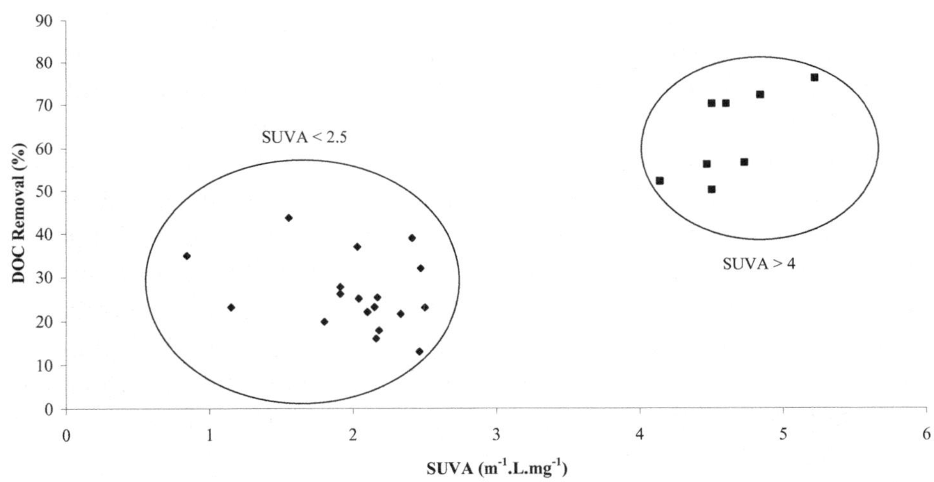

Figure 2.4 Relationship between SUVA and DOC removal focusing on hydrophobic and hydrophilic water sources. (Croué *et al.*, 1993; Owen *et al.*, 1993; Edzwald, 1993; Crozes *et al.*, 1995; Chow *et al.*, 1999; Volk *et al.*, 2000; Bell-Ajy *et al.*, 2000)

Effect of fraction character on removal

Studies on the removal of individual fractions are less well documented. Croué *et al.* (1993a) showed that although high removals of both DOC and THM formation potential

(THMFP) were achieved by coagulation of the humic acid fraction of the organic matter (87%), lower removals were achieved for both the fulvic and hydrophilic acid fractions (55 and 52% respectively). This is expected as, in this case, the humic acid fraction is more reactive in terms of forming THMs (46 μg mg C^{-1} for humic acid compared to 27 μg mg C^{-1} for both fulvic and hydrophilic acid) (Croué *et al.,* 1993a) and is more readily removed due to its lower charge density. Therefore lower doses of coagulant will be required to coagulate equal mass concentrations of humic acid compared to fulvic acids (Kavanaugh, 1978). Marhaba and Pipada (2000) reported similar findings where higher DOC removal efficiencies were achieved on hydrophobic i.e. humic plus fulvic fractions (68%) than on the hydrophilic acid fractions (51%). Bolto *et al.* (2001) reported on the removal of four isolated fractions by coagulation with alum concluding that the easiest fraction to treat was the very hydrophobic fraction (69% UV removal). Contrary to this Chow *et al.* (2000) reported that the easiest fraction to remove by alum coagulation was the hydrophilic charged fraction (~95%) followed by the very hydrophobic fractions (~65%). This is explained by the relatively high SUVA of this fraction compared to the other isolated fractions and hence higher DOC removal.

Another common finding in all the coagulation studies reviewed is that the neutral fractions are the most recalcitrant to removal with coagulation. Marhaba and Pipada (2000) found that only 44% and 52% DOC removal was seen for hydrophilic and hydrophobic neutrals respectively, compared to a 68% and 71% reduction in the hydrophobic acid and base fraction respectively. Similarly Chow *et al.* (2000) observed a 40% DOC reduction for the neutral fractions compared to a 60% reduction for both the very and slightly hydrophobic acids. Bolto *et al.* (2001) observed a 30% removal of UV absorbers for the neutral fraction.

Comparison of the removal of fractions by different authors is difficult unless exactly the same method of fractionation is used. For example Peuravouri *et al.* (2001) compared the fractions obtained using DAX-8 and XAD-8 resins. Although there is very little difference in the major elemental composition of the two fractions obtained, ^{13}C NMR spectroscopy of the fractions showed that the isolate collected from the DAX-8 had a greater content of aliphatics than that collected from the XAD-8. The results that are considered to be comparable in terms of fractionation procedure used are inconclusive in terms of DOC removal indicating that the nature of the fractions are site and probably season specific and cannot be used to predict the efficiency of coagulation on removal of DOC (Table 2.1).

As well as fractions, a number of researchers have looked at removal of MW ranges of NOM. Jiang and Graham (1998) studied the effect of pre-polymerised coagulants on isolated fractions. The advantage of pre-polymerizing coagulants is that the coagulating species formed under normal coagulation can be deteriorated by changes in the water temperature and raw water characteristics. Pre–polymerizing ensures that optimal coagulant species are preformed. They showed that the fractions with the highest MWs (> 3000 Da) were easily removed by both standard and pre-polymerised coagulants (80-90%); however, the fractions with lower MWs (3000-1000 Da) were considerably better removed by pre-polymerised coagulants with up to a 24% increase compared to conventional coagulants in terms of DOC removal. The fractions with lower MWs were more recalcitrant to removal by both types of coagulant. Although slightly higher removal efficiencies (40%) were observed with the pre-polymerised coagulants on the fractions with a MW of 500-1000 Da compared to ~20% for conventional coagulants due to the greater cationic charge on the pre-polymerised coagulant compared to the conventional coagulants.

Changes in the molecular size distribution of a water during treatment by coagulation by means of HPSEC have been reported (Vuorio *et al.*, 1998). The results showed that for the eight distinct molecular size fractions, complete removal of the first fraction (i.e. the fraction with the largest molecular size) is achieved and that the second and third fractions were removed by 92 and 85% respectively. The fourth largest fraction was removed by approximately 50% whilst the remaining four low MW fractions experienced very low removal. Similarly Vartianien *et al.* (1987) looked at 19 water works that employed alum coagulation and showed that for the first six fractions considered previously the removals were 100, 89 and 81% for fractions one, two and three respectively whilst only 66, 40 and 33% reductions were experienced for fractions 4, 5 and 6 respectively. The molecular size range of fractions 1 to 3 is approximately 5000-10000 Da whereas fractions 4 to 6 had an approximate MW distribution of <5000 Da. Similarly Matilainen *et al.* (2002) showed that for raw water (5.5 mg L^{-1} DOC) five distinct peaks are observed by HPSEC. The absolute peak heights for these peaks were monitored throughout the treatment process. It was found that following alum coagulation that peaks 1 and 2 (high MW) were almost completely removed (100 and 93% respectively), however peaks 3 to 5 (medium and low MW) achieved lower reductions (55, 15 and 6% respectively).

Table 2.1
Comparison of reactivity and removal of NOM fractions

Fraction	AKA	Reactivity	DOC (mg L^{-1})	Removal (%)	Reference
HPO-A	-	3 μg TTHM mg^{-1} C	1.2	68 (DOC)	Marhaba and Pipada, 2000
HAF	HPO-A	46 μg $CHCl_3$ mg^{-1} C	5.0	87 (DOC)	Croué *et al.*, 1993a
FAF		27 μg $CHCl_3$ mg^{-1} C	4.9	55 (DOC)	
HPI-A	-	27 μg $CHCl_3$ mg^{-1} C	4.3	52 (DOC)	
HPO-A	VHA	~55 ACE mg^{-1} C	2.4	60 (DOC)	Chow *et al.*, 2000
HPI-A	SHA	~50 ACE mg^{-1} C	1.2	65 (DOC)	
HPO-A	VHA	12.8 μg THM mg^{-1} C	1.7	71 (UV_{254})	Bolto *et al.*, 1999
		13.8 μg THM mg^{-1} C	4	91 (UV_{254})	
HPI-A	SHA	11.4 μg THM mg^{-1} C	0.9	65 (UV_{254})	
		21.9 μg THM mg^{-1} C	0.5	88 (UV_{254})	

(ACE – acetate carbon equivalents)

REMOVAL OF NATURAL ORGANIC MATTER BY CATIONIC POLYMERS

Polymers have been utilized in coagulation/flocculation processes for water purification for at least four decades (Kawamura, 1976). In comparison with alum or iron salts, some of the advantages of using polymers in water treatment are:

- a smaller volume of sludge
- lower coagulant dose requirements
- a reduction in the ionic load of the treated water and the avoidance of aluminum ions
- cost savings of up to 25-30%

The literature indicates that polymers have considerable benefits over inorganic coagulants, mostly arising from the lack of additional solids in the form of metal hydroxide which adds to the burden of the separation process. Because of the lower content of insoluble solids in the reaction mixture, processing is faster; filter runs are longer, the head loss is significantly lower, there is a much smaller sludge volume - usually half - and the sludge has a lower water content (Gregory and Dillon, 2002). In utilizing cationic polymers in lieu of inorganic coagulants the performance is not dependent on pH, and a lower dose of coagulant is applied, in proportion to the turbidity. There is a lower level of dissolved ions in the product water, there are no residual added metal ions such as Al or Mn, and the alkalinity is maintained. Costs can be quite competitive if the dose is not extreme, but costing needs to be holistic, taking into account sludge disposal benefits and the other advantages mentioned (Rout *et al.,* 1999; Nozaic *et al.,* 2001).

Polymers are especially beneficial in coping with the problems of slow settling flocs in low-temperature coagulation or in treating soft colored waters, where they improve settleability and increase the toughness of flocs (Faust and Aly, 1983). The capacity of a treatment facility may be more than doubled with the formation of larger and stronger flocs, the rate of solid and water phase separation can be significantly increased, and the dosage of other chemicals lowered. Also, the range of waters that can be treated is wider.

However, health and environmental issues require attention. Possible negative consequences include a greater sensitivity to incorrect dosage, with turbidity and natural organics removal less efficient in some instances (Nozaic *et al.,* 2001). Residual polymer can interfere with the action of solid adsorbents such as powdered activated carbon, membranes and ion exchange resins. Tagging of polymers to allow sensitive assaying in the product water is the subject of current research (Bennett *et al.,* 2000). Degradation in the presence of chlorine or other disinfectants is a recent aspect that has received attention (Bolto and Levine, 2002). More assimilable organic carbon can be produced that can provide microbial nutrients.

It is well documented that organic polymers may be used as primary coagulants as well as in the more traditional flocculation step of binding already formed small flocs into larger particles (Glaser and Edzwald, 1979; Jackson, 1980; Schlauch, 1981; Amy and Chadik, 1983). The polymer acts as a destabilizing agent via a charge neutralization/ precipitation mechanism, and is also an agent for floc growth (Bolto, 1995). With a few notable exceptions (Leu and Ghosh, 1988), there is not a great deal of published information on the relationship between polymer structure and treatment performance in drinking water production; that is, on the influence of molecular structure on coagulation/flocculation, on the rates of both precipitation and sedimentation, on product water quality and on the solids content of the final sludge. First, some observations on the impurities that have to be removed.

The removal of NOM by cationic polymers has been explored in a number of studies based of simulated waters made from concentrates of humic substances (Bolto *et al.,* 1998, 1999) isolated from various surface waters in southern and south eastern Australia. One ground water from Wanneroo in the western part of the continent was also included in the study. Soil-based Aldrich NOM was included for comparison purposes, as was alum treatment of all the simulated waters. The Aldrich NOM had an average molecular weight of 1990, about double that of the aquatic NOMs (Chow, 1999; Pelekani *et al.,* 1999). The reconstituted waters had DOC levels of *ca.* 5 mg/L. NOM removals were obtained via jar tests and monitored by UV measurements at 254 nm, which were used as a measure of trihalomethane precursors. The optimum dose of polymer was estimated from the UV absorbance data and lay in the range 5-8 mg/L. NOM removal results are given in Table 2.2. The results for the best cationic polymers for treating a particular simulated raw water are shown in bold.

Table 2.2
Percentage removal of UV-absorbing NOM from reconstituted waters by cationic polymers and alum (adapted from Bolto *et al.*, 1998, 1999, 2002)

Coagulant	CD, %	MW	Bell Bay	Hope Valley	Wanneroo	Moorabool	Horsham	Aldrich
PDADMAC-1	100	160K	25	66	35	-	-	**70**
PDADMAC-2	100	260K	49	-	-	47	62	66
PDADMAC-3	100	500K	**68**	**65**	70	49	**70**	-
PDADMAC-4	100	3M	-	-	-	-	65	-
CPAM-1	80	2M	27	52	69	45	60	**73**
CPAM-2	L	H	23	-	31	-	-	18
CPMA	H	H	-	50	**75**	**55**	**71**	-
Chitosan	17*	500K	60	58	-	-	61	60
PEI	6*	10K	49	33	56	-	59	81
Alum	-	-	**88**	**88**	**88**	**69**	**76**	**98**

* At pH 7

Alum clearly performs better than the polymers, only cationic polyacrylamide (CPMA) and PDADMAC approaching its performance in the case of Horsham NOM. In applying organic polymers to the removal of UV-absorbing compounds Bell Bay and Hope Valley NOM was best removed with a high MW and high charge density (CD) polymer in the form of polydiallyldimethylammonium chloride (PDADMAC). A cationic polymethacrylate (CPMA) of high charge but more hydrophobic character was more effective than PDADMAC for Wanneroo and Moorabool. CPMA was the best polymeric coagulant when the water was high in non-polar NOM, as is the case for Wanneroo (to be discussed later). CPMA was also slightly better than PDADMAC for Horsham NOM. For Aldrich samples a CPAM gave better results than PDADMAC. Chitosan and polyethyleneimine (PEI) give surprisingly high removals despite their low CD.

Similar performance has been observed for chitosan in the treatment of typical Norwegian humic waters having color levels of 15-50 platinum color units (PCU) and TOC of 2.3-5.0 mg/L (Eikebrokk and Saltnes, 2002). The best results were obtained at higher chitosan doses, the removals at 7.5 mg/L reaching 79-80% for color and 34-36% for TOC, compared with 94-95% and 62-65% respectively for alum at a dose of 3.1 mg/L as Al.

The so-called polyamines, actually quaternary ammonium polymers made from epichlorohydrin and dimethylamine, are also useful as primary coagulants, with the longer chains being more effective for raw waters of higher turbidity (Lee *et al.*, 2001). Pilot work indicated a 50-80% saving of poly(aluminum chloride) using 0.4-1 mg/L of polymer for waters of average and high turbidity. The removal efficiency for organics was also enhanced. Although organic polymers generally did not perform as well as alum, better results were obtained when particulates are present, so that polymers are best used in conjunction with alum to provide them (Bolto *et al.*, 2001).

Effect of fraction character on removal

The NOM was fractionated using established adsorption procedures (Croué *et al.*, 1993a). Four fractions were obtained, designated as: HPA, hydrophobic acids; TPA, transphilic acids;

Char, charged hydrophilic compounds; Neut, neutral hydrophilic compounds. The results from the various studies are given in Table 2.3, where the data are estimated from measurement of the DOC levels.

Table 2.3
Organic carbon mass in various NOM fractions as percentage DOC of the total NOM (adapted from Bolto *et al.*, 1998, 1999, 2002)

Fraction	Bell Bay	Hope Valley	Wanneroo	Moorabool	Horsham	Aldrich
HPA	47	46	80	34	58	40
TPA	16	21	9	18	14	5
Char	31	32	9	33	11	10
Neut	6	1	2	15	17	45

Since the method of isolating the NOM for some of the waters, namely those from Bell Bay, Hope Valley and Wanneroo, was by coagulation with an iron oxide or adsorption by ion exchange, the results are not representative of the total original NOM of the original water. There is naturally a negligible amount of uncharged material in these isolates. The data vary from water to water, and probably also with the time of year. The above results show that each of the NOMs from surface water sources is similar in its percentage DOC breakdown. The HPA fraction constitutes approximately 45% of the total DOC, the TPA fraction about 20%. The groundwater NOM from Wanneroo has almost double the HPA fraction and half the TPA, and the soil-derived Aldrich NOM much lower TPA at 5%. Of the hydrophilic fractions, the Char fraction can be associated with about 30% of the DOC for three of the surface waters, with that from Horsham being much lower. The Neut fraction comprises a very high 45% for the Aldrich NOM. The waters that have not had their NOM isolated by a membrane technique not do not have a large neutral contribution as they are regeneration effluents from water treatment processes that take out mainly charged species.

The uptake of NOM fractions are presented in Table 2.4, which also includes some alum results for comparison purposes. The treatability varied with the fraction. Alum was best for removal of all fractions for essentially all the waters, but for the HPA fractions, PDADMAC was as good as alum for Hope Valley, and CPMA was nearly as good for Moorabool and Horsham. For the TPA fractions PDADMAC was generally the best, although CPMA and CPAM featured for Hope Valley and Moorabool. PDADMAC as effective as alum for Hope Valley. PDADMAC was the most successful for the Char fractions of Hope Valley and Wanneroo, chitosan for Bell Bay and Horsham, and CPMA for Moorabool. The Neut fractions were a very minor component and there was little effect where data could be obtained, with CPMA apparently outperforming alum for Horsham. These outcomes are summarized in Table 2.5.

PDADMAC is a consistently good performer. The Moorabool case gives striking results, with CPMA is the best polymer for all fractions, in accord with the result for the simulated raw water and the moderately high contents of hydrophobic and transphilic fractions, second only to Wanneroo (Table 2.3). Nuclear magnetic resonance studies have shown that the NOM from Moorabool is less polar overall than that from Horsham (Wong *et al.*, 2002). Moorabool water is also less easily treated by ion exchange than that from Horsham, a further indication of its generally less polar NOM structure (Bolto *et al.*, 2002). Hence it is treated more readily by CPMA , the most hydrophobic of the polymers.

Table 2.4
Percentage removal of NOM fractions by cationic polymers and alum, as measured by UV-absorbance (adapted from Bolto *et al.*, 1998, 1999, 2001)

Polymer	Fraction	Bell Bay	Hope Valley	Wanneroo	Moorabool	Horsham
PDADMAC-1	HPA	58	74	58	46	-
	TPA	44	30	34	35	-
	Char	-	**58**	37	40	-
	Neut	-	-	-	0	-
PDADMAC-3	HPA	65	77	71	55	**57**
	TPA	**66**	59	**69**	39	**51**
	Char	-	54	**74**	29	31
	Neut	-	-	-	4	13
CPAM-1	HPA	-	72	81	58	36
	TPA	-	**62**	61	44	34
	Char	-	39	61	42	16
	Neut	-	-	-	8	12
CPMA	HPA	-	74	**88**	**67**	50
	TPA	-	**61**	64	**51**	47
	Char	-	39	65	**47**	-
	Neut	-	-	-	**12**	**23**
Chitosan	HPA	**75**	67	-	38	49
	TPA	58	49	-	22	35
	Char	**48**	43	-	26	**35**
	Neut	10	-	-	5	0
PEI	HPA	40	49	58	15	26
	TPA	16	25	33	9	23
	Char	-	0	32	0	10
	Neut	**21**	-	-	5	0
Alum	HPA	**88**	77	**91**	**71**	**74**
	TPA	**85**	**66**	**88**	**65**	**71**
	Char	-	**60**	**90**	**83**	**67**
	Neut	**29**	-	-	**24**	**15**

It is interesting to note that weakly basic polymers such as chitosan and PEI give good performance even though they are but slightly charged at neutral pH levels. Thus chitosan is especially effective on Bell Bay HPA and Char, Horsham Char and also Aldrich NOM, despite its low CD. PEI is good for Bell Bay Neut. Such weakly basic polymers may be operating via hydrogen bonding through the free amino groups on the polymer and hydroxyl groups on the NOM.

Table 2.5
Best cationic polymers for removing particular NOM fractions

Fraction	Bell Bay	Hope Valley	Wanneroo	Moorabool	Horsham
HPA	Chitosan	PDADMAC*	CPMA	CPMA	PDADMAC
TPA	PDADMAC	CPAM, CPMA	PDADMAC	CPMA	PDADMAC
Char	Chitosan	PDADMAC	PDADMAC	CPMA	Chitosan
Neut	PEI	-	-	CPMA	CPMA**

* Equivalent to alum performance ** Better than alum performance

The removal of NOM from drinking water supplies can be achieved with cationic polymers, but not as effectively as with alum. For the polymers, the removal of UV-absorbing compounds in Bell Bay and Hope Valley NOM was best achieved with high MW PDADMAC, which has a high CD. CPMA, also of high charge but more hydrophobic in character, was more effective than PDADMAC for Wanneroo and Moorabool. CPMA was the best polymeric coagulant when the water was high in non-polar NOM, as is the case for Wanneroo. CPMA was also slightly better than PDADMAC for Horsham NOM. For Aldrich samples CPAM gave better results than PDADMAC. Hence for waters that are high in non-polar NOM a polyelectrolyte of more hydrophobic character may be better. Chitosan and PEI gave surprisingly high removals despite their low CD. Although organic polymers generally did not perform as well as alum, better results were obtained when particulates were present, and are best used in conjunction with alum to provide them.

Investigation of NOM removal after fractionation into four fractions based on hydrophobic and hydrophilic properties showed that alum was best for hydrophobic and transphilic acids, although PDADMAC was as good for Hope Valley NOM. For hydrophobic acids the top ranking was PDADMAC or CPMA; for transphilic acids it was again PDADMAC, with CPMA and CPAM also active. For hydrophilic charged compounds the best polymers were chitosan or CPAM, both inferior to alum; and for hydrophilic neutral material they were CPMA or PEI, with the former doing better than alum for Horsham NOM. The weakly basic chitosan and PEI, with CDs of 17 and 6% at neutral pH, appear to function by a non-charge mechanism involving hydrogen bonding of the $-OH\cdots\cdots NH_2-$ type. This may open a path towards polymers that are more specific for neutral hydrophilic compounds. Another route to better uptake of neutral compound by polyelectrolytes may require prior treatment of the NOM with a non-ionic surfactant, and removal of the ensuing complex by precipitation with a hydrophobic polymer such as CPMA.

It may be possible in the long term to use results of uptake by various fractions following treatment with different reagents as tools for characterizing NOM.

The reuse of polymer after treatment offers challenges. Regeneration is entirely feasible and inherent in processes like ion exchange, but difficult or costly in coagulation, where in conventional alum treatment the organic contaminants should be removed from an acid extract to restore the efficacy of the alum (Medvedev *et al.,* 1992). In the case of cationic polyelectrolytes, the addition of strong alkali will precipitate PDADMAC and allow its reuse, which fits in with aluminum ore processing where high pH levels are utilized (Keeney, 1987), but is inappropriate in drinking water treatment. A research opportunity could exist however in low-temperature thermal regeneration, based on temperature-sensitive polymers of the poly-*N*-isopropylacrylamide family (Deng *et al.,* 1996).

REMOVAL OF NATURAL ORGANIC MATTER WITH ION EXCHANGE

Quaternary ammonium resins

The use of strongly base anion exchange resins to adsorb NOM from drinking water is well documented (Kölle, 1979; Bursill *et al.,* 1985; Brattebø *et al.,* 1987; Mamchenko, 1993; Symons *et al.,* 1995). Quaternary ammonium resins are normally used in the chloride form, when the following reaction takes place, with R^- representing the charged DOC:

$$\text{Resin-NMe}_3^+\,\text{Cl}^- \;+\; \text{R}^- \;\rightarrow\; \text{Resin-NMe}_3^+\,\text{R}^- \;+\; \text{Cl}^- \qquad (2.2)$$

Regeneration is achieved with an excess of brine or caustic brine, which reverses (2.2). Trihalomethane (THM) precursor removals of up to 81% are reported (Kim and Symons, 1991). In Norway, of 160 water treatment plants designed for NOM removal, 12 use ion exchange (Ødegaard et al., 1999); items of concern at these plants include pretreatment requirements, loss of capacity with time and more frequent regeneration than expected. Ion exchange is claimed as more effective than activated carbon, carbonaceous resins or metal oxides for NOM adsorption (Fettig, 1999). This is especially so in economic terms if on-site regeneration is performed (Heijman *et al.,* 1999). The approach is most competitive for low color waters; it has low installation costs and operation is simple (Hongve *et al.,* 1999). When an anion exchanger is employed to remove the charged organic molecules from raw water, a subsequent alum clarification stage is greatly facilitated. Larger flocs are formed that settle much more rapidly, far less organics are left in the product water, and only a fraction of the original alum dose is required (Bursill *et al.,* 1985).

Research on the adsorption of small organic anions such as benzenesulfonate, naphthalenesulfonate and anthraquinonesulfonate by strongly basic quaternary ammonium resins has shown that the selectivity of the resins increases with an increase in the number of aromatic rings in the organic anion and in the number of carbon atoms around the exchange site (Gustafson and Lirio, 1968). A polystyrene resin was more selective than acrylic species. The high selectivities observed were attributed to a combination of electrostatic interactions and hydrophobic bonding. In the uptake of macromolecular NOM species, it was first thought that molecular adsorption dominated over electrostatic interactions as results with a strong base anion exchanger showed more humate removed than chloride released (Mamchenko, 1993). However, the organic acids present can be ionised to a considerable extent even at low pH, especially in the resin phase, when there is a dramatic increase in the dissociation of all groups. For strong base resins hydrophobic interactions in fact appear to be responsible for only a small proportion of NOM adsorption; one study quantifies this at 7% for a lake water (Symons *et al.,* 1992). Recent work on adsorption of various NOM fractions on styrene resins showed that more highly charged smaller ions were preferred, with size effects being predominant; the retention of organics was postulated to be via a complex mixed mechanism. Non-ion exchange did occur, particularly at low pH and on a weakly basic resin (Croué *et al.,* 1999). The non-removable, presumably uncharged fraction of NOM in natural waters can vary from <10 to about 40% (Fettig, 1999).

In the past it was thought that high MW organics would not fit into the gel phase of a resin. Lightly crosslinked gel resins do better than highly crosslinked ones, and better than highly crosslinked macroporous resins (Gottlieb, 1995). Even for very large organic molecules the relative water content of the gel phase was seen as the only reliable predictor of performance: the performance varied with the resin water content irrespective of whether the resin was gel or macroporous in physical form. The best performer of the resins investigated was a styrene resin of 75% water content.

The results of some later work on NOM removal by a range of structurally diverse resins are summarized in Table 2.6, where MP refers to macroporous (Bolto *et al.,* 2002). The aim was to determine the effects of the physical and chemical structure of the resin on performance. The NOM was from Mount Zero reservoir at Horsham, Eastern Australia, and a soil-derived NOM from the Aldrich Company. It can be seen that there is an increase in NOM removal for resins of higher water content, with the more open structures allowing easier entry of the larger NOM species, in line with earlier work that focused on styrene resins (Gottlieb, 1995). A low degree of crosslinking as in the Type I styrene resin SIR 22P raises the water content of the resin. The results for Type II resins (IRA 410 and 910), which have $-N(C_2H_4OH)Me_2^+$ groups instead of the $-NMe_3^+$ groups found in Type I resins, show improved performance. Having a pendant OH group close to the quaternary nitrogen is a distinct advantage as the resins removed more NOM than would be expected from their water content: an appropriate balance of polar and non-polar regions in the resin structure appears to be required. Polar groups and the degree of crosslinking influence in a positive way the water content of the resin.

Table 2.6
Strongly basic resins tested for NOM removal (adapted from Bolto et al. 2002a)

Resin	*Pores*	*Structure*	*Capacity, meq/mL*	*%Water Content*	*%UV Absorbers Removed*	
					Horsham	*Aldrich*
Amberlite IRA 420	gel	styrene	1.2	43	-	19
Amberlite IRA 410	gel	styrene, Type II	1.3	43	77	32
Lewatit MP 500	MP	styrene	1.6	43	-	20
Amberlite IRA 400	gel	styrene	1.4	45	57	14
Amberlite CG 400	gel	styrene	1.4	45	-	22
Imac HP 555	MP	styrene, $-NEt_3^+$	0.9	49	58	20
Amberlite IRA 402	gel	styrene	1.3	54	-	26
Purolite A520E	MP	styrene, $-NEt_3^+$	1.0	54	-	30
Amberlite IRA 910	MP	styrene, Type II	1.1	55	73	33
Amberlite IRA 401	gel	styrene	0.8	56	-	24
Amberlite IRA 904	MP	styrene	0.7	57	-	43
Amberlite IRA 458	gel	acrylic	1.2	60	69	47
Amberlite IRA 900	MP	styrene	1.0	60	-	42
Amberlite A 26	MP	styrene	1.0	61	-	63
Amberlite IRA 958	MP	acrylic	0.8	69	73	92
Amberlite IRA 938	MP	styrene	0.5	73	-	78
ResinTech SIR 22P	gel	styrene	0.4	75	84	59
CSIRO MASB	MP	methacrylic	0.4	80	89	-
CSIRO PDAA	gel	diallylamine	0.3	90	91	-

Also shown in Table 2.6 are data for the removal of soil-derived Aldrich NOM from synthetic waters. The NOM removals again follow the resin water content, with IRA 938, IRA 958 and A 26 (all MP resins with very large pores) showing higher uptakes relative to water content. Here the pores can be better penetrated by the larger organics present in Aldrich NOM.

The Type II resin IRA 410 also performed better than expected, as for Horsham NOM. The Horsham and Aldrich results show that the size of the NOM molecules is a vital factor. The average molecular weights of the respective UV-absorbing NOMs are 1020 and 1990 respectively (Chow, 1999; Pelekani *et al.,* 1999). Overall, a MP structure is more advantageous than a gel one (MP 500 versus IRA 400, IRA 904 versus IRA 401 and IRA 958 versus IRA 458), as would be expected for diffusion of sizeable organic anions into the resin interior.

For resins of the same water content, an acrylic resin takes out more Aldrich NOM relative to a styrene resin when there is a water content of *ca.* 70% (IRA 958 versus IRA 938 or SIR 22P), but not when there is a lower water content of 60% (IRA 458 versus IRA 900). This suggests the occurrence of hydrophobic/hydrophilic repulsive forces between the styrene resin backbone and highly charged organic ions when there is a 70% water content, which has been postulated before (Croué *et al.,* 1999). It highlights the need for a suitable balance of polar and non-polar regions in the resin.

There is a small particle size effect in the IRA 400 and CG 400 comparison, where the bead sizes used were 200-500 and 5-50 µm respectively. The increase in surface area which results from using such small particles will make for immediate access to more than six times as many surface sites per unit mass as larger resin beads, but this is a small proportion (< 5%) of the internal sites available within the beads. Another aspect is the shorter distance for all sites to be reached, as with larger beads access may be blocked by NOM adsorbed at sites near the resin surface.

Studies of the adsorption of the NOM from Seine River water have been carried out on MP styrene-based resins of various types (Afcharian *et al.,* 1997). Two strong base resins (Lewatit MP 500 and Lewatit S6328A) removed 90 and 93% of the UV-absorbing compounds, while the removals for two non-ionic resins (Lewatit VP OC 1062 and Lewatit VP 1064) were 23 and 28%, and that for a cation exchange resin (Lewatit SP 112) was zero. All the synthetic adsorbents were based on MP polystyrene beads, the strong base resins both being Type I and of comparable water content, as detailed in Table 2.7. The results here are better than those that can be interpolated for Lewatit MP 500 in Table 2.7, which may arise from the apparent higher water content of the resin used in the Seine River study (61-63 versus 43%).

Table 2.7
Removal of NOM from Seine River water by strong base MP styrene resins, from DOC and UV absorption measurements (adapted from Afcharian *et al.,* 1997)

Resin	Capacity, meq/mL	% Water Content	% NOM Removal, from DOC	% NOM Removal, from UV
Lewatit MP 500	1.1	61-63	59	90
Lewatit S6328A	0.8	58-63	73	93

Weakly Basic Resins

Also shown in Table 2.6 are data for a PDAA resin, a crosslinked polydiallylamine (Hodgkin and Eibl, 1988), which has a very basic secondary amino group that is essentially all protonated at pH 7 (McLean *et al.,* 1976). It is notable that all but one of the resins that have a water content of 60% or more, namely IRA 458, IRA 958, MASB and PDAA, possess more polar structures than the Type I styrene resins of low water content used in normal ion exchange applications. This may be because of the presence of hetero-atoms in the bridge between the exchange site and the polymer backbone, a secondary amino group as the exchange site, or else a low ratio of carbon to quaternary nitrogen. The environment around the nitrogen is evidently critical, with hydroxyethyl or the absence of two methyl groups on the N (as with a secondary amino group) being beneficial in NOM removal. The latter behavior fits selectivity theory, in that for weak base resins the charge on the $-NH_2-^+$ exchange site is more concentrated, so the site is more polar than the $-NMe_3^+$ of strong base resins. The $-NMe_3^+$ sites are large in size and have a low charge density, so they have a greater affinity for hydrophobic counter ions (Diamond, 1963; Jensen and Diamond, 1965). Conversely, having a higher charge density, the protonated secondary amino sites have a greater affinity for hydrophilic counter ions. Weakly basic resins have been investigated in other studies (Rook and Evans, 1979; Evans and Maalman, 1979). There are cost advantages in terms of regenerant usage: a strongly basic resin requires salt and alkali well in excess of the stoichiometric amounts, whereas a weakly basic resin has lower chemical requirements in the form of lime and mineral acid at not much above equivalent levels:

$$\text{Resin-NHMe}_2^+\ \text{R}^- + \text{OH}^- \longrightarrow \text{Resin-NMe}_2 + \text{R}^- + \text{H}_2\text{O} \qquad (2.3)$$

$$\text{Resin-NMe}_2 + \text{H}^+ + \text{Cl}^- \longrightarrow \text{Resin-NHMe}_2^+\ \text{Cl}^- \qquad (2.4)$$

The organics are obtained as the calcium salts of humic and fulvic acids. Hence regeneration can be achieved without salt, which can simplify disposal of the waste. This has been done in the past for very highly colored effluents from bleach plants in the paper industry, where the waste is incinerated (Anderson *et al.,* 1974). Weakly basic resins of phenol formaldehyde structure have been tested on various waters. They are effective for organics removal (Boening *et al.,* 1980), but overall their performance was found to be not as good as that of the strongly basic polystyrene resin or its lesser performing acrylic analogue. It is noteworthy that in the removal of color from wood pulp bleach plant liquors a weakly basic phenolic resin gave the best performance of a range of resins studied (Sanks, 1973; Anderson *et al.,* 1982). However, for removing medium MW organics akin to humic acids from tertiary treated sewage a strongly basic polystyrene resin was preferred over all available types (Bolto *et al.,* 1978). Likewise, weakly basic polystyrene resins do not perform as well as their strongly basic counterparts (Anderson and Maier, 1979; Naumczyk *et al.,* 1989).

In a more recent study, the weakly base resins detailed in Table 2.8 were tested for adsorption of NOM at neutral pH (Bolto *et al.,* 2002). The degree of protonation of the resins at neutral pH was estimated from information supplied by the resin manufacturer or from published titration data (Weiss *et al.,* 1966). The removal of NOM by each resin is shown in Table 2.8, together with the resin water content. The maximum removal of 93% of UV-absorbing compounds again accounts for nearly all of the acidic and charged hydrophilic compounds detected in this particular NOM (95%).

Table 2.8
Low basicity resins tested on Horsham NOM (adapted from Bolto et al. 2002a)

Resin	Pores	Structure	Capacity at pH 7, meq/mL	% Sites Charged at pH 7	% Water Content	% UV Absorbers Removed
Reillex 425	MP	pyridine	0.7	0.03	50	53
Amberlite IRA 93	MP	styrene	0.9	3	58	55
Amberlite IRA 68	gel	acrylic	1.2	37	59	58
Amberlite IRA 35	MP	acrylic	1.1	37	69	61
CSIRO MAWB	MP	methacrylic	0.7	40	37	55

Although not as effective as the strong base resins, the very low charged IRA 93 and Reillex 425 have extraordinary uptakes, considering that there are very few charged sites at neutral pH. The more charged IRA 68 and IRA 35 are similar. Water content is more important than basicity level. Thus MAWB, the most basic resin in the series, shows the poorest NOM uptake, as it has a low water content. To explain these results a mechanism other than that shown in equation (2.5) must apply. It is proposed that a non-electrostatic mechanism involving hydrogen bonding of the unprotonated acidic species (carboxylic acid or phenol) in the NOM to the amino groups on the low charged resins is occurring:

(2.5)

A similar mechanism has been postulated in the past for the uptake of *p*-nitrophenol by a weak base resin (Kim *et al.,* 1976). That the mechanisms for high and low charged resins are quite different is further evidenced by the markedly different gradient of the plot of uptake versus water content: 0.19 for the low charged resins listed in Table 2.7 and 0.84 for the high charged ones shown in Table 2.8.

Further information on weak versus strong base resins has recently appeared (Hwang *et al.,* 2001), and is summarized in Table 2.9. The superiority of the strong base resin is clear, although the results are not as impressive as those obtained for the removal of aquatic NOM by the more water swollen resins. Company literature indicates that the resins used in this study have water contents of only 50-64%, versus 75-90% for the better performing resins. The results appear to be about the same as those in Tables 2.5 and 2.6 for resins of similar water content.

Table 2.9
Percentage removal of NOM from four surface waters by MP styrene resins, as determined from DOC measurements (Hwang *et al.*, 2001)

Resin	Type	Capacity, meq/mL	Colorado River	Ribou Reservoir	Seine River	Sacramento-San Joaquin Delta
Dowex MSA 1	strong base	1.0	77	71	57	59
Imac HP 661	weak base	1.4	56	52	47	47

Effect of fraction character on removal

The amount of NOM in natural waters not removed by ion exchange can vary from <10 to about 40% (Fettig, 1999). This is likely to be the uncharged compounds, but further information is desirable. Recent work on adsorption of various NOM fractions on styrene resins showed that more highly charged smaller ions were preferred, with size effects being predominant; the retention of organics was via a complex mixed mechanism: non-ion exchange does occur to a minor extent, particularly at low pH and on a weak base resin (Croué *et al.*, 1999). Further information is now available (Bolto *et al.*, 2002). The NOM from several Australian sources was fractionated using established adsorption procedures (Croué *et al.*, 1993a). Four fractions were obtained, designated as: HPA, hydrophobic acids; TPA, transphilic acids; Char, charged hydrophilic compounds; Neut, neutral hydrophilic compounds. Results are given in Table 2.10, where the data are estimated from measurement of the DOC levels. The method of isolating the NOM for one of the waters, that from Hope Valley, was by ion exchange, so the results are not representative of the total original NOM in this case. There is naturally a negligible amount of uncharged material in the Neut isolate. Uptake of NOM fractions is also included in Table 2.10. The data quoted are for a commercially available gel type styrene resin of high water content. The significant difference in the results for the Horsham and Moorabool NOM is noteworthy.

Table 2.10
NOM fractions in various surface waters, obtained from DOC measurements and expressed as percentage of the total NOM with percentage removal by *ResinTech* SIR 22P shown in parentheses (adapted from Bolto *et al.*, 2002a)

Water Source	NOM Concentrated by	Raw Water	HPA	TPA	Char	Neut
Hope Valley	Ion exchange	-	46	21	32	1
		(98)	-	-	-	-
Moorabool	Reverse osmosis	-	34	18	33	15
		(85)	(97)	(75)	(13)	(24)
Horsham	Reverse osmosis	-	58	14	11	17
		(84)	(92)	(95)	(88)	(10)

Nuclear magnetic resonance studies (NMR) show that Horsham NOM is more polar overall than that from Moorabool (Wong *et al.*, 2002). The larger uptake of NOM from the

Horsham TPA and Char fractions fits the higher carbonyl content relative to Moorabool as shown by NMR. The NOM in the Horsham water may be more oxidized by exposure to sunlight and UV than that from Moorabool. Ion exchange is hence more successful the more polar the charged species.

The results for NOM removal from Suwannee River water, Georgia, as detailed in Table 2.11, are somewhat at variance with other data (Croué *et al.*, 1999). The water content data have been obtained from resin manufacturer literature. Performance is again generally not as effective as with the very water swollen ResinTech SIR 22P, at HPA 32-40% removal versus 92-97%; TPA 40-68% versus 75-95%; Char 42-70% versus 13-88%. However, extraordinarily high removals of hydrophilic neutral compounds have been achieved, at 45-78% versus 15-17%. If these neutral compounds are less polar than those in the previous study, the more crosslinked and more hydrophobic resins used in the Suwannee River study may be responsible.

Table 2.11
Percentage removal of NOM fractions from Suwannee River water by various styrene resins, from DOC measurements (adapted from Croué et al., 1999)

Resin	Pores	Type	Capacity meq/mL	%Water Content	HPA	TPA	Char	Neut	uHA*
Dowex 11	gel	strong base	1.2	52-60	40	68	70	78	8
Dowex MSA 1	MP	strong base	1.0	56-64	38	66	70	78	12
Imac HP 661	MP	weak base	1.4	50-58	32	40	42	45	6

* "Ultrahydrophilic acids," which are colloidal in nature

MIEX® Resin

The magnetic ion exchange (MIEX®) DOC process was developed by Orica Watercare, South Australian Water Corporation and the CSIRO specifically for the removal of DOC from drinking water (Figure 2.5). The resin is a strong base anion exchange resin with a macroporous structure and type 1 quaternary ammonium active sites attached to a magnetic core. The process was developed on a number of recognized premises:

- resins with quarternary ammonia functional groups are more effective (Bolto *et al.*, 2002; Singer and Bilyk, 2002);
- resins with a polyacrylic skeleton are the best for NOM removal (Symons *et al.*, 1995; Gottlieb, 1996; Singer and Bilyk, 2002);
- macroporous resins are more suited to continuous processes than gel resins (Kunin and Yarnell, 1997; Singer and Bilyk, 2002);
- resins need a high specific ion exchange capacity (Brattebo *et al.*, 1987); and
- smaller sized resin particles are more efficient (Meyers, 1995). The MIEX® resin beads have a mean particle size of 150 μm.

As the removal process is ion exchange, only ionised organic matter can be removed. Although the ionization of the active sites is equally efficient over the whole pH range humic substances by their nature of being weak organic acids with carboxylic functionality are only partially ionised below pH 7 and so acidic conditions give less efficient removal.

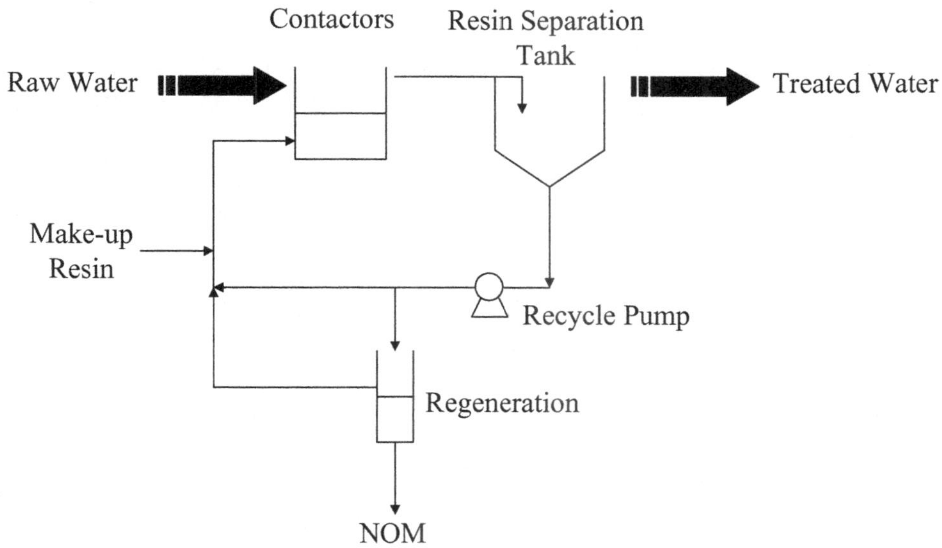

Figure 2.5 Schematic of MIEX® process (adapted from Orica Watercare)

The process involves adsorbing the DOC onto the MIEX® resin in a stirred contactor that disperses the resin beads to allow for maximum surface area. The magnetic part of the resin allows the resin to agglomerate into larger faster settling particles which allow for a recovery rate of greater than 99.9%. Any resin that is carried over is removed in downstream processes (Smith *et al.*, 2001). A further treatment stage is then required as the MIEX® process does not remove suspended matter, this is usually in the form of coagulation or microfiltration (Pelekani *et al.*, 2001; Drikas *et al.*, 2003b).

Kinetic studies for the treatment of waters with MIEX® have shown that equilibrium is generally reached within a short contact time (10-30 minutes) (Lee *et al.*, 2003; Drikas *et al.*, 2003; Budd *et al.*, 2003). Also it has been shown that due to the large surface area relatively low doses of MIEX® are required to achieve maximum removals and there is more resistance to fouling on the surface (Budd *et al.*, 2003).

Because the process is still in its infancy, there is little published information on the bulk removal of NOM by MIEX®. However, published work does show that considerable increases are generally observed in the DOC removal for a range of selected works when MIEX® plus coagulation is employed by comparison with conventional or enhanced coagulation. The removal of UV_{254} is seen to be more pronounced by MIEX® plus coagulation for waters with a low SUVA value i.e. more hydrophilic in nature. This is due to the high SUVA waters having a high hydrophobic content which has been shown to be more readily removed by conventional coagulation. Generally the coagulant dose is greatly reduced following treatment with MIEX® with a reduction of up to 50% observed (Singer and Bilyk, 2002) (Table 2.12).

Table 2.12

Review of published DOC and UV_{254} removal data for coagulation and MIEX® plus coagulated bulk waters (Drikas *et al.*, 2003; Drikas *et al.*, 2003b; Singer and Bilyk, 2002; Delphos *et al.*, 2001; Bourke *et al.*, 2001; Warton *et al.*, 2004)

	SUVA ($L\ mg^{-1}m^{-1}$)	Treatment	Residual DOC ($mg\ L^{-1}$)	DOC Removal (%)	Residual UV_{254} (m^{-1})	UV_{254} Removal (%)	Reference
Mount Pleasant	1.32	Alum	2.6	35	2.2	49	Drikas *et al.*, 2003
		MIEX® +Alum	1.5	53	1.0	77	
Hope Valley	2.4	Alum	4.2	46	7.4	46	Drikas *et al.*, 2003b
		MIEX® +Alum	2.1	64	2.5	82	
Myponga	3.5	Alum	4.7	53	9.8	72	Drikas *et al.*, 2003b
		MIEX® +Alum	2.4	76	3.4	90	
Mantee Co	4.5	Alum	5.3	50	10.6	78	Singer and Bilyk, 2002
		MIEX® +Alum	1.4	87	2.9	94	
Tampa	4.0	Alum	9.3	65	21.9	80	Singer and Bilyk, 2002
		MIEX® +Alum	2.9	86	4.1	96	
Durham	3.5	Alum	3.1	39	4.2	76	Singer and Bilyk, 2002
		MIEX® +Alum	1.2	76	1.4	92	
MWD	3.0	Alum	2	29	4.8	41	Singer and Bilyk, 2002
		MIEX® +Alum	1.1	71	1.6	89	
Hackensack	2.7	Alum	-	-	5.8	45	Singer and Bilyk, 2002
		MIEX® +Alum	-	-	2.0	81	

(continued)

Table 2.12 (continued)

	SUVA (L mg^{-1}m^{-1})	Treatment	Residual DOC (mg L^{-1})	DOC Removal (%)	Residual UV_{254} (m^{-1})	UV_{254} Removal (%)	Reference
Sioux Falls	2.7	Alum	4.9	44	10	26	Singer and Bilyk, 2002
		MIEX® +Alum	2.4	72	3.2	76	
Austin	2.0	Alum	2.6	7	4.1	27	Singer and Bilyk, 2002
		MIEX® +Alum	1.3	54	1.2	79	
Indianapolis	1.9	Alum	-	-	6.8	23	Singer and Bilyk, 2002
		MIEX® +Alum	-	-	5.8	75	
Manchester	1.4	Alum	2	23	2.9	3	Singer and Bilyk, 2002
		MIEX® +Alum	1.4	46	1.6	53	
Chesapeake	3.6 – 4.1	Alum	-	63	-	-	Delphos *et al.*, 2001
		MIEX® +Alum	-	80	-	-	
Olga	~3	Fe	4.7	77	-	-	Bourke *et al.*, 2002
		MIEX® +Fe	3.9	81	-	-	
Wanneroo (winter)	-	Alum	2.2	48	-	-	Warton *et al.*, 2004
		MIEX® +Alum	1.7	60	-	-	
Wanneroo (summer)	-	Alum	2.4	65	-	-	Warton *et al.*, 2004
		MIEX® +Alum	1.8	74			

Similarly the published results for the removal of isolated fractions are limited. However Lee *et al.* (2003) showed that for three isolated fractions of NOM the following removals could be achieved by the MIEX® process alone (Figure 2.6).

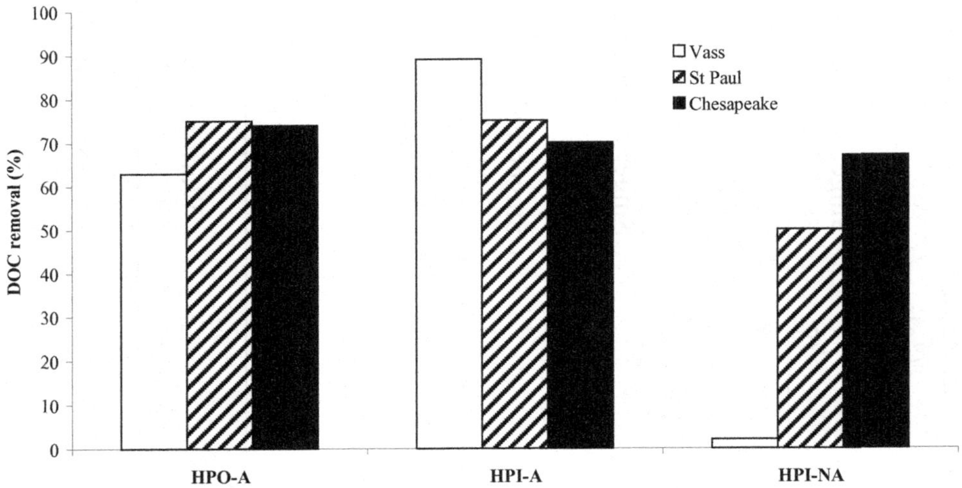

Figure 2.6 DOC removal data for the isolated fractions from 3 US water sources (Adapted from Lee *et al.*, 2003)

The water sources had varying water characteristics with Chesapeake, Vass and St Paul having DOC values of 22.4, 5.1 and 8.0 mg L^{-1} respectively. The data shows that the HPINA fraction is only removed to a certain level, after which no more removal is observed. This may be due to the remaining species being uncharged and consequently unsuitable to removal by the MIEX® process. Furthermore it has been reported, by studying the HPSEC data for the treatment with MIEX®, that the lower MW fractions (e.g., HPI-A and HPINA) were more readily removed than by coagulation (Lee *et al.*, 2003; Hainthaler *et al.*, 2003; Drikas *et al.*, 2003; Drikas *et al.*, 2003b).

Practical Considerations

A practicable NOM removal process can be achieved by using fixed-bed columns (Ødegaard et al., 1999) or fine resin particles such as in the MIEX® process, which makes use of magnetic resin beads (Bourke and Slunjski, 1999). A 2.5 ML/day plant for organics removal by the conventional technology operated for some years in Germany (Kölle, 1984). It utilized a MP styrene resin to treat a groundwater containing 6.5 mg/L of DOC, removing 50% of it in a contact time of only 1.2 minutes, to yield 5000 bed volumes of product water. There was only a 10% fall off in resin capacity after two years of operation, and a five-year resin life was subsequently obtained. However, the plant has reportedly ceased production because of increasing difficulties with the disposal of the spent caustic brine that was used as regenerant (Höll, 1995); the liquor was recycled seven times, with concentration adjustments, before disposal.

Attention to regeneration conditions has allowed a 0.4 ML/day plant at Cape Hatteras, North Carolina, to be operated for 18 months without substantial resin fouling. It had been established that the resin could be more effectively regenerated with chloride than hydroxide solutions as the relative affinity of the resin for chloride is 11 times higher than for hydroxide (Gottlieb, 1995). In practice, making the brine regenerant alkaline converts the carboxylic acid and phenolic groups in the NOM to the fully ionised forms, thus increasing their water solubility at the high concentrations prevailing during regeneration. For example, phenol is soluble to 7.5% in cold water, but as the sodium salt its solubility reaches 20%. The beneficial effect of adding caustic to the brine regenerant is that the peak of organics removed from the resin is after the addition of 1.2 bed volumes (BV) of regenerant instead of 1.6 BV, and their level is 1200 mg/l instead of 800 mg/L (Meyers, 1995). There was then much better performance in the next service cycle, the leakage of organics being halved.

To achieve successful operation it is necessary to remove all the organics after the adsorption cycle, but this is not easy to achieve in practice as the time required would be several hundred hours. To minimize the difficulty of getting out the deeply embedded organic ions, shorter loading cycles are used. The optimum degree of resin loading allowed prior to regeneration is an important parameter that needs to be determined. Thus in sugar decolourisation it is usual to operate at about 50-60% of the maximum throughput, when a typical resin life would be 200 cycles. If the full capacity is utilized the resin life is halved and the average organics removal is 50% instead of 75-80% (McGarvey, 1995). Fouling can be further minimized by using smaller 30-50 mesh beads (360-700 μm) instead of the conventional 16-40 mesh (500-1300 μm), as this provides a greater surface area so that the organic ions penetrate less deeply in the same cycle time. One of the possible unfortunate consequences of using resin beads of high water content may be that because of their softness relative to standard resins, the beads may be subject to greater attrition (McGarvey, 1995). The situation may be further aggravated by the use of smaller beads, which increases the pressure drop.

For removal of charged species (HPA, TPA and Char), ion exchange is better performing than alum. For some low humic waters the hydrophobic fractions are not the main precursors in DBP formation. This is not that common, but when it occurs there can be some difficulty, as the conventional approach using metal salt coagulants may not be adequate. The polar components (TPA, Char and Neut) can give high yields of DBPs, partly because of their aliphatic and carboxylic acid structures (Hwang *et al.,* 2002). Ion exchange can then be very appropriate as it will very efficiently remove essentially all of the TPA and Char fractions as well as the HPA fraction, while alum will give preference to only the larger molecules in these fractions. There is scope for the synthesis of improved resin systems, both strong and weak base. For example, a so-called Type III strong base resin which has hydroxypropyl rather than hydroxyethyl groups attached to the quaternary ammonium nitrogen (Irving *et al.,* 1997) should be tested.

Normal treatment, including ion exchange, is not usually effective in removing neutral hydrophilic compounds, as is illustrated by the data shown above. Probably mainly polysaccharide based, these compounds can be a strong contributor to DBPs in particular waters, or at certain times of the year. Membrane technology is another option for general removal of NOM, although in some waters the neutral fraction can be a potent fouling agent for microfiltration membranes even after coagulation (Carroll *et al.,* 2000). There is hence scope for new adsorbents that are specific for non-charged hydrophilic compounds.

The regeneration of loaded quaternary ammonium resins with strong brine and sometimes accompanying alkali can be a concern, as the excess regenerant can cause waste disposal problems. The regeneration of weakly basic polyamine resins on the other hand can be achieved without salt, the chemical requirements being stoicheometric amounts of lime and mineral acid. There is need for an improved method that minimizes disposal difficulties: something like the use of hot water as achieved in the inorganic realm in desalination with weakly basic resins (Bolto, 1996). A research opportunity could exist in thermal regeneration, using temperature-sensitive polymers of the poly-*N*-isopropylacrylamide family (Deng *et al.,* 1996). The use of light is even more futuristic, although it can be used to desorb fulvic acid from iron oxide surfaces (Bjorklund *et al.,* 2001).

REMOVAL OF NATURAL ORGANIC MATTER BY ADSORBENTS

Raw water processing to produce drinking water normally involves physicochemical procedures, usually based on coagulation and flocculation of suspended solids and colloids, but adsorption of soluble material by solid substrates such as ion-exchange resins (Bolto, 2003, incorporated in Parsons *et al.*, 2003), and clays, metal oxides and activated carbon, can also be employed. In this review the focus is on the use of the latter adsorbents, some of which can also be used in modes that involve entrapment of turbidity, especially if used in conjunction with polyelectrolytes. The earlier report outlined the nature of some of the natural impurities in drinking water sources. The data surveyed here are essentially on equilibrium situations rather than kinetics, except for activated carbon.

Clays

The addition of clay to purify turbid river water has been practised for hundreds of years in the Sudan (Madsen and Schlundt, 1989). Bentonite, a three-layered aluminosilicate, will remove particulates, including microbes. It has been employed for flocculating suspended solids in conventional water works (Cox, 1964). The aim in adding clay, as with other adsorbents, is to convert NOM to an insoluble form. The adsorption of NOM by clays has been postulated to take place by a ligand exchange between the carboxylate groups on the organic molecules and the OH groups bound to a metal on the surface of the solid (Kaiser and Zech, 1998):

$$\text{>M-OH} \quad + \quad RCO_2^- \quad \rightarrow \quad \text{>M-}O_2CR \quad + \quad OH^- \qquad (2.6)$$

There are other analogous interpretations such as that involving hydrogen bonding (Schulthess and Huang, 1991), but as the organic acids are more than half ionised under normal water treatment conditions the reaction in equation (2.5) seems the more pertinent. Another proposed mechanism is via hydrophobic interactions, for which there is evidence in the case of clays, but not for metal oxides (Jardine *et al.*, 1982; Baham and Sposito, 1994).

The coating of charged organics on a clay has a strong impact on the amount of coagulant required (Gibbs, 1983). The charge on the particles is controlled by the adsorbed layer of NOM. As some of the organics have polyelectrolyte character, there will be an increase in the normally negative charge on the particles because some of the carboxylate groups on the NOM are not bound to the metal. The charge on the particles is also influenced by salinity and the divalent cation content of the water. The surface potential is an important parameter influencing colloid stability and adsorption behavior, and in natural systems is invariably negative, irrespective of the nature of the primary particle (Beckett and Le, 1990). The potential decreases in magnitude as the ionic strength of the water increases because of charge shielding (Beckett, 1990), so that less coagulant is required at higher salinities. As drinking water salinity levels are strictly limited, ionic strength effects will not be dwelt on in any detail in this review.

On addition of kaolinite, another aluminosilicate, to humic acid solutions, the organic material is mostly adsorbed onto the clay surface. Kaolinite has been found to have a much stronger affinity for soil-derived humic acid than montmorillonite, an aluminum magnesium silicate that contains sodium and calcium counterions (Parazak *et al.*, 1988). The interaction is believed to involve polyvalent metals at the clay surface, and since kaolinite has a higher aluminum content than montmorillonite, it adsorbs humic acid more strongly. Also, montmorillonite has a high negative charge, while that of kaolinite is a factor of ten lower, and is dependent on pH. Hence there will be a strong electrostatic repulsion between montmorillonite and the NOM that will decrease with decreasing pH. On adding humic acid to clays there is very little change in the electrophoretic or zeta potential, as documented for kaolinite at pH 8, the IEP of which is pH 4-5 (Sposito, 1992). This has been confirmed in later measurements of zeta potentials following the adsorption of NOM from Lake Hohloh in Germany onto kaolinite or montmorillonite, which causes only a small increase in negative charge. The increase is larger for montmorillonite at low pH levels, but is much less than occurs with a metal oxide such as alumina (Specht *et al.*, 2000). Substances of higher MW showed the greatest affinity towards the clay minerals in the latter work. Another study has confirmed that this, with greater adsorption of a 15 mg/L solution of soil-derived humic acid over that of a 8.4 mg/L solution of Dismal Swamp fulvic acid, at 55 and 25% removal respectively at neutral pH (Schulthess and Huang, 1991). There was a much greater removal with aluminum oxide, at 90 and 85%, as there was also with kaolinite, where the effect was about equal removal at ~85% for each. The removal was reversed for mordenite, another aluminosilicate that has an IEP of 1.3 (Elliott and Huang, 1981). The relative difference was small though, at 10 versus 15% removal. A similar performance resulted with silica.

Other clays that have been investigated include gibbsite, another aluminosilicate, and imogolite, a calcium aluminosilicate. Ligand exchange plays a less important role in adsorption of soil-derived NOM on gibbsite (Parfitt *et al.*, 1977), the higher adsorption capacity being ascribed to the larger surface area (900 versus 100 m^2/g), and hydrogen bonding of un-ionised organic acids to the clay.

A recent paper discusses the adsorption of humic substances by tobermorite, a fibrous calcium silicate of formula $Ca_5Si_6O_{16}(OH)_2.2H_2O$ (Kaneco *et al.*, 2003). The adsorbent had pore sizes of 50-2000 μm and a surface area of 19 m^2/g. A column test procedure showed 60-62% removal of fulvic acid (20 mg/L, obtained from brown forest soil) and 97% removal of humic acid (30 mg/L, obtained from peat soil) from synthetic waters over the pH range 5-9. The adsorption was 40-50% and 30-50% respectively greater than that observed with adsorbents based on sodium aluminosilicate, such as zeolite and 5Å molecular sieve, despite the zeolite having a similar surface area and the molecular sieve a surface area 20 times larger; the pore sizes though were very much smaller, at 3×10^{-4} and 5×10^{-4} μm. Nevertheless, the adsorbents with small pore sizes adsorbed humic acid more easily than fulvic acid, indicating that adsorption was mainly taking place on the exterior surface of the particles. It arose from specific interactions between the tobermorite and carboxylic and phenolic groups in the NOM, as well as from electrostatic considerations as has been noted elsewhere (Specht *et al.*, 2000; Gu *et al.*, 1995; Vermeer *et al.*, 1998).

Another calcium silicate, xonotlite, has been tested for the same soil-derived humic substances removal (Katsumata *et al.*, 2003). At neutral pH, the optimum for performance over the pH range 5-9, 35% removal of fulvic acid from a 20 mg/L solution was achieved in 24 h, and 80% removal of humic acid (15 mg/L solution). The xonotlite was present at 5 g/L. The heat of adsorption for humic acid was 330 kJ/mol, indicating a chemical adsorption mechanism. Not only was the larger sized humic acid removed more effectively than the smaller fulvic acid, MW fraction studies showed that the humic acid of larger size (measured as 10-100 kDa) was preferentially adsorbed. This was attributed to the presence of more functional groups in the humic acid, which would result in multi-point attachment of one NOM molecule to many sites on the adsorbent. The pore size of the xonotlite was 50-2000 μm, which would allow entry of these larger species.

Silica

The addition of silica to a Tasmanian water and solutions of humic and fulvic acids at pH 6 and 8 caused the zeta potential of the silica to become at least 10% more negative, showing that the silica becomes coated with the organics (Dixon and Kolarik, 1990). The binding of like-charged particles in this way is not unusual, a specific interaction between them being advocated (Buleva and Petkanchin, 1999). This probably involves hydrogen bonding from the silica surface hydroxyl group to the carbonyl oxygen in the carboxylic acid group of the humic substances, analogous to the interaction shown in (2). Substantiation of such an effect comes from a report that shows that the adsorption of soil-derived humic acid on sand is much increased on lowering the pH: the amount adsorbed at pH 8 is doubled at pH 3.5 (Petrovic *et al.*, 1999).

In a fundamental investigation of the role of humic acid in the transport of negatively charged colloids through porous media, the adsorption of Suwannee River humic acid on 0.2 mm silica beads and sulfate latex colloids of size 98 nm was examined (Franchi and O'Melia, 2003). The organics adsorbed on both the latex and the beads reduced the deposition of suspended particles and enhanced the re-entrainment of deposited particles in the porous media. The effects were considered to arise from the additional electrostatic and steric contributions to the repulsive interaction energy between the latex and the glass surfaces. It was proposed that at low ionic strength the major attachment mechanism was reversible deposition in "shallow minima", independent of the presence of humic acid. At higher ionic strengths deposition was regarded as a combination of two mechanisms: capture in the "primary well" and capture in the "secondary minimum".

Clay/Silica and Polymer Combinations

A silica or clay polyelectrolyte mix can give a synergistic improvement in NOM removal when the clay alone is effective. Thus the removal is only partial when solutions of aquatic NOM free of suspended solids are treated with long chain cationic polymers of high charge content (Bolto *et al.*, 1998, 1999), whereas in direct filtration with a floating filter, where additional finely divided solids in the form of kaolinite are present as well, much better removals are obtained (Kancharla *et al.*, 1997). This behavior parallels that reported for treatment of pond water NOM with the polyelectrolyte chitosan (Murcott and Harleman, 1993). There chitosan alone performed poorly in terms of turbidity and colour removal, but a chitosan/bentonite mix gave an improvement in colour removal. Bentonite itself was not effective. Likewise, tertiary treatment of sewage effluent with cationic polymers is successful if the turbidity is above 1NTU, being raised if necessary by the addition of clay (Isais and Al Rammah, 1999). The combination is more useful than metal salt coagulants, which produce a voluminous sludge that results in shorter filter runs.

Studies of a Bell Bay, Tasmania, water of TOC 9 mg/L treated with a cationic polyacrylamide (CPAM) or PDADMAC at neutral pH have shown that there is more effective true colour removal when 20 mg/L of fine clay or silica particles is added (Bolto *et al.*, 1998, 2001). With 7 mg/L of PDADMAC, a polymer of MW 120,000 and high CD, colour removal is increased from 44 to 60% by adding the silica. Removal of UV-absorbing compounds, used as an effective measure of NOM compounds that form disinfection by-products, is less enhanced, changing from 23 to 30% on addition of silica. The results are surprising in that silica is the only one of the additives that has a strong negative charge under the test conditions. The iso electric point (IEP) of silica is at pH 2-3, so a higher pH results in an increase in negative charge (Sposito, 1992). The addition of a cationic polymer must result in a strong bridging between the anionic forms of the two species, thus binding any free NOM to the loaded silica. It has recently been shown (Huck *et al.*, 2002) that high doses of alum are effective for *Cryptosporidium* removal, but the addition of silica had no noticeable effect.

Clays differ from silica and metal oxides in that they have two types of charged sites: a permanent negative charge arising from isomorphous substitution of metal ions within the aluminosilicate layers, which is expressed at the surface of these layers making up the clay, and a pH-variable charge on the edges of the layers due to >Al-OH groups and their interactions with protons and hydroxyl ions from the water (Rengasamy and Churchman, 1999).

For Bell Bay water of initial TOC 4-5 mg/L, PDADMAC gives a change in true colour removal from 44 to 55% when ball clay is initially present, and remains essentially unchanged at 41% in the case of bentonite. Removal of UV-absorbing compounds is raised insignificantly from 23 to 28% with ball clay, and to 26% with bentonite (Bolto *et al.*, 1998). These effects are minor, and except for the colour removal in the Bell Bay ball clay case, not much above the experimental error. They are in accordance with the higher aluminium content of ball clay, which is mainly a disordered kaolinite with sheared layers of high surface area (Grim, 1968). The best results are obtained with a final filter of 0.22 μm instead of the usual 0.45 μm: true colour removal with clay present then increases from 66 to 80%, but removal of UV-absorbing compounds is less affected, increasing only slightly from 60 to 64%. Raising the CD from 8 to 100% gives better results in treating water containing ball clay: colour removal is raised from 55 to 63% and the removal of UV-absorbing compounds from 28 to 40%. As the more highly charged CPAM is the more effective, hydrophobic interaction in a patch charge manner between polymer molecules adsorbed onto adjacent particles does not appear to occur. This contrasts with previous studies with NOM from soil where increasing the CD of CPAMs did not decrease performance, and the mechanism involving hydrophobic interaction seemed to apply (Parazak *et al.*, 1988).

A wider range of clays has been explored with reconstituted water from Hope Valley, South Australia (Bolto *et al.*, 2001). It includes palygorskite (a magnesium aluminosilicate), illite (a clay containing aluminium, silicon, iron, magnesium and potassium), Arumpo bentonite and Wyoming bentonite DMDOAB, a commercial product made from bentonite that had been treated with the long chain quaternary ammonium salt dioctadecyldimethylammonium bromide, $(C_{18}H_{37})_2NMe_2Br$. The cation exchange capacities were palygorskite 37, illite 18, and Arumpo bentonite 66-75 cmol+/kg. These contrast with kaolinite at 4-11 cmol+/kg (Churchman *et al.*, 1996). The results in Table 2.13 show that colour removal with a PDADMAC of MW 260,000 is enhanced by adding 20 mg/L of a clay, the best outcome being with palygorskite and illite.

The uptake of UV-absorbing compounds is generally not enhanced, except for the significant improvement obtained with illite. The best performance is thus with the clays of lowest cation exchange capacity, or those having the lowest negative charge. The surface treatment of Wyoming bentonite does not give rise to any advantage, even though the surface charge of the clay has been essentially neutralised by the long chain quaternary ammonium salt. The resulting hydrophobic surface is evidently detrimental to the adsorption of NOM, which is either charged, or hydrophilic if neutral.

Table 2.13

Removal of NOM from reconstituted Hope Valley water by clays (20 mg/L) and pDADMAC (4-5 mg/L) at neutral pH (Bolto et al., 2001. Reprinted from *Water Research.* 35. Bolto, B.A., Dixon, D.R., Eldridge, R.J. and King, S.J. Cationic polymer and clay or metal oxide combinations for NOM removal. 2669-2676. Copyright (2001), with permission from Elsevier.

Clay	Color Removed %	UV Absorbers Removed %
None	65	49
Palygorskite	87	44
Illite	81	68
Arumpo bentonite	71	50
Wyoming bentonite DMDOAB	76	52

This is contrary to the results obtained with montmorillonite modified with a cationic surfactant in the form of a hexadecyltrimethylammonium salt, $C_{16}H_{33}NMe_3X$, where an improvement in the uptake of humic acid from water containing 20-400 mg/L of a soil-derived humic acid was obtained (Önkal-Engin *et al.*, 2000). The result was ascribed to two effects: changes in the clay structure such that the pore diameter was increased, and a decrease in the surface area because of the pores being filled with surfactant. The organophilic nature of the modified clay resulted in a higher adsorption capacity for humic acid. However, such soil-derived material is known to be more hydrophobic than aquatic humic substances, and therefore to be more amenable to adhering to a less polar surface (Bolto *et al.*, 2002).

Hydrotalcite, a magnesium aluminium hydroxide hydrate that is an anion- rather than a cation- exchange clay, has also been tested using a water from Horsham, Victoria, Australia. With 5 mg/L of PDADMAC true colour removal is increased from 81 to 92% by adding 20 mg/L of the clay (Bolto *et al.*, 2001). The removal of UV-absorbing compounds is barely changed, from 65 to 67%. When used as an NOM removal reagent in its own right, 20 mg/L of hydrotalcite removes 15% of colour and 6% of UV absorbers. But when 333 mg/L of hydrotalcite is added to the strongly hydrophobic acids (SHA) fraction of Horsham NOM, it removes 57% of the UV-absorbing compounds, showing its preference for such non-polar material. Iron-containing versions of hydrotalcite are as effective as those based on aluminium for soil-derived NOM (Seida and Nakano, 2000). The latter work included studies of analogous synthetic hydrotalcites where the adsorption capacity increased with an increasing M^{2+}/M^{3+} ratio in the cases of Mg^{2+}/Al^{3+} and Mg^{2+}/Fe^{3+}, but this is reversed for Ca^{2+}/Fe^{3+} types. Organics removal is postulated to occur by ion exchange and by adsorption onto the hydroxy groups of the adsorbents, the latter as shown in (2). However, hydrotalcite and hydrotalcite-like adsorbents are less effective than the surfactant-modified montmorillonite in removing soil-derived humic acid from water (Önkal-Engin *et al.*, 2000).

Other polymers have been tested with ball clay for the removal of the SHA fraction of aquatic NOM obtained from Moorabool, Victoria, which comprised 34% of the total NOM (Bolto *et al.*, 2001). The TOC of the test solution was 5 mg/L, to which was added 20 mg/L of the clay. The effect of pH is shown in Table 2.14 on the following page, for PDADMAC, polyethyleneimine (PEI) and chitosan. The ionisation levels or CDs are calculated from available basicity data for PEI (Bloys Van Treslong and Staverman, 1974) and chitosan (Muzzarelli, 1977). PDADMAC, being a quaternary ammonium polymer, does not vary its charge with pH, but its performance in removing the SHA fraction is nevertheless slightly dependent on pH. Added clay has a greater effect on colour than on UV-adsorbing compounds. Neutral pH level gives the better result, which is a reflection of the nature of the colour bodies as a function of pH: they will be more charged at the higher pH and hence will complex more easily with the polymer. With added clay, the negative charge of which increases with increasing pH, the polymer will be consumed in coping with the additional solids.

The variation in cationic character of the weakly basic polymers dominates their performance: they have a higher charge at acidic pH, when the polymers alone give greater SHA removal. For PEI, the presence of clay has a significant effect at neutral pH, but has no effect under acidic conditions. For chitosan, clay addition shows no advantage. The interaction between chitosan and coloured species appears to be much greater than for PEI. This has been ascribed to hydrogen bonding of NOM to the large number of electron-rich hydroxyl groups in the polysaccharide. Such groups do not exist in PEI, which has essentially the same cationic charge as chitosan at pH 4.5. The fact that colour removal is greater at the lower pH, when there are more undissociated carboxylic acid groups in the NOM, suggests that the interactions unique to chitosan.

Table 2.14
Removal of the SHA fraction of Moorabool NOM by polymers at varying pH, with and without 20 mg/L of ball clay present (Bolto et al., 2001. Reprinted from *Water Research.* 35. Bolto, B.A., Dixon, D.R., Eldridge, R.J. and King, S.J. Cationic polymer and clay or metal oxide combinations for NOM removal. 2669-2676. Copyright (2001), with permission from Elsevier.

pH	Polymer (CD)	Optimum Doses, mg/L		Colour Removed, %		UV Absorbers Removed, %	
		No Clay	Clay	No Clay	Clay	No Clay	Clay
4.5	PDADMAC (100%)	3.0	2.0	50	80	42	44
7.0	PDADMAC (100%)	3.1	4.0	86	100	55	54
4.5	PEI (95%)	0.5	0.4	60	60	43	43
7.0	PEI (6%)	3.9	1.0	8	80	15	35
4.5	Chitosan (98%)	1.5	2.0	100	60	40	43
7.0	Chitosan (17%)	2.2	5.0	60	67	38	35

Metal Oxides

A similar mechanism holds for organic acids uptake by metal oxides, which are stronger adsorbers of NOM as there are more metal ions accessible to interact with the NOM, via its chelating ligands of carboxylic acid and phenolic groups. There are no counteracting anionic sites such as the silicate groups in clays that tend to reduce this interaction. Iron oxides in particular have received considerable attention (Parfitt *et al.*, 1977; Tipping and Cooke, 1982; Kolarik, 1983; van Velsen *et al.*, 1991; Gu *et al.*, 1995; Korshin *et al.*, 1997; Kim and Walker, 2001). The main mechanism is again seen as ligand exchange as in (1), at least for iron and aluminium species, as has been determined for example in detailed studies of the uptake of aquatic NOM by the ferric oxide haematite, α-Fe_2O_3 (Gu *et al.*, 1995). A study of the binding of Suwannee River fulvic acid by goethite (α-FeOOH) has shown that the higher MW organics are removed preferentially, and the more strongly adsorbing compounds had stronger acidic functional groups (Wang *et al.*, 1997). The binding has been described in terms of a heterogeneous surface complexation model called the ligand and charge distribution model (Filius *et al.*, 2003). Based mainly on spectroscopic evidence, surface speciation studies showed that almost all the hydroxyl groups of the adsorbed fulvic acid are involved in outer sphere complexation, and that their carboxylate groups form inner and outer sphere complexes. The carboxlate groups remaining un-coordinated are predominantly de-protonated. The carboxylate groups are important for binding the fulvic acid at low pH, whereas the hydroxyl groups are more important at high pH. Inner sphere complexes are important at low pH; outer sphere complexes are more important at high pH. The model accurately predicts the co-adsorption of protons and can simultaneously describe the concentration, pH, and salt dependency of the fulvic acid adsorption. It can be extended to humic acids, but is less suitable for describing the adsorption of larger molecules such as polyelectrolytes.

Fractionation of pond NOM has shown that with haematite the larger-sized hydrophobic fraction is preferentially adsorbed over the smaller-sized hydrophilic fraction (Gu *et al.*, 1995). This increased uptake could arise from the stronger binding that occurs because of multi-point attachment of the larger NOM species. The same effect is observed with soil-derived humic acid on haematite, where the large humic acid species actually displace the small fulvic acid molecules (Vermeer and Koopal, 1998). The overall uptake decreases with increasing pH (Mahiuddin *et al.*, 1992). In another example based on Suwannee River humic substances (Murphy *et al.*, 1990), the humic acid fraction was adsorbed by haematite more effectively than the fulvic acid one (80 and 64% removal at pH 6 respectively). An interesting comparison is made with kaolinite, which is a less potent adsorbant (47 and 17% respectively).

The removal of NOM by oxides in a river system is reported (McKnight *et al.*, 1992). An acidic, metal-enriched stream, on meeting a pristine stream, precipitated hydrous iron and aluminium oxides which resulted in a 40% decrease in the NOM content of the water, accompanied by fractionation and behaviour consistent with a ligand exchange mechanism.

A recently proposed process makes use of sand particles coated with iron oxide (Chang *et al.*, 1997). The capacity is limited by the amount of oxide that can be deposited on the media (~5% by weight), a process carried out by using $FeCl_3$ solutions and raising the pH, followed by heat treatment at 110°C for 14 h. For a typical lake NOM concentration of 3-5 mg/L, the system needed regenerating with 0.1N NaOH every few days. Regeneration is achieved by circulating the alkali through the bed for 10-20 minutes. A slow sand filter version that has not required regeneration after 16 months has also been tested (McMeen and Benjamin, 1997).

The analogous removal of residual DOC from secondary effluent by iron oxide particles has been studied (Choo and Kang, 2003). Ferrihydrite, goethite and haematite of particle size <10 μm were compared, and ferrihydrite, made by neutralising a ferric chloride solution was the most effective at taking out UV absorbing organics. At a dose of 100 mg/L the respective removals from an effluent containing 10.5 mg/L of DOC were 33, 9 and 8% at neutral pH. It was postulated that both the hydrophobicity and coordination ability of the organics played an important role, with phenolic compounds being selectively removed. This comparison is useful in that it puts a focus on the surface properties of the oxide, which are summarised in Table 2.15. The oxide with the highest IEP should be the most effective in terms of coordination, in that it will have the most positive surface charge under neutral conditions.

Table 2.15
Isoelectric points of various iron oxides

Iron Oxide	Structure	IEP at pH	Reference
Haematite	α-Fe_2O_3	8.0-8.5	Sposito, 1992
Maghaemite	γ-Fe_2O_3	6.7	Stumm and Morgan, 1981
Goethite	α-FeOOH	7.8	Stumm and Morgan, 1981
Ferrihydrite	Amorphous $Fe(OH)_3$	8.5	Stumm and Morgan, 1981
Magnetite	Fe_3O_4	6.5	Stumm and Morgan, 1981

Metal Oxide and Polymer Combinations

The effect of adding ferric oxide, magnetite, manganese dioxide, alumina or magnesium oxide in conjunction with varying doses of PDADMAC on the removal of NOM from a water obtained from Wanneroo in Western Australia has been studied (Bolto *et al.*, 2001).

Ferric oxide

With 2 g/L of ferric oxide, lowering the pH to 4.5 gave good performance, with an 85% removal of UV-absorbing compounds when 2 mg/L of polymer was added as well, versus 67% without polymer and 40% with polymer alone. 84% removal was possible with a higher polymer dose of 4 mg/L at neutral pH. With no polymer the removal was 67%. Ferric oxide has an IEP at pH 8.0-8.5 (Sposito, 1992), so the positive charge on the oxide is enhanced under acidic conditions. Similar results have been obtained for a soil-derived humic acid with PDADMAC and ferric oxide, when the polymer reversed the negative charge that exists on the oxide under alkaline conditions (pH >9), and more humic acid was adsorbed (Kim and Walker, 2001). There was an increasing positive charge produced by the polymer in the pH range 3.5-10.

Magnetite

Magnetite, a mixed ferric/ferrous oxide, performs nearly as well as ferric oxide on aquatic NOM, despite its lower IEPat pH 6.5 (Kolarik, 1983). However, the assistance of the polymer was more marked, with an 80% removal of UV-absorbing compounds with a PDADMAC dose of 5 mg/L, versus 17% with no polymer. 5 mg/L of polymer alone gave a 67% removal. Colour removal was even better, at 96% for the combination, 23% for no added polymer and 78% for polymer alone. There appeared to be an additive effect, almost as though different compounds are removed by the oxide and by the polymer. The poor result for magnetite alone relative to ferric oxide is a reflection of the lower charge on the iron. The effect with polymer assisting is of a magnitude such that, with the ease of handling rapidly reacting microparticles with magnetic properties, a large-scale process is viable (Maloney *et al.*, 1989), the magnetite being regenerated, by treating it with alkali, and reused.

Manganese dioxide

Manganese dioxide at the 2 g/L level has a negligible influence on NOM removal at neutral pH, compared to the use of polymer alone. The α-form of manganese dioxide has an IEP at pH 1.5-2.5 (Sposito, 1992), so that it is quite negatively charged under neutral conditions. The lack of uptake shows that, in contrast with silica, there is no specific interaction between the MnO_2 and the NOM, the like charges inhibiting attachment of the organics to the oxide. Work with β-MnO_2 at neutral pH has shown that soil-derived humic acids are adsorbed, as is tannic acid, but the latter to a lesser extent (Bernard *et al.*, 1997). The β-form has different surface charge properties to the α-form, and an IEP at the higher pH of 7.2 (Stumm and Morgan, 1981) or 7.3 (Healy *et al.*, 1966), so that there is less repulsion of the NOM under the treatment conditions. A 60% removal of the UV-absorbing compounds in the soil-derived humic acid sample was obtained with 5 mg/L doses of the adsorbent. Again, such NOM is known to be more hydrophobic and of higher MW than the aquatic version, so that it is removed more readily (Bolto *et al.*, 2002).

Alumina

Alumina at neutral pH gave 74% removal of UV-absorbing compounds from Wanneroo water using a polymer dose of 4 mg/L, so it is less effective than the iron oxides. The removal was only 28% in the absence of polymer. Alumina has an IEP at pH 8-9 (Sposito, 1992), and exhibits a very large change in zeta potential (from +12 to -6 mV) on exposure to humic acid from brown coal at pH 8 (Tombácz *et al.*, 1999). It does not bind the NOM as effectively as ferric oxide, indicating a greater specific interaction between the organics from Wanneroo water and iron, compared with aluminium. The superior performance achieved parallels published results on uptake of soil-derived humic acid, which wss 30% greater for ferric oxide compared to aluminium oxide (Parfitt *et al.*, 1977).

Magnesium oxide

The addition of 0.5 g/L of magnesium oxide to Wanneroo water gave excellent removal of both colour (96%) and UV-absorbing compounds (92%) which was little affected by adding polymer. With only a 50 mg/L oxide dose the respective removals were 30% and 33%. With PDADMAC at 4 mg/L removals improved to 87% and 72%. In the treatment of water from Moorabool, a higher dose of 2 g/L of the oxide was necessary to achieve complete removal of colour when the removal of UV-absorbing compounds was 68%. For a water from Horsham, Victoria, the removals were 94% for colour and 84% for the UV-absorbing compounds, with no improvement resulting from the addition of 2 mg/L of polymer. The polymer assists to a small degree in colour and UV absorber removal from the SHA fractions of both waters. Its effect was particularly pronounced for the weakly hydrophobic acids (WHA) fraction of the Moorabool water, with a doubling of the removals. There was no discernible effect of polymer on the Horsham hydrophilic neutral fraction. It appears that the treatability of the three waters with magnesium oxide was dependent on the proportion of hydrophobic compounds in the NOM, which was highest for Wanneroo at 89% (Bolto *et al.*, 1999). A dose of 0.5 g/L was needed there whereas, for Horsham and Moorabool, which had respective contents of these very highly coloured compounds of 62% and 52%, a higher dose of 2 g/L was required. The pH level of 10 arising from this treatment contributed to better performance because of the greater ionisation of the carboxylic acid and phenolic groups in the NOM.

Aluminium Hydroxide

The simplest way of making use of hydrated aluminium oxide or hydroxide particles for joint treatment with polymer is to dose the raw water with alum plus polymer, as is often done in practice, and which is well documented in the literature. Thus a dose of 2 mg/L of cationic polyamine (a epichlorohydrin/dimethylamine polymer) of low MW but high CD gives a positive effect when used in conjunction with the optimum dose of metal salt coagulant on various US waters (Amy and Chadik, 1982). Shown in Table 2.16, the results are in terms of the removal of precursors for trihalomethanes (THMs) formed on chlorination. The improvement was generally greater in the case of iron, and was observed at lower metal salt doses when the flocs would still be of opposite charge to that of the polymer.

Table 2.16
Percentage removals of THM precursors by metal salts and 2 mg/L of polyamine (adapted from Amy and Chadik, 1982)

Water Source	At Alum Optimum Dose	Alum & Polymer	At Iron Optimum Dose	Iron & Polymer
Biscayne Aquifer	53	60	48	65
Edisto River	60	74	73	82
Mississippi River	58	67	55	64
James River	26	49	30	32
Ilwaco Reservoir	68	74	77	82
Daytona Beach Aquifer	63	58	40	42

Removal of color and UV-absorbing compounds using alum is shown in Table 2.17. For Moorabool water the optimum alum dose was 35 mg/L, when 72% of the UV-absorbing compounds was removed. With 20 mg/L of alum and 0.5 mg/L of PDADMAC of MW 260,000 the same removal was possible. By increasing the polymer dose to 1.5 mg/L the alum dose could be lowered to 10 mg/L with almost the same performance, at 69% versus 72% removal. Colour removal was complete for all but the lowest alum dose situation.

In the case of Hope Valley, where the optimum alum dose was 20 mg/L, 87% removal of the UV-absorbing compounds was achievable. By using 1 mg/L of the same PDADMAC the removal was increased to 92%, and by halving the alum dose and adding 2-4 mg/L of polymer removals of 78-81% were achieved. Colour removal was improved by the polymer.

For Wanneroo the optimum alum dose was 30 mg/L, giving 90% removal of UV-absorbing compounds. Here the alum dose could be reduced to 10 mg/L by adding 1 mg/L of PDADMAC, when the same removal performance was obtained. With 3 mg/L of polymer the alum dose could be lowered to only 5 mg/L. Again, colour removal was greater, and was enhanced by polymer dosing. It is significant that the ease of NOM removal is in parallel with the content of very hydrophobic compounds in the waters; namely, Wanneroo > Hope Valley > Moorabool.

Table 2.17
Removal of colour and UV-absorbing compounds by alum/pDADMAC (Bolto et al., 2001. Reprinted from *Water Research*. 35. Bolto, B.A., Dixon, D.R., Eldridge, R.J. and King, S.J. Cationic polymer and clay or metal oxide combinations for NOM removal. 2669-2676. Copyright 2001, with permission from Elsevier)

Water	**Alum mg/L**	**PDADMAC Mg/L**	**Removal %** Color	UV Absorb
Moorabool	35	0	100	72
	20	0.5	100	72
	20	1.5	100	69
	5	1.0	86	58
	0	4.0	-	51
Hope Valley	20	0	95	87
	20	1, 2, 3 or 4	all 100	92-95
	10	0	0	0
	10	0.2	38	37
	10	0.4	84	75
	10	0.6	76	65
	10	0.8	84	71
	10	1, 2, 3 or 4	87-92	74-80
	5	4	83	69
	0	4	46	31
Wanneroo	30	0	-	90
	20	1	96	88
	10	0	86	78
	10	1 or 2	both 100	90-95
	5	3	95	85
	0	1	0	41
	0	2	62	55
	0	3	76	68

Activated Carbon

Widely used in water treatment, activated carbon is universally recognized as an effective adsorbent for natural and synthetic pollutants, such as taste and odor compounds, algal toxins, pesticides and industrial chemicals (Njam *et al.*, 1991; Owen *et al.*, 1998; Newcombe, 1999). It is also in common usage for decolorizing industrial liquors and food products, notably in the sugar industry. NOM is certainly adsorbed by activated carbon, the effectiveness ranging from poor to very good. The extent of removal depends on the type and amount of carbon and on the ionic strength, pH and DOC levels. In particular, Ca^{++} and Mg^{++} ions can enhance the adsorption of humic substances because of the formation of carbon-metal-NOM complexes that involve acidic groups on both the carbon and the NOM (Weber *et al.*, 1981). A more recent study of the calcium effect on three coal-based carbons has shown that it depends on the type of NOM (Li *et al.*, 2002). With a coal-derived humic acid the adsorption capacity increases markedly with increasing Ca^{++} dosage, but the capacity for a natural river water NOM is much less affected, while that of NOM left after conventional coagulation is not affected. It was also noted that carbons that had more pores in the size range 3-10 nm exhibited higher capacities. There is a vast literature on the topic, and only a brief incursion will be made here, with a focus on more recent publications. Laboratory and full-scale data are available for a range of waters, showing better than 90% TOC removal initially to about 50% removal after 125 days of operation (Owen *et al.*, 1998). Usually present at much higher concentrations than micropollutants that may also be present, the NOM can interfere with the adsorption of compounds such as taste and odor compounds, pesticides and algal toxins (Li *et al.*, 2003a, 2003b). Hence considerable effort has been put into understanding the mechanism of interaction of activated carbon with this range of compounds, allowing the adsorbent to be utilized more effectively for a particular application (Newcombe *et al.*, 1997).

In the absence of strong electrostatic effects adsorption occurs by a pore-filling mode in which available pore space is filled. This would apply at low pH when the NOM and carbon have little charge and the NOM would be in a shrunken state. The carbons are all negative or neutral in surface charge at neutral pH. Tests on ten activated carbons derived from coal, peat, wood, lignite or coconut found them to have, depending on the amount of DOC, up to eight times the uptake of NOM from Myponga Reservoir, South Australia, at pH 3 as occurs at pH 7, when the NOM has a substantial anionic charge. The smaller-sized NOM molecules are preferentially taken up, there being very little competition between the larger NOM fraction and 2-methylisoborneol for example (Newcombe *et al.*, 1997). For carbons made from low rank coal, the pore size distribution is one of the most important parameters in adsorption of river water NOM (Othman *et al.*, 2000). As the MW of the NOM increases the time required for diffusion into the pores increases. In one example with an aquatic NOM, the fulvic acid fraction required 17 h to come to equilibrium, whereas the humic acid fraction needed about two days (Njam *et al.*, 1991). In addition to charge and size effects, the hydrophobicity/hydrophilicity balance and the presence of NOM groups capable of hydrogen bonding to the carbon are parameters that can influence adsorption (Newcombe, 1999). Activated carbon is a more complex material than the other adsorbents considered here, so interactions at the molecular level are less well defined; the lack of a thorough understanding of the molecular structure of the NOM is extended to the realm of the adsorbent itself. The mechanism of interaction is hence less certain, and anomalies are more difficult to explain.

At the practical level, granular activated carbon (GAC) has been found to be effective for NOM and trihalomethanes (THMs) removal at a water treatment plant in Ankara (Çapar and Yetiş, 2002). Regeneration is usually performed off site, with the regenerated carbon being put to a lower grade use. The key costs are for carbon replacement or regeneration, with transport being another important factor. The main determinant for carbon life is physical strength, and coconut carbon seems to have an advantage over other types. Other ways of restoring carbon effectiveness are in use: steam treatment, washing with hot solvent or the use of chemicals, usually hot alkali (Degrémont, 1991; McLaughlin, 1995). Surfactant washing could be feasible. Biological regeneration is another possibility, but it is not yet used on an industrial scale. Prior oxidation would be an aid as it would produce reaction products that are more rapidly biodegraded. An alkali wash after oxidation may be adequate, much as has been attempted in ion exchange (Bolto *et al.*, 1978). There is a report of full-scale studies of water treatment showing thermally reactivated GAC to be more effective than virgin GAC in a six-month trial (Moore *et al.*, 2001). This was ascribed to the reactivated material having mostly mesopores and fewer micropores than the virgin GAC.

In a comparative study of ion exchange and GAC it has been observed that the carbon will remove 93% of the UV-absorbing compounds from Seine River water, compared with 90% and 93% for two anion exchange resins, 23% and 28% for two non-ionic resins, and nil for a cation exchange resin, all based on macroporous polystyrene beads (Afcharian *et al.*, 1997). However, for this surface water GAC gives a preferential adsorption of THM precursors not seen for the other adsorbents. Preferential adsorption is also seen for fluorescing fulvic acid (a Suwannee River sample) and humic acid (derived from soil) on activated carbon, relative to the non-fluorescing variants of these compounds (Schmita and Wells, 2002). This was ascribed to the innate hydrophobic nature and macromolecular size of the humic substances, rendering them amenable to adsorption on activated carbon. The effect is not surprising as fluorescence correlates with both aromatic content and molecular size. Earlier work had shown that adsorption has a significant dependence on the adsorbent to solution mass ratio; as the ratio is increased, progressively smaller organic molecules can be adsorbed (Summers and Roberts, 1988). Adsorption occurred at neutral pH, and some desorption at pH 11. The result was attributed to the carbon (Calgon F-300) having a positive surface charge at pH 7, so that anionic material was readily taken up, and a negative surface charge at the higher pH, resulting in the repulsion of the organic anions. With non-ionic polymers, large molecules adsorb in preference to small ones, and the adsorption capacity increases with MW.

A powdered activated carbon (PAC) approach is also effective if the carbon is introduced at the coagulation stage, when more than 70% of the NOM can be removed from a lake water (Ljubas *et al.*, 1999). Here the carbon is used on a throw away basis, the residue appearing in the sludge. There is an economic limit to the amount that can be dosed to the system, even though PAC is two to three times less expensive than GAC. The maximum dose is usually about 25 mg/L, depending on circumstances. Thus studies reported in a review of the area (Njam *et al.*, 1991) record that in one case a PAC dose of 100 mg/L reduced the THM formation potential (THMFP) from 53 to 25 μg/L; in another a 90% reduction could be achieved, but with a dose as high as 500 mg/L; a slight reduction of 7.5-29% was possible in another case; other workers have found PAC to be ineffective for this duty. That performance varies with the type of carbon is exemplified in one instance where 25 and 21 mg/L doses of Aqua Nuchar and Nuchar SA (both from Westvaco Corporation) gave a 4% and 16% reduction respectively of water having a THMFP of 362 μg/L. Other examples are given in Table 2.18 for joint metal salt coagulant/PAC treatment (Amy and Chadik, 1982). They clearly show that the reduction of the THMFP depends largely on the type of carbon being used and the source of the water.

Table 2.18
Removal of THM precursors by alum or iron salt coagulants in conjunction with 50 mg/L of PAC (adapted from Amy and Chadik, 1982)

Water	Coagulant Type	Dose mg/L	THMFP Reduction % Corborendum 20	THMFP Reduction % Hydrodarco HDH
Daytona Beach Aquifer	Al	30	50	28
	Fe	30	61	11
Ilwaco Reservoir	Al	2.5	69	63
	Fe	2.5	0	0
	Fe	2.5	84	82

A new nanoporous carbon has been reported that has rapid adsorption properties and a high capacity for removing soil-derived humic acids (Han *et al.*, 2003). Silica sol particles were incorporated as templates in a resorcinol polymer that was crosslinked with formaldehyde and carbonised. The product had 16 times the capacity of conventional activated carbons. An activated carbon fibre made into a cloth couples the processes of adsorption and ultrafiltration (Pignon *et al.*, 2000). It was efficient for microorganics such as phenol, but not for humic substances.

Adsorbents Summary

The calcium silicates tobermorite and xonotlite appear to be the best performers of the clays, with xonotlite removing 97% and 80% of a soil-derived humic acid from synthetic solutions. The adsorbents generally seem to have a preference for the larger sized organic anions. When the addition of clay alone has only a small effect, the presence of the particles, as would normally occur in natural waters, gives a synergistic improvement in the removal of aquatic NOM if cationic polymers are also added. With PDADMAC and clay consisting mainly of kaolinite there is a small gain, with 16% better removal of colour in one case, but there is little change in the removal of UV-absorbing species. Illite, another clay of low cation exchange capacity or low net negative surface charge, is also effective, but bentonite has little influence. In one instance hydrophobic surface treatment of bentonite was to no avail, but in another example with montmorillonite there was an improvement in the uptake of soil-derived NOM.

The best NOM removal by metal oxides is with ferric oxide, when the uptake of NOM is further enhanced by 29% at neutral pH on adding polymer. Effectiveness is in the order Fe_2O_3 > Fe_3O_4 > Al_2O_3 > MnO_2. Magnesium oxide at a much lower dose is nearly as useful as ferric oxide, but of course it raises the pH level significantly. Treatability with magnesium oxide is dependent on the proportion of hydrophobic compounds in the NOM, which was highest for Wanneroo in the cases studied. For Horsham and Moorabool a higher dose was required. Because of the ease of handling very small and hence rapidly reacting magnetite microparticles once they are magnetised, a cost-effective large-scale operating system incorporating a cationic polymer as an aid has been devised and is use in several countries. The solid reagent is regenerated and reused after alkali treatment.

Mixes of alum and polymer are a commonly used and convenient practical way of having suitably reactive particles present, with waters containing the more hydrophobic compounds being the easiest to treat. For one of the waters studied, from Wanneroo, a 67% reduction in the alum dose was possible by adding 1 mg/L of PDADMAC, to give even better performance than alum alone at the optimum dose. For Hope Valley no reduction in alum requirement occurred, but better performance was achieved with the addition of 1 mg/L of polymer, the removal of UV absorbers rising from 87% to 92%. By halving the alum dose and adding 2-4 mg/L of PDADMAC removals of 78-81% were possible. For Moorabool the alum dose could be lowered by 43% with the addition of 0.5 mg/L of polymer. Better performance is again in line with a higher content of hydrophobic material.

Adsorption on activated carbon is effective for NOM removal, taking out as much organic material as an anion exchange system, and showing a preference for THM precursor compounds in the case of surface waters. There is much experience in its use, but it is an expensive process, with regeneration and replacement of attrition losses being the crucial issues. Employed in sparing quantities in the PAC mode, it can be a practical approach for low DOC waters where it can be used on a throw away basis, particularly for seasonal problems. To target neutral hydrophilic NOM, prior treatment with a non-ionic surfactant could facilitate removal of the ensuing complex by conventional alum coagulation, precipitation by a hydrophobic polyelectrolyte, or adsorption on a non-polar adsorbent, activated carbon or a hydrophobic membrane.

COMPARISON OF NATURAL ORGANIC MATTER REMOVAL BY DIFFERENT PROCESSES

The removal of UV absorbing compounds from simulated waters using NOM from Hope Valley, Horsham and Moorabool, as well as the soil-derived Aldrich material, have been measured for a number of water treatment techniques. Data for the best performing commercial anion exchangers (Bolto e*t al.,* 2002a), cationic polyelectrolytes (Bolto *et al.,* 1999) and alum are compared in Figure 2.7 (Bolto *et al.,* 2002b). The resins used were a gel styrene resin, ResinTech SIR 22P, and a macroporous (MP) acrylic resin, Amberlite IRA 958, in batch tests at 3 g/L of resin over 16 h. The cationic polymers were a pDADMAC, Magnafloc LT 510, and a cationic polymethacrylate, Diaclear MK 4000, utilised at the optimum dosage of 5-8 mg/L and 7-8 mg/L respectively. The optimum dosage for alum was 30-60 mg/L.

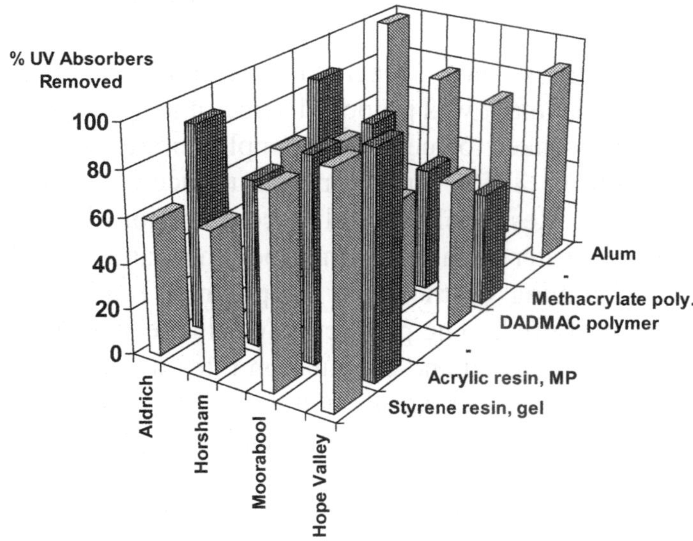

Figure 2.7 NOM removal from various waters by resins, polymers and alum (Bolto *et al.*, 2002b. Reprinted from *Water Research.* 36. Removal of THM precursors by coagulation or adsorption processes. Bolto, B.A., Dixon, D.R., Eldridge, R.J., King, S.J. and Linge, K. 5066-5073. Copyright 2002, with permission from Elsevier)

The average apparent molecular weights of the respective NOMs were 1200, 1000, 1000 and 2000 respectively (Chow, 1999), as determined by size exclusion chromatography (Chow *et al.,* 2000). It can be seen that Aldrich NOM, which had the largest NOM molecules, is the one most effectively removed by alum coagulation. It is the least readily adsorbed by the styrene resin, but the methacrylate polymer is significantly better performing than PDADMAC. In precipitation with a cationic polymer, the organic anions are better removed by the more hydrophobic reagent, but in adsorption by ion exchange the more hydrophilic resin is the better performer. Ion exchange is best for Hope Valley and Moorabool, but alum coagulation is slightly better for Horsham. The acrylic resin is generally better performing than the styrene resin on all samples, and particularly for Aldrich. The relative behaviour observed with the Moorabool and Horsham organics is not likely to be related to molecular size, but rather to the polarity of the species present in each case. Thus the cationic polymethacrylate is more effective, relative to PDADMAC, for Moorabool than for Horsham, as would be expected from the less polar overall character of Moorabool NOM.

The DOC in Seine River water has been adsorbed by ion exchange and GAC in a comparative study (Afcharian *et al.,* 1997). The GAC removed 93% of the UV-absorbing compounds, versus 90 and 93% for two anion exchange resins (Lewatit MP 500 and Lewatit S 6328A), 23 and 28% for two non-ionic resins (Lewatit VP OC 1062 and Lewatit VP 1064), and nil for a cation exchange resin (Lewatit SP 112). All the synthetic adsorbents were based on macroporous polystyrene beads, the anion exchangers being Type I strong base resins of comparable water content. However, the GAC exhibited preferential adsorption of THM precursors that the other adsorbents did not.

NATURAL ORGANIC MATTER REMOVAL BY COMBINATIONS OF PROCESSES

Resin/coagulant combinations have received attention recently. Although the combination of a trap resin with coagulant was proposed some time ago (Bursill *et al.,* 1985), the practical testing has only recently taken place with examples of resin and alum (Singer and Bilyk, 2002; Drikas *et al.,* 2003). THM formation was reduced by 60-90% and the DOC content by 64-74% respectively. Resin/polyelectrolyte combinations have been tested in both possible ways: first treating the water with resin, then polymer, and vice versa (Bolto *et al.,* 2002b). Because of potential fouling of the resin by polymer (unlikely as they have like charges), the first procedure is preferred. The production of sludge if an inorganic coagulant is employed last will then be less pronounced and coagulation rates will be much faster (Bursill *et al.,* 1985). The Horsham reconstituted water was treated with 3 g/L of ResinTech SIR 22P and a cationic polymethacrylate in the form of 2 mg/L of Magnafloc LT 510 or 4 mg/L of Diaclear MK 4000 (Table 2.19).

The UV and colour removals were substantially increased by adding polyelectrolyte after resin treatment, to essentially complete removal of UV absorbing compounds, and certainly colour. This order of addition was slightly better than when exposure to the resin was the second step. Thus the reagents were taking out quite different organic compounds. The preference of resins for the smaller compounds and of coagulants for the larger ones is well documented (Johannsen *et al.,* 1993; Lefebvre and Legube, 1993).

Table 2.19

Combination treatment of Horsham water (Bolto *et al.,* 2002b. Reprinted from *Water Research.* 36. Removal of THM precursors by coagulation or adsorption processes. Bolto, B.A., Dixon, D.R., Eldridge, R.J., King, S.J. and Linge, K. 5066-5073. Copyright 2002, with permission from Elsevier)

Reagents	% UV-Absorbing Compounds Removed	% Colour Removed
Resin SIR 22P	74	42
Resin + LT 510	93	100
MK 4000	68	79
Resin + MK 4000	91	100
MK 4000 + Resin	95	94

REMOVAL OF NATURAL ORGANIC MATTER FRACTIONS BY DIFFERENT PROCESSES

The uptake of NOM fractions of Moorabool and Horsham waters by the same acrylic and styrene resins is compared with coagulation results in Figures 2.8 and 2.9. The UV absorbing compounds in the VHA and SHA fractions of both waters were more effectively removed by the resins, which were similar in their behaviour, than alum. Alum was better in this regard than the polyelectrolytes. For Moorabool the Char fraction was more easily removed by alum than by the resins, with the polymer results lying in between. However, for Horsham the resins were better than alum. For both waters the resin and alum results were similar for the Neut fraction, the polymers being less effective for Moorabool, but about the same as the other reagents for Horsham. The accuracy of the measurements is very poor because of the low UV absorption by the neutral fractions. In both cases the methacrylate polymer was better than pDADMAC. Hence to account for the different performances overall, polarity effects must be taken into account.

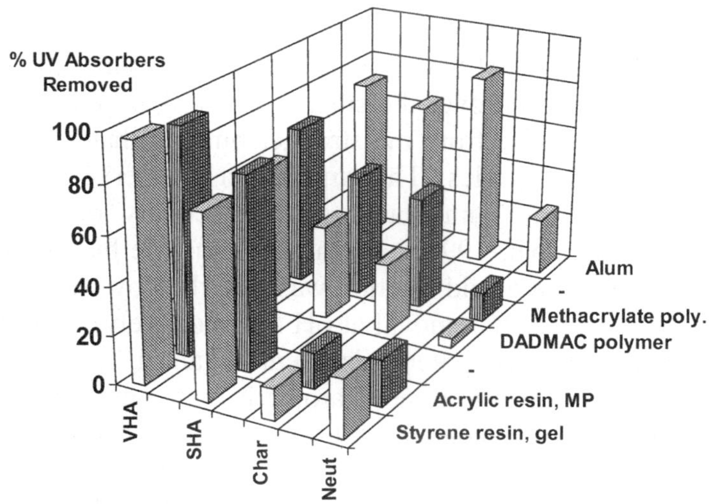

Figure 2.8 Removal of Moorabool NOM fractions by different processes (Bolto *et al.*, 2002b. Reprinted from *Water Research.* 36. Removal of THM precursors by coagulation or adsorption processes. Bolto, B.A., Dixon, D.R., Eldridge, R.J., King, S.J. and Linge, K. 5066-5073. Copyright 2002, with permission from Elsevier)

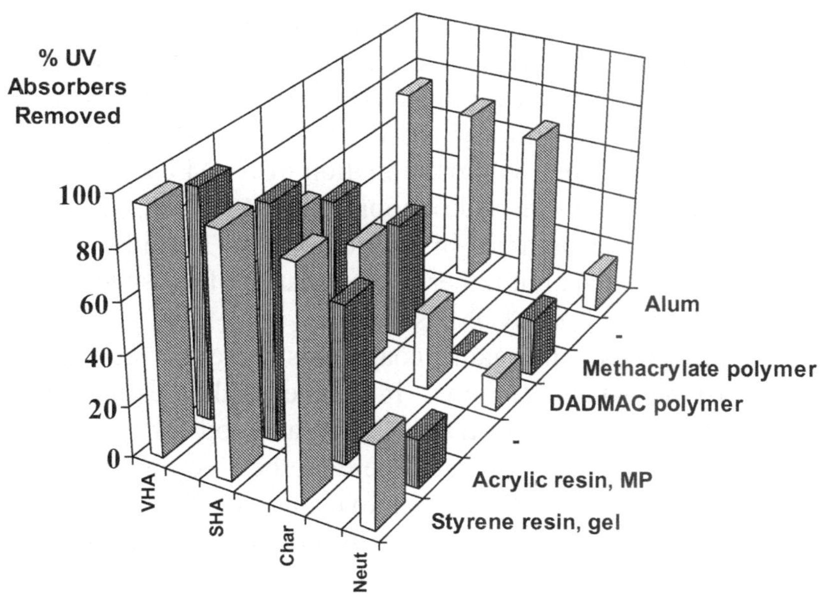

Figure 2.9 Removal of Horsham NOM fractions by different processes (Bolto *et al.*, 2002b. Reprinted from *Water Research.* 36. Removal of THM precursors by coagulation or adsorption processes. Bolto, B.A., Dixon, D.R., Eldridge, R.J., King, S.J. and Linge, K. 5066-5073. Copyright 2002, with permission from Elsevier)

Nuclear magnetic resonance studies (NMR) have shown that the Horsham NOM, despite being higher in the amounts of hydrophobic compounds present, had more polar groups overall than Moorabool NOM (Wong *et al.,* 2002). The relative uptake of NOM from the SHA fractions fits the higher carboxyl content for Horsham than Moorabool that was shown by NMR; the same holds for the Char fractions. This explains why these Horsham fractions were better removed by ion exchange than those of Moorabool.

Fractionation of pond NOM has shown that with haematite the larger-sized hydrophobic fraction is preferentially adsorbed over the smaller-sized hydrophilic fraction (Gu *et al.,* 1995). This increased uptake could arise from the stronger binding that occurs because of multi-point attachment of the larger NOM species. The same effect is observed with soil-derived humic acid on haematite, where the large humic acid species actually displace the small fulvic acid molecules (Vermeer and Koopal, 1998). In another example based on Suwannee River NOM (Murphy *et al.,* 1990), the humic acid fraction was adsorbed by haematite more effectively than the fulvic acid one (80% and 64% removal at pH 6 respectively). An interesting comparison is made with kaolinite, which is a less potent adsorbent (47 and 17% respectively).

Iron oxide has been shown to remove aromatic compounds preferentially (Korshin *et al.,* 1997). Adsorption on iron-oxide coated sand takes out acidic fractions of NOM and does not remove neutral hydrophilic compounds to any significant extent. Spectroscopic analysis by ^{13}C NMR confirmed that the adsorbed NOM was rich in aromatic and carboxylic acid moieties.

CONCLUSIONS

In coagulation, polymers such as high molecular weight polydiallyldimethylammonium chloride (PDADMAC) and cationic polyacrylamides of high charge are nearly as effective as inorganic coagulants, color removals being 86-100% of those obtained with alum. Ion exchange using the best commercially available resins designed for this purpose, a gel polystyrene and a macroporous acrylic resin, is more effective than alum treatment for two of the natural waters studied, but inferior for a third. The resins are overall superior to cationic polymers. There is an interesting contrast in behavior, with adsorption of NOM by ion exchange resins being more successful the more hydrophilic the resin, but with precipitation of NOM by a cationic polymer, organic anions being better removed by a more hydrophobic reagent. Alum was very definitely the best reagent for removal of the larger organics in Aldrich NOM. For adsorption on resins, outcomes were not related to molecular size, but rather to the polarity of the species present.

Similar tests have been applied to NOM separated into four fractions based on hydrophobic and hydrophilic properties. Inorganic coagulants were not as effective as ion exchange for the elimination of ionic NOM fractions. It was better than cationic polymers for removal of humic and fulvic acids, although PDADMAC was as good for one water. For the removal of charged compounds alum then PDADMAC were the best performers for that water. Unequivocal evidence was obtained that coagulants remove material that is not adsorbed by resins, and vice versa. A combination of coagulation with alum or a cationic polymer, and adsorption by an anion exchanger, removed essentially all of the NOM. The preference of the coagulants was for the larger, more hydrophobic molecules, and of resins for smaller highly charged hydrophilic molecules.

Activated carbon can be very useful for taking up UV-absorbing compounds, being as effective as anion exchange resins in one study. Also, it exhibits a preferential adsorption of THM precursors that ion exchangers do not. This has been ascribed to the innate hydrophobic nature and macromolecular size of the humic substances, rendering them amenable to adsorption on activated carbon. Adsorption by activated carbon can be an expensive process, with physical loss by abrasion and regeneration being the crucial issues, but employed in sparing quantities in the powdered activated carbon mode it can be a very practical approach for low DOC waters where it can be used on a throw-away basis.

The effectiveness of selected processes for NOM fraction removal is summarized in Table 2.20, where some of the data are interpolated from the general behavior of the reagent. It can be seen that very few reagents can adsorb the neutral hydrophilic fraction, and combinations of the processes considered would be of little avail for its removal. Mainly polysaccharide based, these compounds can be a strong contributor to DBPs in particular waters or at certain times of the year, as in some eutrophic supplies. Membrane technology is another option for general removal of NOM although, for some waters, the neutral hydrophilic fraction can be a fouling agent for microfiltration membranes even after coagulation.

Table 2.20
Comparison of process effectiveness

Reagent	VHA	SHA	Char	Neut	Preferred Species
Inorganic Coagulant	+++	++	++	+	High MW, charged, hydrophobic
Cationic polymer	+++	++	+		High MW, charged, hydrophobic
Anion exchanger	+++	+++	++		Charged, nearly MW independent, hydrophilic
Clay	+++	+	+	+	High MW, charged, hydrophobic
Iron oxide	+++	+	+	+	High MW, charged, hydrophobic
Activated carbon	++	+++	+	+	Low MW, hydrophobic

There is hence scope for new adsorbents that are specific for non-charged hydrophilic compounds. Chitosan, an amino polysaccharide, can function as a soluble cationic coagulant for these compounds, as can polyethyleneimine (PEI). It is postulated that they act via hydrogen bonding as in $-OH \cdots NH_2-$ (Bolto *et al.,* 1998). A charge mechanism is unlikely as the polymers have only 17 and 6% ionisation respectively at neutral pH. A crosslinked chitosan or PEI resin could therefore be efficient for removal of neutral hydrophilic compounds by adsorption.

Prior treatment with a surfactant, probably non-ionic, then removal of the ensuing complex by conventional alum coagulation, precipitation by a hydrophobic polyelectrolyte, or adsorption on a non-polar adsorbent or activated carbon or a hydrophobic membrane are possibilities (Bolto, 2002). It may be necessary to include a biological treatment step, and if necessary preceded by oxidation as in an ozone/biological activated carbon sequence. A further alternative, after oxidation of the compounds with ozone to convert them to carboxylic acids, is removal of these reaction products by ion exchange (Meltzer and Jornitz, 2002).

CHAPTER 3
MONITORING OF ELEVATED CONTENT WATERS

INTRODUCTION

Periods of high organic loading are often linked to heavy rainfall or snow melt waters and there is a need for techniques to monitor both load and character of the organics. Here we have looked at what information can be gained from monitoring different water quality parameters to see if plant operating data can be used to define the nature of NOM. The ideal monitoring tool would be an online tool that could be used as a surrogate for both NOM concentration and give a fairly accurate indication of the levels of THM expected in distribution and any likely impact on treatment.

NOM can be characterised by a wide range of analytical techniques on either bulk or fractionated samples. The literature identified a number of key water quality parameters to be included in the monitoring programme and relationships between the character and the potential to form THMs. Parameters identified include DOC (mg L^{-1}), ultraviolet absorbance at 254 nm (UV_{254} (m^{-1})) and THMs formed (μg L^{-1}). THM data is often reported as THM formation potential (THMFP) where the THMs formed are divided by the DOC of the sample to give the μg of THMs formed per mg of carbon in the sample (μg THM mg^{-1} C). The key relationships found in the literature are:

- The higher the DOC, the higher the UV_{254}
- The higher the DOC, the higher the concentration of THMs produced
- The higher the SUVA, the higher the THM-FP (SUVA, is defined as the UV absorbance of a given sample determined at 254 nm and divided by the DOC concentration of the solution (expressed in m^{-1}. L mg^{-1} C).

For example data from 88 water sources from 10 references (Allgeier and Summers 1995, Amy *et al.* 1987, Collins *et al.* 1986, Croué *et al.* 1993a, Nokes *et al.* 1999, Ratnaweera *et al.* 1999, Siddiqui *et al.* 2000, Vilge-Ritter *et al.* 1999, Volk *et al.* 2000 and White *et al.* 1997) was collated and a good linear relationship between DOC and UV_{254} was observed with an R^2 value of 0.93 (Figure 3.1).

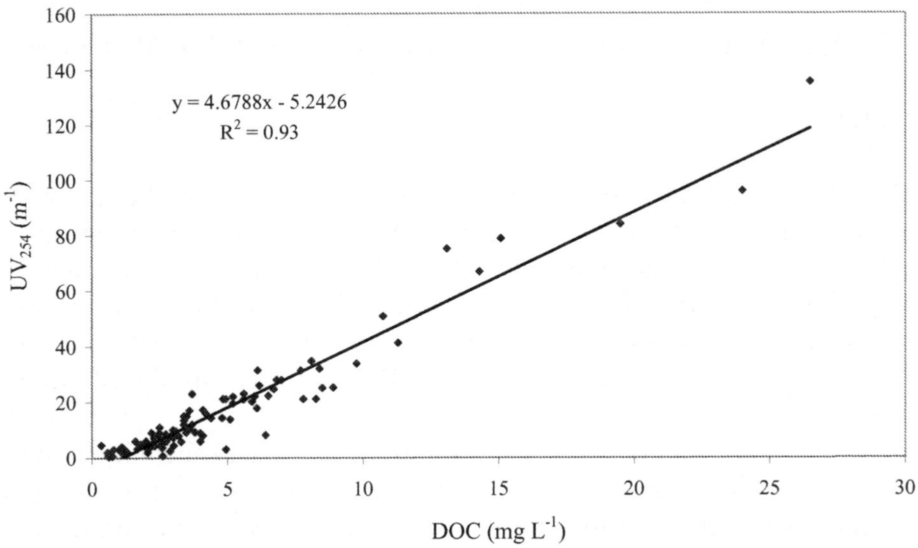

Figure 3.1 Relationship between DOC and UV_{254} for a range of 88 source waters (adapted from Allgeier and Summers 1995, Amy *et al.* 1987, Collins *et al.* 1986, Croué *et al.* 1993a, Nokes *et al.* 1999, Ratnaweera *et al.* 1999, Siddiqui *et al.* 2000, Vilge-Ritter *et al.* 1999, Volk *et al.* 2000 and White *et al.* 1997)

Goslan (2003) investigate the relationship between DOC and THMs formed using data from 55 different source waters and showed that as DOC increases so do the THMs formed (Figure 3.2). The correlation found though (R^2 = 0.55) was weak. The relationship between DOC and THMs formed was dependent on the water having a high humic component. Excellent correlations have been reported between DOC and THMFP for a single water but when waters from different sources are included, the correlations are not surprisingly as good. This is because waters from different sources tend to have different specific THM yields as determined by their particular watershed characteristics (Reckhow and Singer 1990).

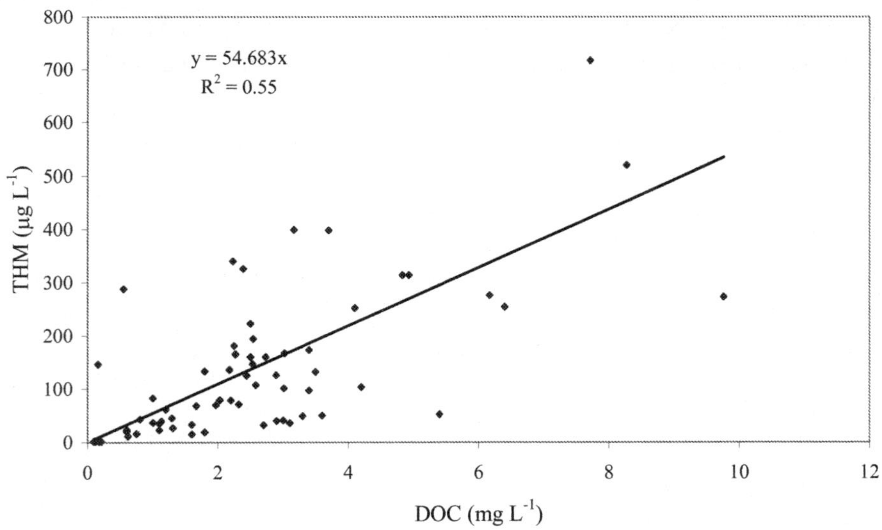

Figure 3.2 Relationship between DOC and THM for a range of 55 source waters (adapted from Goslan 2003)

A useful indicator of NOM character is SUVA (expressed in m^{-1}. L mg^{-1} C). Edzwald and Tobiason (1999) first defined guidelines for SUVA and showed a good correlation between the value and the nature of the organic material (Table 3.1). SUVA has also been shown to correlate well with the aromatic content of NOM (Krasner *et al.* 1996, Croué *et al.* 1999). The more aromatic the NOM the more hydrophobic it will be leading to a water with a higher SUVA.

Table 3.1
Guidelines for the nature of NOM and expected DOC removal by coagulation (Edzwald and Tobiason 1999)

SUVA	Composition	Coagulation	DOC Removal
>4	Mostly aquatic humics. High hydrophobicity, High MW	NOM controls. Good DOC removal	>50% for Alum >50% for Ferric
2-4	Mixture of aquatic humics and other NOM. Mixture of hydrophobic and hydrophilic NOM, mixture of MWs	NOM influences. DOC removals OK	25-50% for Alum Little greater for Ferric
<2	Mostly Non-Humics. Low hydrophobicity. Low MW	NOM has little influence. Poor DOC removal	<25% for Alum Little greater for Ferric

A plot of THM-FP against SUVA data from the literature (40 bulk waters from 6 references, Allgeier and Summers 1995, Amy *et al.* 1987, Afcharian *et al.* 1997, Collins *et al.* 1986, Kitis *et al.* 2001) was prepared (Figure 3.3) and the correlation found was poor (R^2 = 0.47). This is in contrast with Reckhow *et al.* (1990) who reported an R^2 value of 0.78. However the reported value was for separated NOM fractions at one site, not bulk waters.

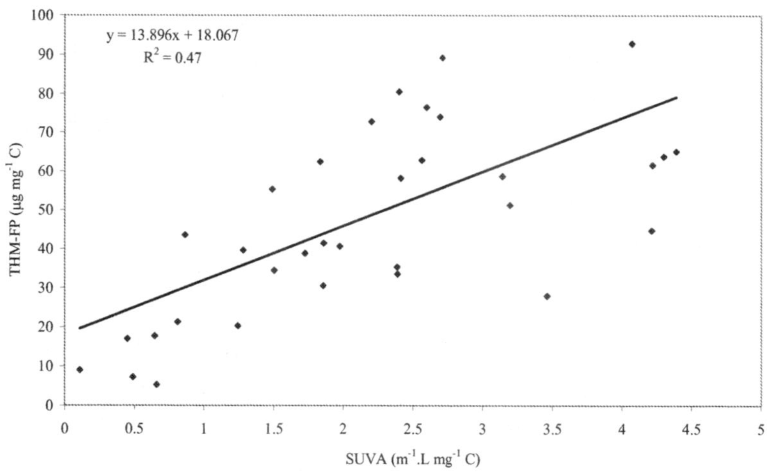

Figure 3.3 Relationship between SUVA and THM-FP for a range of 40 waters (adapted from Allgeier and Summers 1995, Amy *et al.* 1987, Afcharian *et al.* 1997, Collins *et al.* 1986, Kitis *et al.* 2001)

The literature on NOM characterisation nearly always focuses on dissolved organic matter as in practice, particulate organic matter (POM) is separated from the aqueous phase by filtration through 0.45 μm pore filters. Once the POM has been removed, the choice of which separation method to use for isolation/concentration/fractionation will depend on the objectives of the study, the equipment available and the DOC concentration of the water.

Typically the bulk organic matter can be separated on the basis of either molecular charge or size and a range of options for concentration, isolation and fractionation are outlined in Figure 3.4. The options for separating the NOM on the basis of adsorption/charge are resin fractionation, chromatography and less often, mineral adsorption. To separate the NOM on the basis of the size of the molecules, the options include membrane fractionation, flow field-flow fractionation (FFFF) and again chromatography. Chromatography is included in both categories as size exclusion chromatography columns can be used to separate organic matter on the basis of molecular size whilst normal chromatographic columns can be used to separate the organic matter on the basis of polarity.

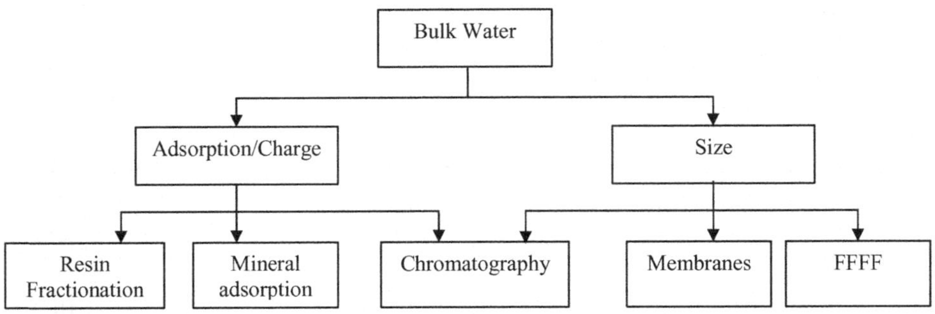

Figure 3.4 Options for isolation, concentration and fractionation of NOM

Resin fractionation

Amberlite XAD non-ionic resins have been used extensively as adsorbents for organic solutes from water over the last 30 years. These resins have a large surface area and sorption of organic acids is determined by the solute's aqueous solubility and the solution pH (Cheng 1977). At low pH, weak acids are protonated and absorbed on the resin. At high pH, weak acids are ionised and desorption is favoured. Differences in resin pore size, surface area and chemical composition result in different capacity factors for the same solute on each resin. The properties of the XAD resins are shown (Table 3.2).

Table 3.2
Resin Properties (adapted from Aiken *et al.* 1979)

XAD Resin	Pore Size (Å)	Surface Area ($m^2 g^{-1}$)	Polymer Composition	Polarity
1	200	100	Styrene divinylbenzene	None
2	90	330	Styrene divinylbenzene	None
4	50	750	Styrene divinylbenzene	None
7	80	450	Acrylic ester	Slight
8	250	140	Acrylic ester	Slight

The use of these resins was explored and validated throughout the late 1970s to the 1990s (Kim *et al.* 1976, Van Rossum and Webb 1978, Leenheer 1981, Thurman and Malcolm 1981, Blok *et al.* 1983, Kunte and Pfeiffer 1985) and reviews carried out (Aiken 1985, 1988, Giabbai *et al.* 1987, Ben-Poorat *et al.* 1987, Le Cloirec *et al.* 1990). Aiken (1988) concluded that the use of XAD resins for the isolation of humic substances from water are the most efficient and allow processing of large amounts of water. His critical review investigated all parts of the isolation procedure including initial filtration of the sample, flow characteristics of the column, sample preservation and the advantages and disadvantages of the use of nonionic macroporous sorbents. The evidence gathered in the reviews culminated in a study outlining the factors to be considered when isolating and characterising aquatic humic substances using Amberlite resins (Malcolm 1991). The main points are summarised below:

- Proper and complete cleaning of resins is vital to ensure that there are no impurities in the resin that may contaminate the isolated organic material.
- Resin columns do not have an infinite capacity for solute sorption and the sample volume and resin volume should reflect this.
- The amount of concentrated material that can be produced depends on the money and time available.
- With the correct precautions when handling and storing resins, quality results can be produced.
- Excessive resin bleed precluded the use of XAD-7 for concentration of humic substances from water.
- Testing of XAD-1 discontinued as its manufacture was stopped.
- XAD-8 exhibited the highest sorptive capacity for humic solutes and yielded 100 percent recovery of sorbed humic substances compared to XAD-2 and XAD-4 where the recovery of humic solutes ranged from 75 - 85 percent.

Once Amberlite XAD-8 was established as the most suitable resin for humic substance absorption and the validity of the fractions had been accepted, two defined methods were developed based on the methods of Leenheer (1981) and Thurman and Malcolm (1981). The first quantitatively evaluated the use of Amberlite XAD-8 and XAD-4 resins in tandem to separate hydrophobic acids and neutrals and hydrophilic acid (Malcolm and MacCarthy 1992). The second also used the XAD-8/XAD-4 two-column array to separate the organic matter into its hydrophobic and hydrophilic components (Aiken *et al.* 1992). Due to its small pore size and large surface area compared to XAD-8, XAD-4 was deemed more suitable for adsorbing non-humic organic substances (i.e. hydrophilic organic matter with a smaller molecular weight). The isolated materials are operationally defined based on the fractionation procedure. Operational definitions of the fractions obtained using the XAD-8/XAD-4 two column are as follows (Aiken *et al.* 1992):

- *Hydrophobic acid fraction* - that portion of the DOC that sorbs on a column of XAD-8 at pH 2 and are eluted at pH 13. This fraction can contain aliphatic carboxylic acids of 5-9 carbons, one- and two- ring aromatic carboxylic acids, one- and two- ring phenols and aquatic humic substances.
- *Hydrophilic acid fraction* - that portion of the DOC contained in the XAD-8 effluent at pH 2 that sorbs on a column of XAD-4 resin and are eluted at pH 13.

This fraction can contain polyfunctional organic acids and aliphatic acids of five or fewer carbons.

From these definitions it can be seen that the hydrophobic acids are large complex organic molecules that are aromatic in character whereas the hydrophilic acids are much smaller and simpler aliphatic molecules.

Size Fractionation

High Performance Liquid Chromatography (HPLC) is a method of measuring molecular size that has been extensively applied to humic and fulvic acids. HPLC uses a column for separation of species. A solvent (mobile phase) is pumped through the column (that is packed with the stationary phase) and the column effluent passes through a detector, often UV, where the eluted species are detected. HPLC methods can be based on adsorption or partitioning as well as ion-exchange. The mobile phase is pumped through the column at high pressure with pumps. HPLC is sometimes distinguished by the relative polarity of the mobile and stationary phases. Normal-phase chromatography is when the mobile phase is a non-polar solvent and the stationary phase is a highly polar material. With reverse-phase chromatography, the mobile phase is a relatively polar solvent and the stationary phase is non-polar (Sawyer *et al.* 1994).

Reverse-phase HPLC of aquatic humic substances has not produced well resolved component chromatograms. Broad trailing peaks indicative of solute-solute or solute-sorbent, direct phase interactions are produced by reverse-phase chromatography. Normal-phase chromatography with an aqueous mobile phase has been found to fractionate aquatic humic substances according to the nature of their polar functional group content (Leenheer 1985). An attempt to find the structure of aquatic fulvic acid using reverse-phase HPLC was made but only 70% of the sample was recovered from the column (Saleh *et al.* 1989). This indicates that some of the fulvic acid was adsorbed onto the column stationary phase and was not removed by the mobile phase flowing through the column.

The introduction of rigid, aqueous compatible Size Exclusion Chromatography (SEC) materials, that can be silica- or polymer-based, have made it possible to characterise the molecular size of humic substances. The separation by SEC is based on differential permeation of molecules of various size into a porous matrix. As the sample traverses the column, the small compounds permeate the matrix pores to a greater degree than the larger components and are retained longer. The elution therefore depends on molecular size, with the larger materials eluting first and the smallest last.

There are some important variables with HPSEC used for separation of NOM – the mobile phase, column packing (stationary phase) and use of standards for molecular size calibration. The recommended mobile phase is distilled water for separation of non-ionic compounds. This does not work for humic substances that are repelled by ionic sites on the stationary phase because of their negative charge in aqueous solutions. Fulvic acid will elute in the void volume of the column and the fraction of stationary phase available for penetration will be zero. The ionic repulsion can be reduced by increasing the ionic strength of the mobile phase by the addition of neutral salts or a buffer. Many types of column packing have been tried and tested but the most popular are TSK gels that are silica based (Peuravuori and Pihlaja 1997a). Another popular column material is Protein-pak that is also silica based (Zhou *et al.* 2000).

The standards are also important in order that the size of the NOM can be determined. Some studies do not use standards but use HPSEC entirely comparatively (Vuorio *et al.* 1998, Frimmel and Abbt-Braun 1999, Hesse *et al.* 1999, Specht *et al.* 2000, Makharadze *et al.* 2000). Other standards used are polystyrene sodium sulphonates, dextrans and globular proteins. The standards should have a similar structure to the NOM to obtain as close results to the actual size of NOM as possible.

The review has highlighted the use of UV_{254}, DOC/TOC, HPSEC and fractionation for monitoring and these have been used to follow seasonal changes in NOM throughout this project to investigate the information provided.

MATERIALS AND METHODS

Resin fractionation

Bulk waters were fractionated by XAD resin adsorption techniques into their hydrophobic acid (HPO-A) and hydrophilic acid (HPI-A) fractions using a method adapted from Malcolm & McCarthy (1992). A schematic of the procedure is shown in Figure 3.5. The resins used were Amberlite XAD-7HP resin and Amberlite XAD-4 resin (Rohm & Haas, PA, USA). Amberlite XAD-7HP is an acrylic ester polymer. It has been used in place of XAD-8 resin as its manufacture has been discontinued. Amberlite XAD-4 is a styrene divinylbenzene polymer. Bio-Rad AG-MP-50 resin (Bio-Rad Laboratories Ltd., Herts, UK), a non-macroporous cation exchange resin, was used to hydrogen saturate the fractions produced.

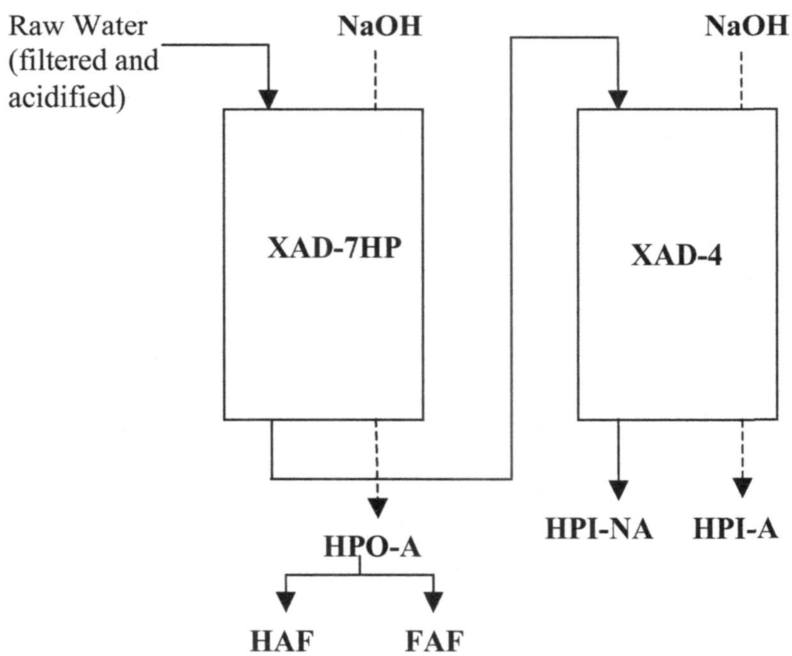

Figure 3.5 Schematic of Fractionation Process

Resin preparation

Amberlite XAD-7HP Resin - Amberlite XAD-7HP resin (1.5 L) was slurried with NaOH (0.1 M, 1.5 – 2 L) and the fines decanted off, this was then repeated. The resin was stored in Methanol (1.5 – 2 L) for 24 hours. The resin was sequentially Soxhlet extracted for 48 hours each with Methanol, Acetonitrile and Methanol again (~ 1.8 L). The resin was packed into the column and rinsed with reverse osmosis (RO) water until the column effluent DOC was < 2 mg/l. The column was rinsed with 2.5 bed volumes (BV) NaOH (0.1 M) followed by 2.5 BV HCl (0.1 M) to remove impurities.

Amberlite XAD-4 Resin - As Amberlite XAD-7HP Resin preparation. Bio-Rad AG-MP-50 Resin - The resin was Soxhlet extracted for 24 hours with Methanol. The resin was then slurried with RO water and packed into the column to be used. NH_4OH (3 M) was pumped through the column until breakthrough of ammonia was observed. The resin was hydrogen saturated by pumping four bed volumes of HCl (2 M) through the column. The resin was rinsed with RO water until the specific conductance of the column effluent was the same as the influent water. The resin was stored in Methanol. Each column was wrapped in aluminium foil to prevent algal growth.

Fractionation procedure

The following resin quantities and sample volumes are based on the fractionation of natural reservoir water. Raw inlet water (75 L) from Albert Reservoir was passed through a 1 μm pre-filter capsule and a 0.45 μm filter capsule (both from Whatman plc., Maidstone, Kent, UK) and acidified to pH 2 using HCl. All of the acidified filtered water was put through the XAD-7HP/XAD-4 column pair (resin volume was 1200 ml in each column). The effluent from both columns contained the non-acid hydrophilic fraction (HPINA).

The XAD-7HP column was back eluted with NaOH (0.1 M, 1800 ml). The eluate was acidified to pH 2 and passed through a 60 ml XAD-7HP column. This was the hydrophobic acid fraction (HPOA). The XAD-4 column was back eluted with NaOH (0.1 M, 1800 ml). The eluate was acidified to pH 2 and passed through a 60 ml XAD-4 column. This was the hydrophilic acid fraction (HPIA).

The pH of the HPOA was adjusted to 1 (± 0.2) by adding concentrated HCl (12.1 M), and left to settle for 24 hours and centrifuged. The supernatant (fulvic acid fraction – FAF) was decanted. The residue (humic acid fraction – HAF) was dissolved in the minimum required volume of NaOH (0.1 M, ~50 ml). The HAF was hydrogen saturated by passing it through a 5 ml column of Bio-Rad AG-MP-50 resin and rinsed with RO water (5 ml). The FAF was passed through a 5 ml column of Bio-Rad AG-MP-50 resin and rinsed with RO water (5 ml).

The HPI-A was pumped through a 5 ml column of Bio-Rad AG-MP-50 resin and rinsed with RO water (5 ml).

The fractions produced consisted of FAF, HAF, HPIA and HPINA. The recovery of the DOC was quantified by measuring the influent DOC of the water and the DOC and volume of the fractions produced. Unextracted material was assumed to be made up of the hydrophobic neutral fraction (HPON) as well as unrecovered material. This was not quantified.

High performance size exclusion chromatography (HPSEC)

Bulk waters were analysed at their natural DOC concentration and filtered with a 0.45 μm filter capsule (Whatman plc., Maidstone, Kent, UK) before analysis. Fractions were diluted to 1 mg L^{-1}. HPSEC was carried using an HPLC (Shimadzu VP Series, Shimadzu, Milton Keynes, UK) with UV detection set to 254 nm. The mobile phase was 0.01 M sodium acetate at a flow rate of 1 ml min^{-1}. The column was a TSK – gel G3000SW 7.5 mm (ID) × 30 cm and the guard column was TSK gel 7.5 mm (ID) × 7.5 cm (Tosoh Biosep GmbH, Stuttgart, Germany). For each sample a chromatogram of UV absorbance (arbitrary units) against time (minutes) was produced. Part way through the study a new column and guard column were purchased. This column was a BIOSEP-SEC-S3000 7.8 mm (ID) × 30 cm and the guard column was a 'Security Guard' fitted with a GFC-3000 disc 4.0 mm (ID) × 3.0 mm (Phenomenex UK, Cheshire, UK). The column packing in all columns was identical. Due mainly to the smaller guard column dimensions from Phenomenex compared with Tosoh Biosep GmbH, the retention times produced were different for each column. However the mechanism of separation and number of peaks produced by each column were the same. Calibration of the HPSEC is described in Appendix 1.

Laboratory Measurements

DOC was measured using a Shimadzu TOC-5000A analyser (Shimadzu, Milton Keynes, UK). TOC was calculated by measuring the total carbon (TC) and the inorganic carbon (IC) and subtracting the IC from the TC. The TC standard was made by dissolving 2.125 g potassium hydrogen phthalate in 1 L RO water. The IC standard was made by dissolving 1.750 g sodium hydrogen carbonate in 500 mL RO water and adding it to 2.205 g sodium carbonate in 500 mL RO water. Each standard had a concentration of 1000 mg L^{-1} and working standards were diluted to the appropriate concentration with RO water. The machine was calibrated regularly. The analyser took up to five replicates and reported an average of three given that the coefficient of variance was not greater than 2%. The analyser was recalibrated if the value of the standards were not within 10% of the expected value.

Ultraviolet (UV) absorbance at 254 nm was measured using a Jenway 6505 UV/VIS spectrophotometer (Patterson Scientific Ltd., Luton, UK). The machine was calibrated using RO water as a blank.

SEASONAL MONITORING

UV_{254}

There were a number of parameters that could have been used to monitor THM formation in distribution including TOC, DOC, UV_{254} and colour (absorbance at 400nm). Banks and Wilson (2002) investigated the use of UV_{254} on a number of water treatment works in the UK and were successful in using it as a surrogate for THM precursors. A number of on-line UV monitors are now installed on UK water treatment works and (i) linked to models predicting the formation of THMs at key points in distribution and (ii) produce UV_{254} targets at each works. Data from Albert WTW is shown below (Figure 3.6) and this shows significant seasonal changes in UV_{254} especially in the autumn months where raw water UV increases substantially following the autumn rains. This would indicate an increase in the concentration of aromatic and more hydrophobic material entering the works. No increase was observed in 2003 which was a dry year in the UK but the impact of this can be seen for 2004 with UV increasing from 3 to 9 cm^{-1}.

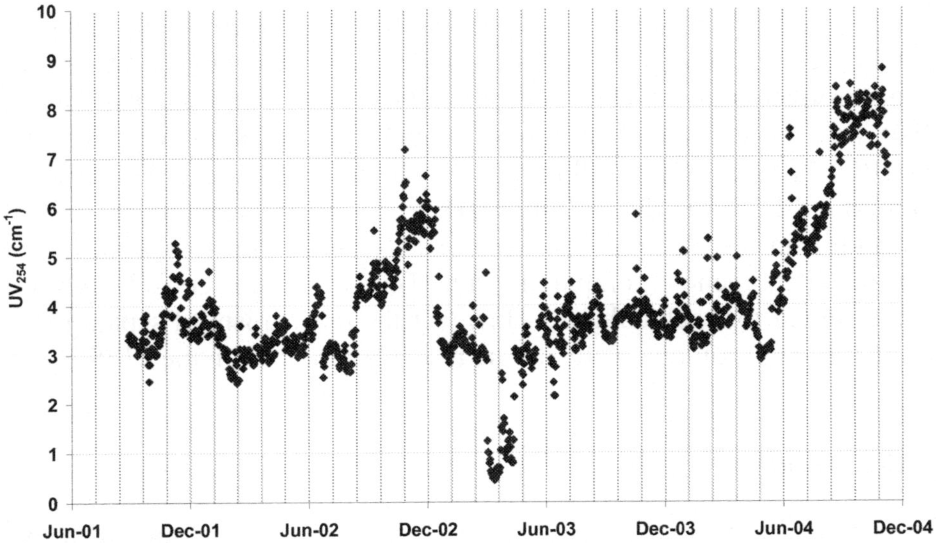

Figure 3.6 Seasonal changes in Albert raw water UV_{254}.

Organic content

At FCWTF, organic levels increased over a relatively short period during the spring snowmelt period from April to June 2004 (Figure 3.7). The DOC concentrations in the Poudre river increased from 4.4 $mg.L^{-1}$ to 7.4 $mg.L^{-1}$ DOC between the 4th and the 13th of May 2004 with a corresponding increase in TOC from 4.5 $mg.L^{-1}$ to 7.7 $mg.L^{-1}$ over the same period. Comparison of 2003 and 2004 data also reveals some annual variation with the TOC concentration in 2003 reaching a peak of 11.5 $mg.L^{-1}$. At Albert WTW, DOC concentrations also showed both seasonal and annual variation (Figure 3.7). In 2003 a low of 4.3 $mg.L^{-1}$ DOC was recorded in September, rising to 8.6 $mg.L^{-1}$ following heavy rains at the beginning of October. However, in 2004 concentrations remained above 8 $mg.L^{-1}$ for the majority of the year and rose to 13.8 $mg.L^{-1}$ following heavy rains in August 2004. Similar concentrations were recorded for Bamford WTW over the same period (September-December 2003), where the DOC concentrations increased from 4.5 $mg.L^{-1}$ to 10.2 $mg.L^{-1}$.

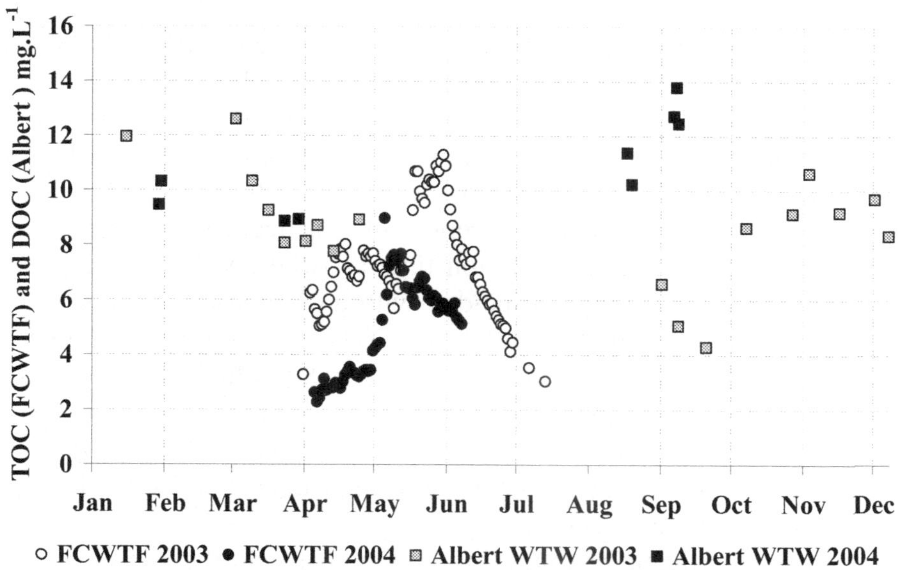

Figure 3.7 **Seasonal changes in organic content of Albert reservoir (Albert WTW) and the Poudre River (FWWTF).**

HPSEC

Raw water samples with high and low organic concentrations for both the UK (Albert and Bamford WTWs) and US sites (Poudre river at FCWTF), were analyzed using HPSEC (Figure 3.8).

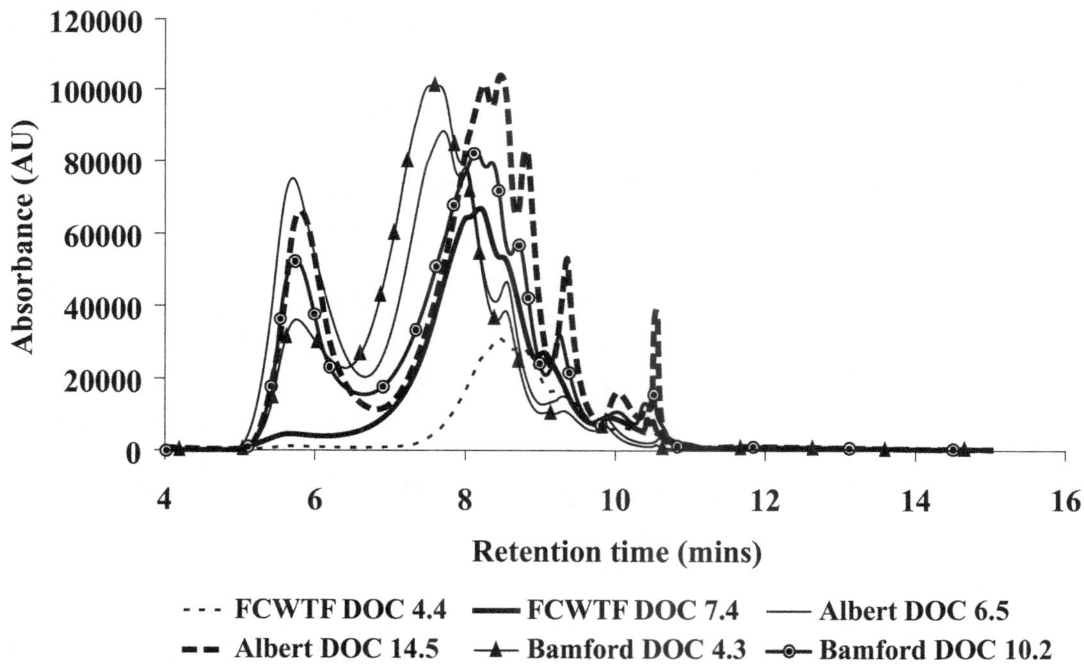

Figure 3.8 **HPSEC chromatograms for high and low DOC raw water samples at FCWTF, Albert and Bamford WTWs**

Results show an increase in the higher MW components for FCWTF when the DOC increased from 4.4 to 7.4 mg.L^{-1}. This was demonstrated by a shift in position of the main peak from an elution time of 8.5 to 8.1 minutes corresponding to an increase in size from <3 kDa to 3-4 kDa respectively (Appendix 1). Samples from both Albert and Bamford WTWs also showed a shift in HPSEC peaks when the DOC concentration increased from 4.3 and 6.5 to 10.2 and 14.5 mg.L^{-1} respectively, although in this case it represented a decrease in MW. For example, the main peak shift from 7.8 to 8.5 minutes elution time at Albert WTW corresponds to a change from 4-5 kDa to 3-4 kDa. This change coincided with an increase in peaks at the later elution times of 8.8, 9.4, 10.1 and 10.6 minutes (3-0.5 kDa). The Albert and Bamford HPSEC profiles also identified a peak at approximately 5.8 minutes (>5 kDa), which was not present with either of the FCWTF samples. This suggests that, whilst both sites experienced an increase in DOC, the species involved were different. It appears that DOC increases resulting from snow melt runoff are generally in the higher MW range whilst increases in the DOC of the upland peat catchment waters cover a much wider range of MW.

There are also significant changes in the intensity of the peaks seen in a HPSEC (Figure 3.9). Albert chromatographs exhibit seven distinct peaks (see Appendix 1 for the assigned molecular weight range). Over a 12 month period peaks 2 and 3 (corresponding to a MW greater than 4 kDa) are consistently the highest peaking in the autumn of 2002, staying consistent until February 2003 and then falling until autumn 2004. The low molecular weight peaks are significantly lower but also show some seasonal variability.

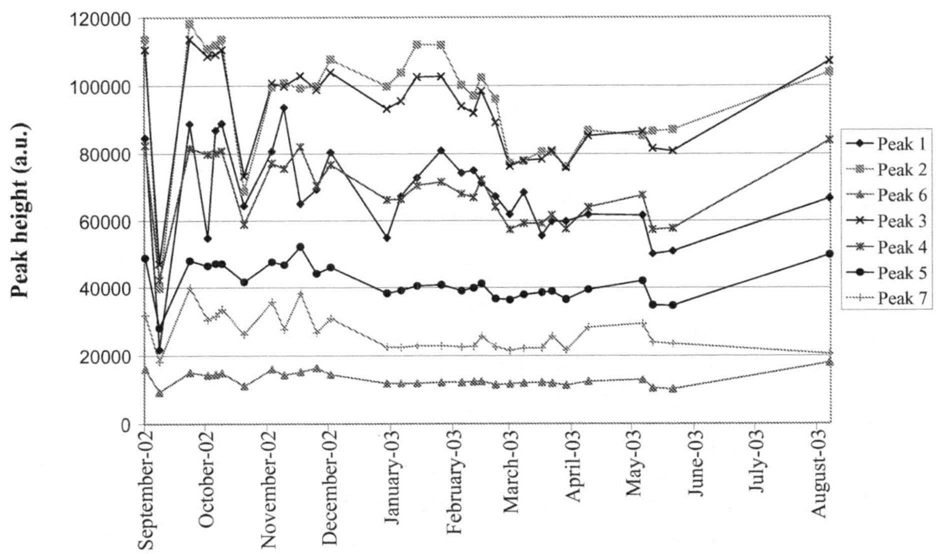

Figure 3.9 Seasonal changes in HPSEC peak height for Albert raw water.

Fractionation

The three raw waters have been fractionated periodically throughout the project. This has revealed that the increase in organics at FCWTF was due to an increase in the hydrophobic content of the water (Figure 3.10). There was no HAF present when the DOC was 4.4 mg.L^{-1}, although when the DOC increased to 7.4 mg.L^{-1}, HAF contributed 1.7 mg.L^{-1}. The total hydrophobic fraction (HAF and FAF) at FCWTF increased from 2.8 mg.L^{-1} to 5.5 mg.L^{-1} whereas the hydrophilic fractions (HPIA and HPINA) only increased from 1.6 to 1.9 mg.L^{-1}.

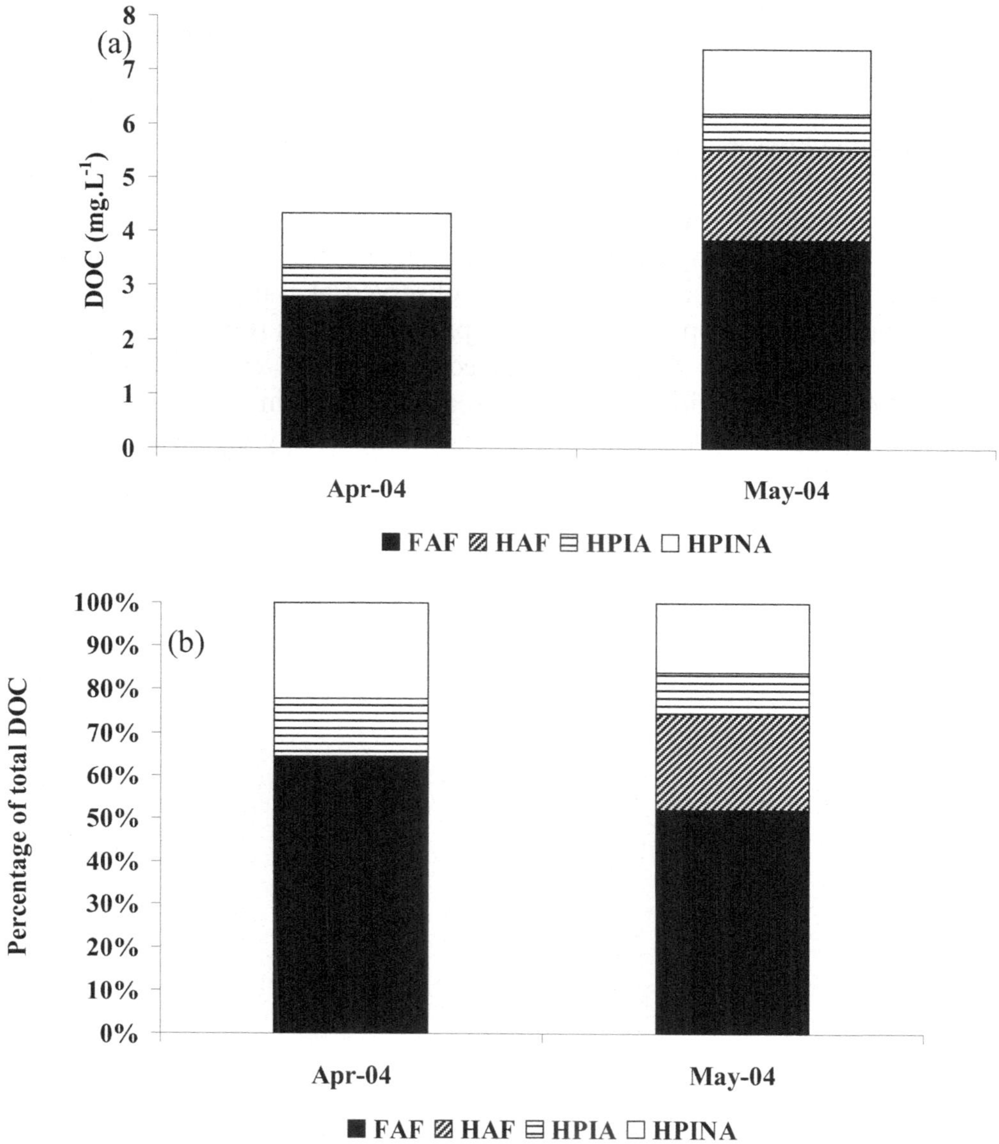

Figure 3.10 Raw water NOM fraction make-up at FCWTF expressed as (a) total DOC concentrations and b) as a percentage of the total DOC concentration.

Fractionation of the raw water at Albert WTW during the experimental periods (Figure 3.11), confirmed that up to 80% of the organic matter was associated with the fulvic and humic acid fractions. Throughout the majority of the sampling period, excluding the August 2004 sample, the DOC concentration of the hydrophilic (combined) fraction remained relatively stable at 2.7 ± 0.8 mg.L^{-1}. In contrast, the hydrophobic fractions, HAF and FAF, demonstrated slightly more variability, in terms of total DOC, with a concentration throughout the sampling period of 2.6 ± 1.3 mg.L^{-1} and 3.4 ± 1.0 mg.L^{-1} respectively. The variation in the hydrophobic fraction concentrations were the major components of change in the fractional make up of the water. To illustrate, in October 2003, the relative contribution of each fraction to the total DOC of the water was 27%, 31% and 41% for the FAF, HAF and hydrophilic fractions respectively. Whereas, in January 2004, the relative fraction amounts were 57% for FAF, 14% for HAF and 25% for the hydrophilic fractions; demonstrating a greater FAF contribution and a decreased HAF concentration, although still resulting in an overall increase from the combined hydrophobic fractions.

A focused fractionation study on Albert WTW during the autumn of 2003 revealed an increase in both the hydrophobic and hydrophilic NOM fractions coinciding with the first period of heavy rainfall at the end of the summer. The hydrophobic fractions (HAF and FAF), increased 20%, from a combined total of 4.50 $mg.L^{-1}$ to 5.38 $mg.L^{-1}$ during the period of September to October 2003. Whereas the hydrophilic fractions (HPIA and HPINA) increased 80% from 2.10 $mg.L^{-1}$ to 3.82 $mg.L^{-1}$. Similar increases were observed at Bamford (Figure 3.12), at 44% and 75% for the combined hydrophobic and hydrophilic fractions respectively. The most significant increase in the hydrophobic fractions occurred towards the end of the sampling period. For instance, the FAF concentration at Albert increased 80% from 2.51 $mg.L^{-1}$ to 4.50 $mg.L^{-1}$ from October to December, when the total DOC concentration only increased from 9.2 $mg.L^{-1}$ to 9.7 $mg.L^{-1}$. Again, similar results were observed at Bamford, with a 97% increase in FAF over the same period.

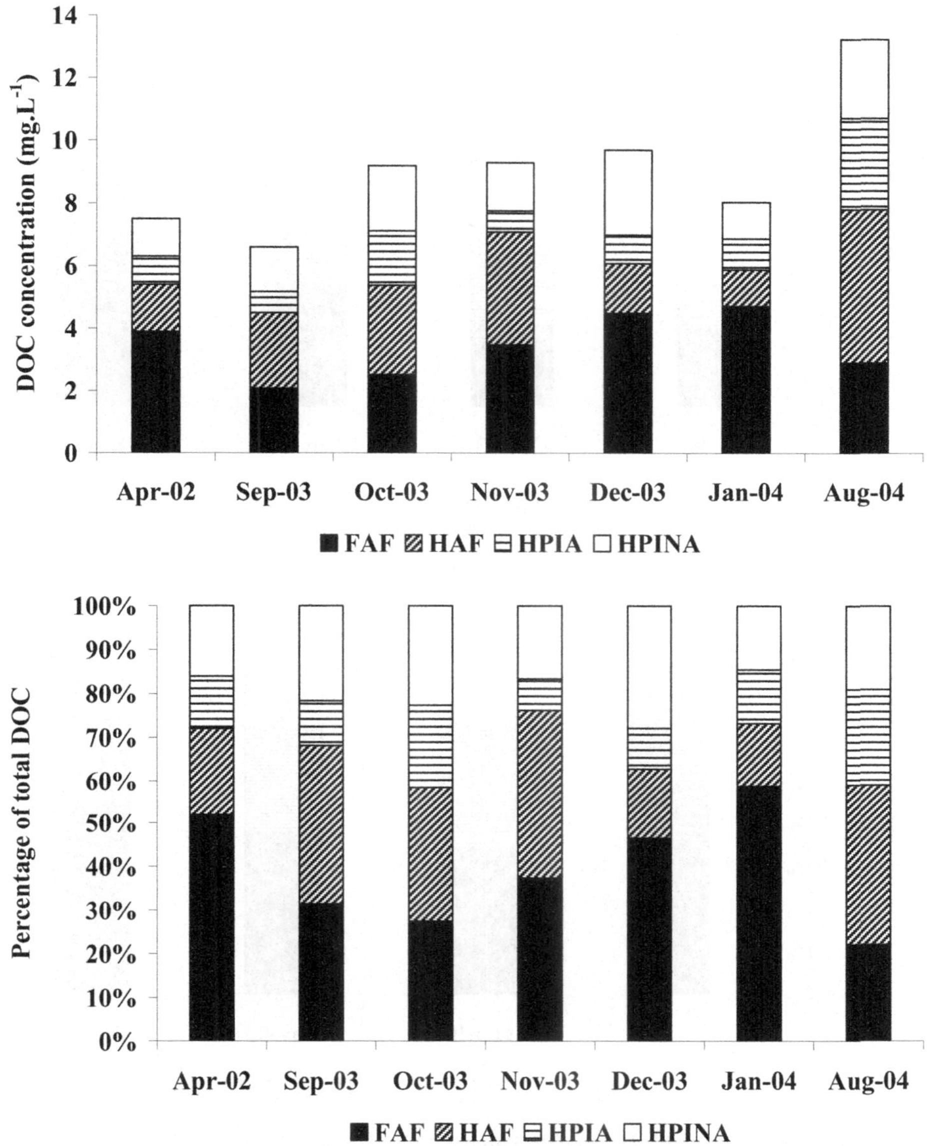

Figure 3.11 Raw water NOM fraction make-up at Albert WTW expressed as (a) total DOC concentrations and (b) as a percentage of the total DOC concentration.

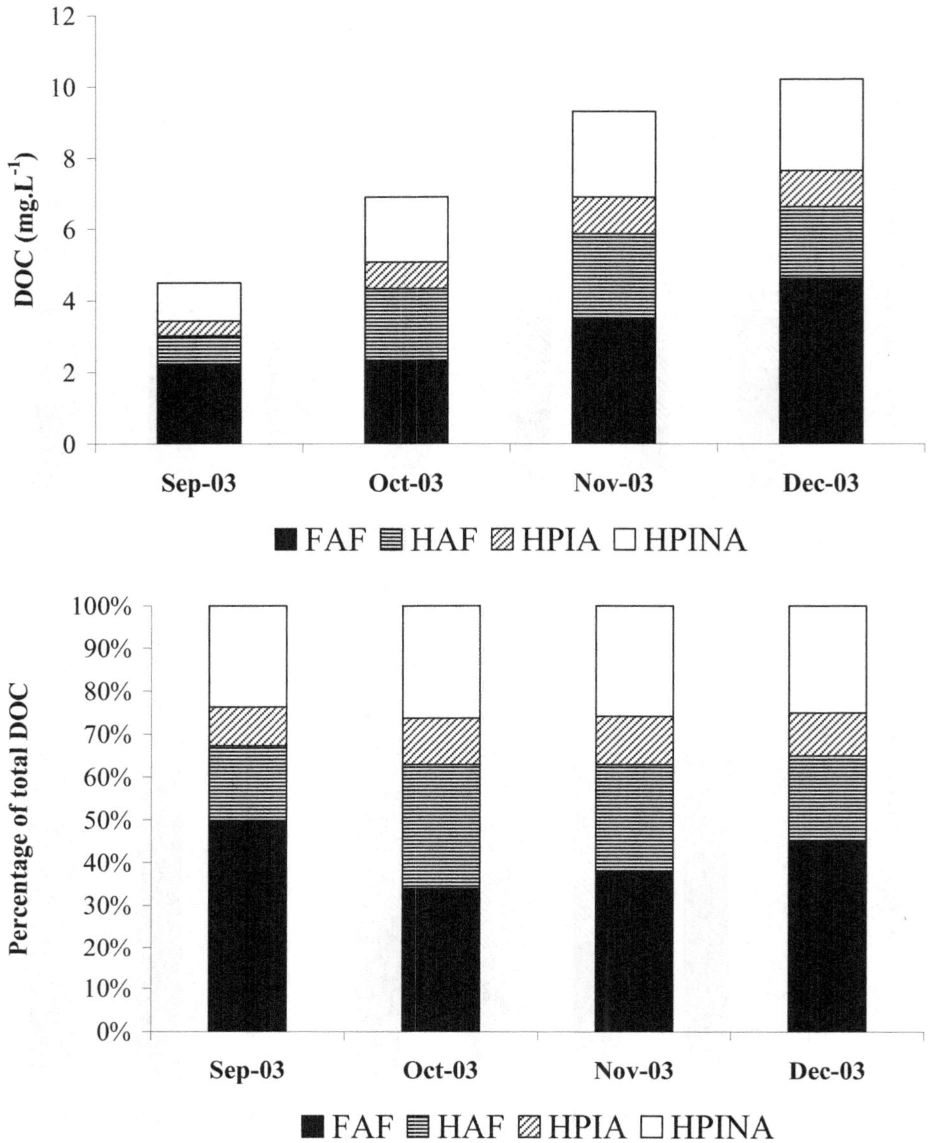

Figure 3.12 Raw water NOM fraction make-up at Bamford WTW expressed as (a) total DOC concentrations and (b) as a percentage of the total DOC concentration.

It is worth noting that increases in organics take place in sequence. The increase is initially in the hydrophilic fractions followed, after a time delay, by an increase in hydrophobic fractions. The increase in hydrophobic fractions occurs when the total organic content has become more stable so that the change, which can have a significant impact on coagulant dose requirements, may well go unnoticed. This effect was also noticed following the first heavy rainfall after summer 2004. Comparisons between January and August fractionation data at Albert WTW confirmed conclusions drawn from HPSEC plots previously. This revealed not only a 30% increase in hydrophobic material from 5.9 mg.L^{-1} to 7.8mg.L^{-1}, but also a 150% increase in hydrophilic material from 2.2 to 5.4 mg.L^{-1}, as the total DOC increased from 8.3 to 13.3 mg.L^{-1}.

Comparison with published data reveals similar levels and variation in DOC across different seasons for river and lake systems from both Europe (Ratnaweera *et al.*, 1999) and the US (Aiken *et al.*, 2004). For instance, the DOC from an intake from the Santa Ana River varied between a low of 2.2 $mg.L^{-1}$ and a high of 9 $mg.L^{-1}$ between June 2000 and May 2002 (Aiken *et al.*, 2004). Fractionation of waters from different sources and locations reports much greater variability than observed in the current study. Analysis from around 100 different surface source waters indicated that typically about 56% of the material is hydrophobic in character (Leenheer and Huffman, 1976). Similarly Croué *et al.*, (1993a), reported that water taken from the Apremont reservoir was 51 % hydrophobic and is higher than the 38 % reported for the Colorado river in September 2000 (Aiken *et al.*, 2004). The range of values (57-79%) in the current study are similar to values reported for highly colored waters in Canada, Scandinavia and Russia where the hydrophobic content typically varies between 60 and 80 % of the total DOC (Malcolm, 1985). The observed differences are consistent with the view that as the DOC of a water increases the majority of the additional organic matter is likely to be hydrophobic in character (Malcolm, 1985) and hence high DOC waters tend to be more hydrophobic in character.

With regards to NOM fraction make-up during the period of elevated organics, FCWTF raw water contained a higher proportion of hydrophobic NOM (74% in May 2004 during runoff) when compared to Albert reservoir samples (59% in August 2004). Carlson *et al.* (1994) also observed an increase in the hydrophobic content of the water at FCWTF, from 33% to 50% during snowmelt run-off. The total TOC concentration was reported to be significantly lower in 1994, at approximately 4.8 $mg.L^{-1}$ compared to 7.7 $mg.L^{-1}$ during the 2004 run-off period. Carlson *et al.* (1996) state that NOM due to snow run-off is also mainly composed of high MW hydrophobic material, such as humic and fulvic acids. Therefore elevated organic levels are likely to coincide with a higher hydrophobic proportion of NOM (Malcolm, 1985). At both the UK sites this was not the case as the increased organic concentrations were due to an increase in both the hydrophobic and hydrophilic content of the water. Scott *et al.* (1998), looked at the seasonal variations in NOM obtained from a UK upland peat catchment system. They observed variations in the hydrophilic content of the water of between 20 and 80%, with the higher proportions occurring during the summer months. The observed variations throughout the year are often explained in relation to a microbially driven mechanism of DOC release (Scott *et al.*, 2001). The microbes within the soil matrices are more active during warmer, relative drier period in the summer / early autumn but the impact of this is not seen at the treatment works until sufficient rainfall/snow melt occurs to flush the released materials into the intake system. After which time the soil becomes colder and more water laden and aerobic microbial activity significantly decreases. The increases in hydrophilic content following the initial periods of heavy rain during the late summer and autumn were attributed to its relative ease of dissolution, whereas the hydrophobic material is leached more slowly.

Overall each of the parameters evaluated (UV_{254}, DOC/TOC, HPSEC and fractionation) provide useful information on either the load or the character of NOM entering a water treatment works. UV_{254} is the only real option for on-line monitoring and it has been successfully shown to identify seasonal trends and can be related to THM formation. DOC is primarily an off line measurement of organic load and much of the same information can be found with UV_{254} with out the delay in measurement. HPSEC is a very useful technique for monitoring changes in the profile of organics entering and although it requires more specialised equipment once famililar with the interpretation it can be a powerful tool for optimising treatment processes. Here fractionation has been able to show how increases in organics from the two sites vary considerably in both composition and character. Snowmelt contributes an increase mainly in hydrophobic fractions, whilst the UK upland sites show an increase in both hydrophobic and hydrophilic fractions. At the UK sites, hydrophilic fractions increase following the first flush of heavy rainfall at the end of the summer whilst the increase in hydrophobic fractions is delayed and may, therefore, go unnoticed except by UV_{254} monitoring.

CHAPTER 4
TREATMENT PROCESSES

OVERVIEW

This Chapter presents the results of investigations at both bench scale and pilot plant scale into the optimization of conventional coagulation and flocculation processes for the treatment of waters with elevated organic content. It also reviews the results of bench scale trials into the use of the MIEX® ion exchange process.

TREATMENT BY COAGULATION

Optimum coagulation is defined as the coagulation conditions of dosage and pH that achieve a minimum particle count and turbidity, maximum TOC, DOC and DBP removal, and minimum coagulant residual (Edzwald and Tobiason, 1999). Given that the coagulation of NOM is predominately charge controlled for both soluble and colloidal material then the electrical character of the NOM is likely to be a key parameter in describing the process. Characterisation in this regard relates to either surface potential (expressed as zeta potential or electrophoretic mobility) or the total surface charge (measured as colloidal charge or charge density) (Ratnaweera *et al.*, 1999). The application of charge measurement for the diagnosis and control of water and wastewater treatment processes is not new, although the basic understanding of how charge controls performance has not changed considerably. Zeta potential indicates the specific surface potential of the colloidal system and is commonly related to the stability of the colloidal materials under any given set of water characteristics. However, it does not necessarily provide a proportional indication of the amount of coagulants required for destabilization. In contrast, the total surface charge of the raw water is proportional to the theoretical coagulant demand providing a charge neutralisation mechanism prevails.

The mechanism of NOM coagulation is thought to be charge neutralisation for colloidal matter and charge complexation for soluble compounds. Additional organic removal also occurs due to adsorption of NOM onto newly formed NOM-coagulant precipitates (Randtke, 1988). Consequently, optimisation of the coagulation process occurs under acidic conditions between the iso electric point (IEP) of the coagulant and the NOM. This is typically around pH 4.5-5 for iron and pH 5-6 for aluminum based coagulants (Amirtharajah and O'Melia, 1990). In reality, optimized charge conditions occur over a range of values, rather than specifically at the IEP of the system, such that operational zeta potential windows exist within which residuals are minimised. Several studies have reported conditions favourable for particle removal with zeta potential values ranging between +10 and –10 mV (McCurdy et al., 2004; Tseng et al., 2000, Carlson et al., 1996). Ratnaweera et al. (1999), suggest that colloid destabilisation occurs before complete neutralisation of surface charge, hence the optimum zeta potential may not necessarily be zero.

Raw water characteristics are known to vary both temporally and spatially, often experiencing rapid changes during periods of elevated organics (Carlson *et al.,* 1994; Owen et al., 1993). Changes in land management, such as the disturbance of peat or vegetation damage, can lead to increased decomposition and the production of loosely bound organic material (Worrall and Burt; 2004 Niskavaara et al., 1997). The accumulated organics are then flushed into the aquatic environment during the first rainfall following a dry period (Hurst et al., 2004; Goslan et al., 2002). Mountain regions are also affected by a first flush phenomenon, when snowmelt waters percolate though undisturbed forest floors containing organic material, resulting in high DOC concentrations. Water treatment is often controlled on bulk water parameters such as UV_{254} absorbance or residual turbidity levels. Previous research has shown that such an approach may fail to detect changes in the NOM composition and character of the water, and the change in charge and associated coagulant demand.

This section presents the results of investigations at both bench scale and pilot plant scale into the optimization of conventional coagulation and flocculation processes and extensive use is made of zeta potential measurements to establish whether this could be used as an optimization tool.

Raw water sources

The raw waters used during the coagulation experiments were taken from both UK and US water sources experiencing elevated organic concentrations. The UK sites were two moorland WTWs in the north of England; Albert WTW, situated in Halifax, and Bamford WTW situated in Derbyshire. These are located in moorland peat catchment areas producing colored water throughout the year, although still subject to rapid changes and increases in organic levels following periods of heavy rain. Table 4.1 summarizes the characteristics of the two sources, which are typically of low turbidity (1.4-10.9 NTU), low alkalinity (20 $mg.L^{-1}$ $CaCO_3$) and high DOC (4.5-10.2 $mg.L^{-1}$). The third water source was taken from the Poudre river intake at Fort Collins water treatment plant facility (FCWTF), Colorado, USA. This is also a low turbidity (0.4-8.3 NTU), low alkalinity (11-35 $mg.L^{-1}$ $CaCO_3$) water. Fort Collins is situated along the front range of the Rocky mountains and is only subject to rapidly changing organic levels during the spring snowmelt runoff.

Table 4.1
Raw water characteristics during sampling periods.

Date	Source	Scale	UV_{254} (absorbance)	DOC ($mg.L^{-1}$)	Turbidity (NTU)	pH	SUVA ($m^{-1}.L.mg^{-1}$)
Apr- 02	Albert	Bench	38.1	7.5	No data	5.9	5.1
Sept-Dec 03	Albert	Bench	45.0-63.6	6.6-9.7	5.9-10.9	5.5-6.5	4.6-7.4
Sept-Dec 03	Bamford	Bench	38.5-45.9	4.5-10.2	1.4-4.2	5.5-6.5	3.6-8.7
Jan-04	Albert	Pilot	37.6-43.6	8.7-10.6	5.4-5.5	6.1-6.2	4.1-4.3
Mar-04	Albert	Pilot	36.9-39.5	6.5-9.4	5.4-5.5	5.9-6.0	4.2-5.7
Apr-Jun 04	FCWTF	Pilot	9.9-22.5	2.0-7.7	0.5-8.3	7.1-7.6	2.1-3.2
Aug-Sept 04	Albert	Pilot	50.4-52.5	11.4-13.8	5.1-5.3	6.0-6.1	3.8-4.4

Analytical Equipment

Experimental work was undertaken in the UK at Cranfield University and in the US at the FCWTF, and therefore the equipment used to undertake the analysis did vary slightly, although wherever possible, similar instruments were employed.

- The turbidity of the samples was measured in the UK using a HACH2100 turbidity meter (Camlab, UK) and a HACH2100AN bench top turbidimeter at FCWTF.
- DOC was measured at Cranfield using a Shimadzu TOC-5000A analyzer (Milton Keynes, UK). At FCWTF, the TOC and DOC was measured using a 800 Sievers TOC analyzer and standard method 5310C.
- UV absorbance at a wavelength of 254nm was measured at Cranfield using a Jenway 6505 UV/Vis spectrophotometer (Patterson Scientific Ltd., Luton, UK). At FCWTF, this was measured using a HACHDR/4000U spectrophotometer.
- High performance size exclusion chromatography (HPSEC) analysis was performed using a high performance liquid chromatography (HPLC) device (Shimadzu VP series, Shimadzu, Milton Keynes, UK). The method used was a TSK gel GW3000SW 7.5mm internal diameter, 30cm in length. The guard column was also TSK, 7.5mm internal diameter, 7.5cm in length (Tosoh Biosep GmbH, Stuttgart, Germany). UV detection was set at a wavelength of 254nm with a 0.01M sodium acetate mobile phase with a flowrate of $1 ml.min^{-1}$. From the results for each sample, a chromatogram, of UV_{254} absorbance units against time in minutes, was produced. The HPSEC calibration was developed at Cranfield University (Goslan, 2003), allowing comparison with previous work on similar water sources carried out at Cranfield.
- The zeta potential was measured at Cranfield with a zetasizer 2000HSA and with a zetasizer 2000 at FCWTF (Malvern Instruments, US). The surface charge was measured with the zetasizer at 20°C. The zetasizer yields estimates and standard deviations of the zeta potential directly, and all measurements were conducted in triplicate. The general accuracy of the zetasizer was determined using standard solutions (Malvern Instruments, UK). The zeta potential was initially measured at varying stages of the jar test procedure from the end of the rapid mix to the final settlement, and was found to show no significant variation.

Raw Water

Charge Density

The charge density of both the fractions and the raw water samples was determined by an adjusted form of the method used by Kam and Gregory (2001), which monitored the change in charge as a function of PDADMAC dose. This was shown to produce a clearer end point determination than traditional ion titration methods. The PDADMAC, with a MW of 100,000-200,000, was obtained from Sigma-Aldrich as a 20% by weight aqueous solution. Varying amounts of a 0.1% PDADMAC solution were added to a 100 mL solution of deionized water with 2.5 mg of the appropriate NOM fraction adjusted to pH 7 with NaOH, and the zeta potential measured until the point of zero charge or iso electric point (IEP) had been established. The process was then repeated twice. The hydrophilic fractions were analyzed at lower concentrations and with a more dilute PDADMAC concentration due to availability issues. The charge density of the PDADMAC was determined to be 6.2 $meq.g^{-1}$ from the manufacturers information, and was similar to the value reported by Kam and Gregory (1999), of 5.78 $meq.g^{-1}$, determined experimentally by colloid titration. This value (6.2 $meq.g^{-1}$), was then used to calculate the charge density of the NOM in terms of $meq.g^{-1}$. With regards to the raw water samples, the same method was followed using one liter of raw water, of a known DOC concentration, and the charge density determined in terms of both meq.g DOC^{-1} and $meq.L^{-1}$.

Charge density measurements of the raw water samples were shown to increase as the organic concentrations increased. For example, in relation to water fractionated from samples collected from FCWTF, The charge density of the Poudre river raw water increased from 0.009 $meq.L^{-1}$ to 0.029 $meq.L^{-1}$, compared to 0.023 $meq.L^{-1}$ at Albert WTW in January (Table 4.2). This corresponds to an increase from 2.0 meq.g DOC^{-1} to 3.9 meq.g DOC^{-1} for the Poudre river. Collins Green (1997) also calculated the colloid charge capacity (CCC) of blended Horsetooth reservoir and Poudre river water at FCWTF, using a fluorescence polarization colloid titration method. The CCC was found to increase from 2 to 2.5 meq.g DOC^{-1}, during the 1997 spring snowmelt run-off period. This value is lower than calculated here for the 2004 run-off period, which was solely for Poudre river water, According to Carlson *et al.* (1994), the blended water studied by Collins Green (1997) has been shown to contain a lower hydrophobic content which would explain the reduced charge. Collins *et al.* (1986), also achieved similar values when using a potentiometric titration method to determine the carboxylic acidity of the hydrophobic organic matter for a number of raw waters and fulvic acid components. For instance, charges of 0.035 $meq.L^{-1}$ and 0.030 $meq.L^{-1}$ were determined for the Grasse River and Floridan aquifer samples respectively.

Table 4.2
Charge density of raw water samples and NOM fractions

Sample date	Location	HAF	FAF	HPIA	HPINA	Raw water
Apr-02	Albert	6.8	4.2	0.06	Nm	-
Jan-04	Albert	4.3	4.5	1.5	Nm	2.7 (0.023 $meq.L^{-1}$)
Apr-04	FCWTF	-	-	-	Nm	2.7 (0.009 $meq.L^{-1}$)
May-04	FCWTF	-	3.7	-	Nm	3.8 (0.029 $meq.L^{-1}$)
Aug-04	Albert	3.7	3.5	1.2	Nm	3.7 (0.049 $meq.L^{-1}$)

Note- values are in meq.g DOC^{-1}.
Nm- not measurable

NOM fractions

Charge density measurements of the raw water fractions revealed the majority of the total colloidal charge to exist in the hydrophobic fractions. The charge density for the FAF fractions of both FCWTF and Albert WTW are similar, between 3.5 meq.g DOC^{-1} and 4.5 meq.g DOC^{-1} (Table 4.3). Scott *et al.* (2001) reported values of carboxylic acidity for humic and fulvic acids of 2.1 $meq.g^{-1}$ and 2.9 $meq.g^{-1}$ respectively. Whereas Kam and Gregory (2001) reported slightly higher values of 5-5.6 meq.g DOC^{-1} for commercial humic acid. Similar values have been reported elsewhere with humic and fulvic waters typically exhibiting charge densities between 5-15 meq.g DOC^{-1} (Table 4.3).

Table 4.3
Previously determined humic substance charge densities

Substance	Method	Charge density (meq.g^{-1})	Reference
Fulvic acid	Potentiometric titration	13.5 ±1.4 (hydrophobic carboxylic acidity)	Collins *et al.*, 1986.
Grasse River water	Potentiometric titration	8.0 ±1.2 (hydrophobic carboxylic acidity)	Collins *et al.*, 1986.
Humic and fulvic acids	pH titration	5-10	Tipping, 1993
Fulvic acids	Colloid titration	10-15	Edzwald, 1993
Humic acids	Colloid titration	5-10	Edzwald, 1993
Organic carbon	Alkametric titration	12 (at max charge density) 6.1 at pH 4	Tiller and O'Melia, 1993
Humic acid	Colloid titration	5-5.6	Kam and Gregory, 2001
Humic acid	Colloid titration	1.5 ± 0.04	Mikkelsen, 2003

The charge density of the hydrophilic fractions was lower, at 0.06-1.5 meq.g DOC^{-1} for instance, for the HPIA fraction. In all cases, the charge density for the non acid fraction (HPINA) was below the level of detection. Edzwald (1993) reported similar findings for both hydrophilic bases and neutrals and found that the acid fraction was at least an order of magnitude smaller than the hydrophobic fractions.

There is a significant difference between the charge density result for the April 2002 HPIA, at 0.06 meq.g DOC^{-1}, and both the January and August 2004 HPIA samples at 1.5 and 1.2 meq.g DOC^{-1} respectively, all obtained from the fractionation of Albert raw water. In this case, HPSEC traces of the HPIA fraction clearly demonstrate its variability, with respect to both molecular size and UV_{254} absorbance (Figure 4.1). January and August 2004 samples show the existence of a primary peak, eluted at approximately 7 minutes and corresponding to a molecular size of approximately 5 Kilo Daltons (Goslan, 2003), which was not present in the April 2002 sample. The area under the graph also increased by 278% in August 2004 compared to the April 2002 sample when the corresponding HPIA charge load increased by approximately 2 orders of magnitude.

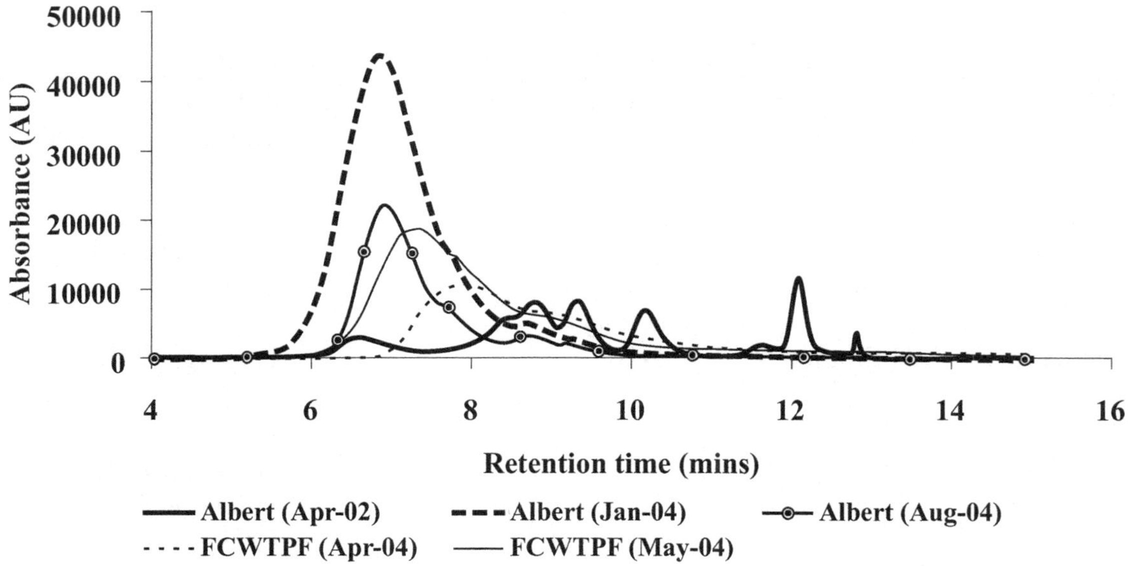

Figure 4.1 HPSEC chromatograms for the HPIA fraction

Once again the results indicate changing composition of the NOM fractions within a total DOC concentration which can lead to varying coagulant dose requirements. These changes cannot be detected by simple DOC measurement.

Conversion of the charge densities to charge concentration reveals that the largest proportion of the total load comes from the FAF fraction. To illustrate, in the case of the water collected in April 2002, the total charge of the individual fractions was 0.013, 0.026 and 0.08 x 10^{-3} meq.L^{-1} for the HAF, FAF and HPIA samples respectively. To establish how much impact variation in the FAF fraction may have on the total charge, the charge density of the FAF fraction was monitored throughout the sampling period (Figure 4.2).

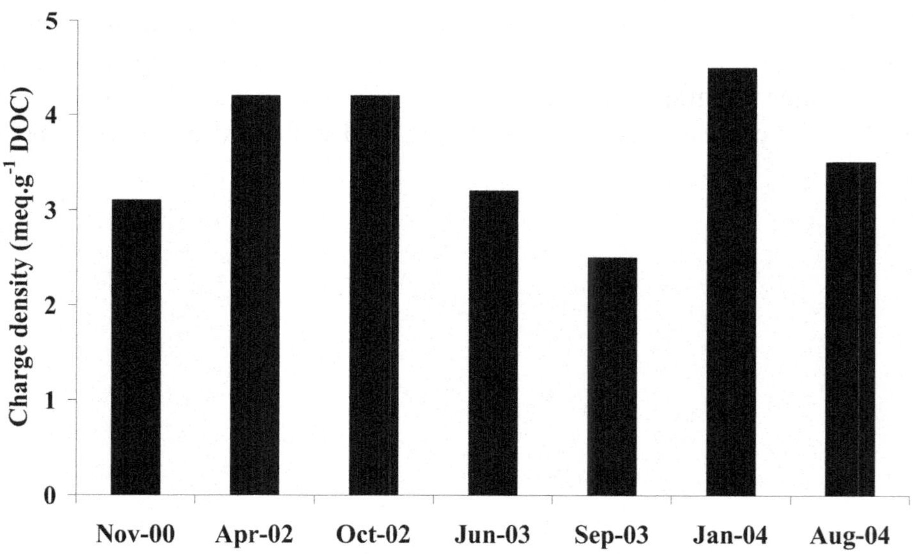

Figure 4.2 Charge density variation of the Albert WTW FAF fraction

The charge density varied between 2.5 meq.g DOC^{-1} in September 2003 to 4.5 meq.g DOC^{-1} in January 2004 representing a variation of 45 % compared to the maximum recorded value. Combining the change in charge density and DOC concentration for the same period reveals that the total charge of the FAF fraction varied between 0.006 and 0.022 $meq.L^{-1}$ throughout the sampling period. In April 2002 the charge density calculated from the FAF concentration was 0.016 meq/L and this increased to 0.021 $meq.L^{-1}$ in January 2004, an increase of 31%. Corresponding DOC concentrations were 7.5 $mg.L^{-1}$ and 8.3 $mg.L^{-1}$ respectively, an increase of only 10%. The differences in carboxylic acidity or charge density will therefore influence the coagulant demand and the water treatment process as a higher charge is associated with the larger MW fractions (Collins *et al.*, 1986; Ratnaweera *et al.*, 1999).

Effect of pH on surface charge

In relation to the specific colloidal charge, the zeta potential of the raw water was -15 ±4 mV throughout the sampling period. Alteration of the pH of the water revealed a stable zeta potential value between pH levels of 4 and 8 and an IEP of 1.7-2.2 (Figure 4.3). Similar results were obtained when comparing both Albert and Bamford raw water samples from September 2003, when the IEP was between a pH of 2 -2.1 for both sites (Figure 4.4).

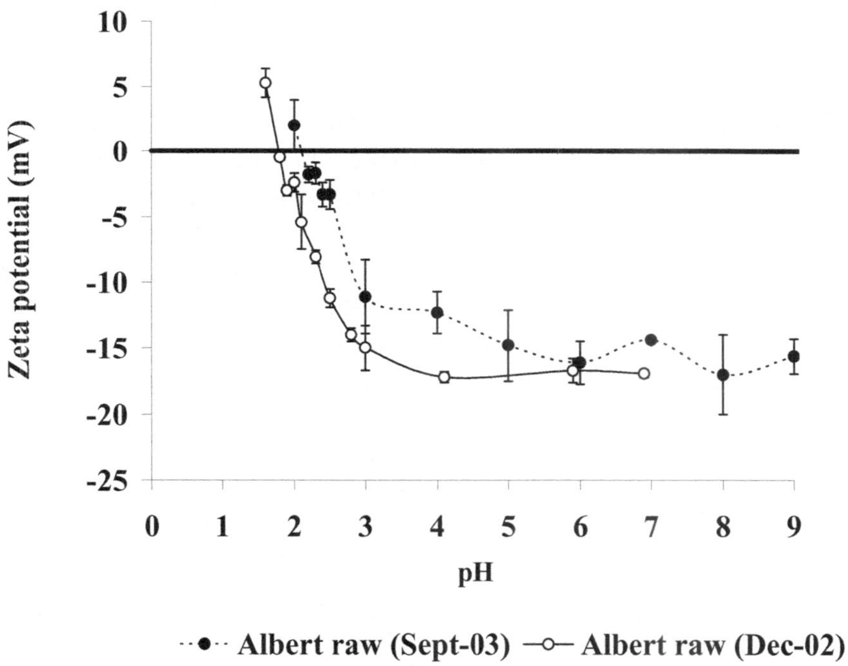

Figure 4.3 Zeta potential of Albert raw water samples with varying pH.

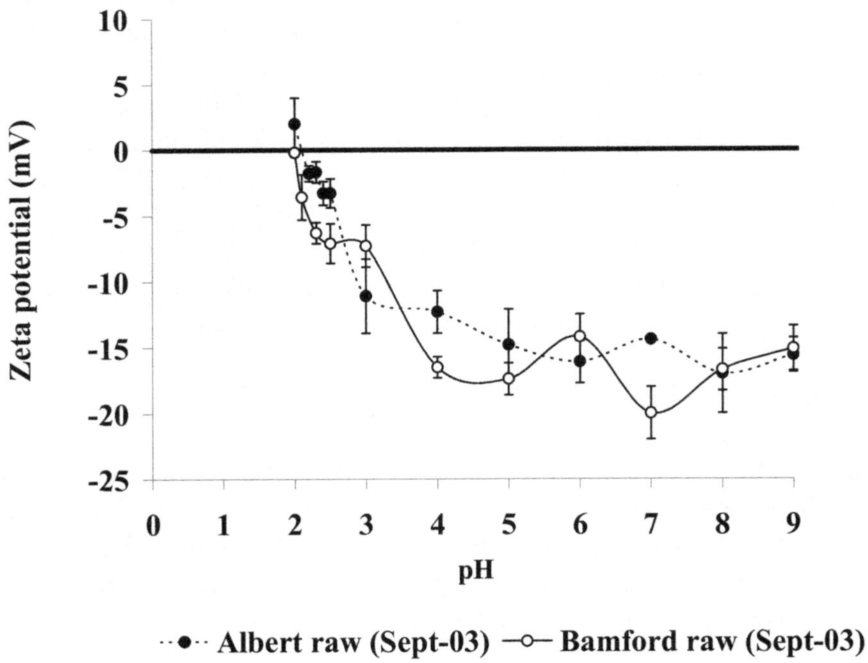

Figure 4.4 Zeta potential of Albert and Bamford raw water samples (Sept-03) with varying pH.

Within the study, both the raw water zeta potential and its pH profile remained fairly constant irrespective of changes in DOC or fractional make up. The results indicate that the charge of the colloids within the raw water can be characterized by an acidic dissociation site model (Hunter, 1981). This is expected as the main functional groups identified within the structures of NOM are predominately carboxylic and phenolic groups with pK_a in the range of 3-6 and 8-10 respectively (Edwards *et al.*, 1996; Gu *et al.*, 1995; Kretzschamr and Sticher, 1997; Tiller and O'Melia, 1993). The stable response to pH until a value of around 4 indicates that for the current water source the carboxylic groups are likely to dominate the charge response of the water.

Comparative studies, varying the pH between 1 and for the individual fractions extracted from April 2002 raw water, revealed a clear difference between hydrophobic and hydrophilic fractions (Figure 4.5). The two hydrophobic fractions have similar profiles to the raw water such that the zeta potential values post pH 4 were -13 ± 3.7 mV, - 17 ± 0.3 mV, and -25 ± 2.2 mV for the FAF, raw water and HAF respectively. In contrast, the zeta potential of the hydrophilic fraction decreased consistently across all the pH values tested. For instance, the zeta potential of the HPIA and HPNIA fractions decreased from 2.1 ± 0.2 mV at pH 2 for both fractions to -4 ± 3.2 mV and -7 ± 2 mV at pH 9 for the HPIA and HPNIA fractions respectively (Figure 4.5). The corresponding IEP of all the fractions was 1.5, 1.6, 1.7, 4 and 4.7 for HAF, FAF, raw water, HPIA and HPNIA respectively. Previous work on the charge of NOM fractions has shown that the pK_a of the FAF fraction to be around 2 ± 0.3 and can be attributed to the carboxyl groups (Leenheer *et al.*, 1995a; 1995b). Whereas the hydrophilic fractions are composed of weaker acidic groups with correspondingly much higher values for their IEP Comparing the different fractions to the raw waters shows as expected that the hydrophobic materials dominates the specific colloidal charge character of the water.

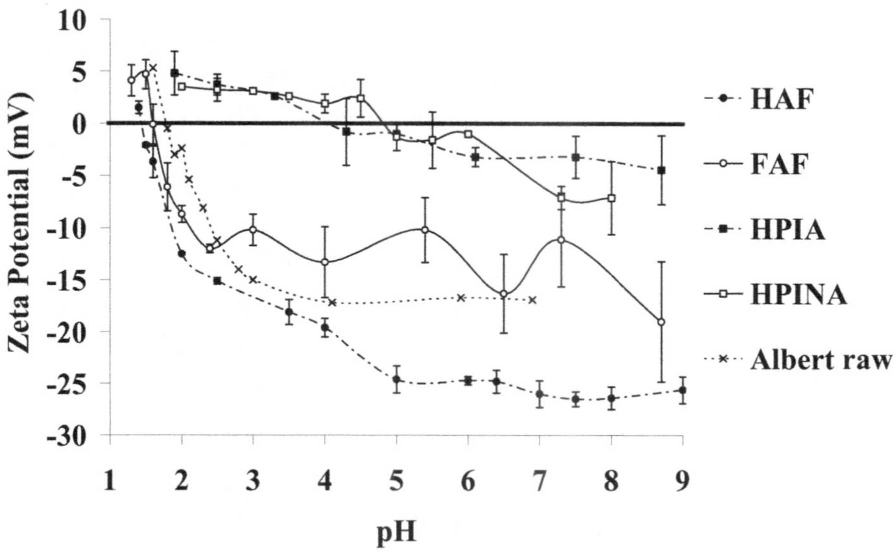

Figure 4.5 Zeta potential of April 2002 NOM fractions with varying pH.

On the basis of the results of the study, a number of the raw water characteristics can be expected to influence the required coagulant dose. In particular the charge density of the hydrophobic fraction is up to an order of magnitude greater than that of the hydrophilic (HPIA) fraction, and variations in the hydrophilic fraction, which are often very significant, may therefore go unnoticed. The increase in MW of this fraction can have a significant impact on the coagulant demand. The pH profile of the HPINA fraction shows it to have minimal or slightly positive zeta potential at operational coagulation pH but its charge density is below the current limit of detection.

Bench- scale trials

Method

NOM Fractions extracted from Albert WTW during April 2002 were used to investigate the NOM coagulant interactions. Experiments were conducted by adding 20 mg of sodium hydrogen carbonate to one liter of deionized water containing the required amount of NOM fraction and a magnetic stirrer bar. The beaker was then placed on a magnetic stirrer and 2 mg of required coagulant, as the metal, was added in order to achieve the required NOM:coagulant ratio. Portions of the solution were then removed and adjusted to varying pH values in the range 2-9 and the resulting zeta potential recorded. The increase in ionic strength as a result of a reduction to pH 2 did not influence the zeta potential. In addition, the zeta potential of the original solutions were also measured over a two hour time period and also remained stable.

Sampling for the bench scale coagulation experiments trials took place during the period of September to December 2003 using raw water from Albert and Bamford WTWs. Raw water samples were stored at 5°C prior to jar testing, and subsequent jar testing, using a Phipps and Bird PB-900 six paddle jar tester, was undertaken at room temperature (20°C). The initial rapid mix was for 2 minutes at 200 rpm whilst adding the required dose of ferric sulfate, 5-25 $mg.L^{-1}$ as Fe (Ferripol xl, EA West, UK). Sodium hydroxide was then added, until the required coagulation pH, whilst stirring for an additional 1.5 minutes at 200 rpm. The jars were then stirred for 15 minutes at 30 rpm and left to settle for a further 20 minutes before taking the samples, which were extracted from the centre of the jar. The zeta potential was initially measured at varying stages of the jar test procedure from the end of the rapid mix to the final settlement, and was found to show no significant variation. The surface charge was measured with a Zetasizer at 25°C.

NOM-Coagulant Interactions

Bench scale tests were carried out using raw water samples and NOM fractions (Apr-02) from Albert WTW to investigate how raw water characteristics and, in particular, the properties of the NOM fractions would affect their interactions with coagulants (M^+). The interactions were monitored as changes in the IEP of the sample at various DOC to coagulant mass ratios and, in all cases, the IEP of the system was found to decrease as the DOC: M^+ ratio increased. In December 2002, the IEP of the systems decreased from 8 to 2.8 as the DOC:Fe ratio increased from 0 to 2.4 (Figure 4.6). The exact relationship between DOC:M^+ ratio and IEP was different for each sampling period and ranged between the extremes of December 2002 and September 2003.

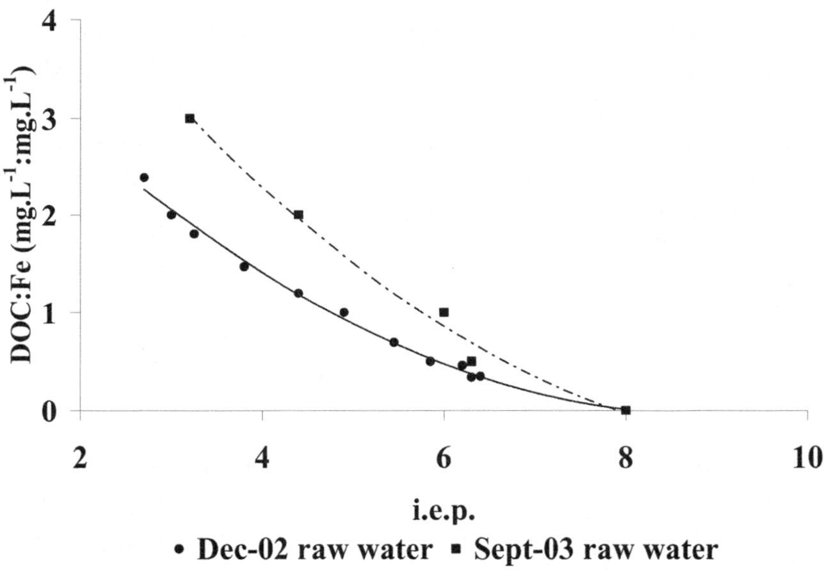

Figure 4.6 IEP and DOC: coagulant ratios for Albert raw water.

As an example, at a DOC: Fe ratio of 2:1, the IEP was 3 in December 2002 compared to 4.2 in September 2003. Importantly, the existence of a variation in the relationship between IEP and mass ratio between different water samples indicates that the changes must be due to character (Tseng *et al.*, 2000) and can be explained in terms of the variation in total charge of the water from one period to the next. For example, in September 2003 the samples were less influenced by the organic matter than in December 2002, when coagulating at the same mass ratio. The differences coincide with the approximate minimum and maximum in the charge density of the FAF fraction described previously.

Calculation of the neutralizing charge provided by adding ferric, based on speciation data (Jiang and Graham, 1998), reveals the coagulant charge density of the added ferric to be 35.5, 30.7, 12.6 and 1.7 $meq.g^{-1}$ Fe at pH values of 4, 5, 6 and 7 respectively. The decrease in available charge as pH increases gives a clear indication why coagulation must proceed under acidic conditions; the charge neutralizing capacity of ferric is 20.8 times greater at pH 4 compared to pH 7. Comparison between mass and charge ratios under equivalent conditions reveals that at the IEP there is a broad agreement between the two. Importantly, the change of mass ratio at any given pH from one water sample to the next indicates that stoichiometric ratios are not appropriate for predicting dose requirements. Instead a simple charge density measurement appears to provide a more consistent indication of dose requirement.

NOM fractions

The interactions of the individual NOM fractions with iron based coagulants were investigated further by increasing the concentrations of organics obtained from the April 2002 fractionation, dosed with 2 $mg.L^{-1}$ (as metal) of coagulant (Figure 4.7). The FAF fraction results in the largest shift in IEP from pH 8 with no NOM present (i.e. ferric hydroxide) to pH 3 for a 2:1 FAF: Fe ratio. This was followed by the HAF and HPIA fractions at pH 3.8 and 4.8 respectively at the same mass ratio. The HPINA did shift the IEP although increasing the DOC concentration further did not have an additive effect. Therefore, this implies only a certain concentration of the HPINA combined with the iron which once exceeded did not affect the IEP of the system. Importantly, the FAF fraction appears to be critical in determining the resultant charge properties of the NOM-coagulant system and hence the dose requirements. Interestingly, complimentary work on floc structural properties has indicated that variation in the FAF component generates the greatest degree of change in the size and strength of the resultant flocs (Jefferson *et al.*, 2004).

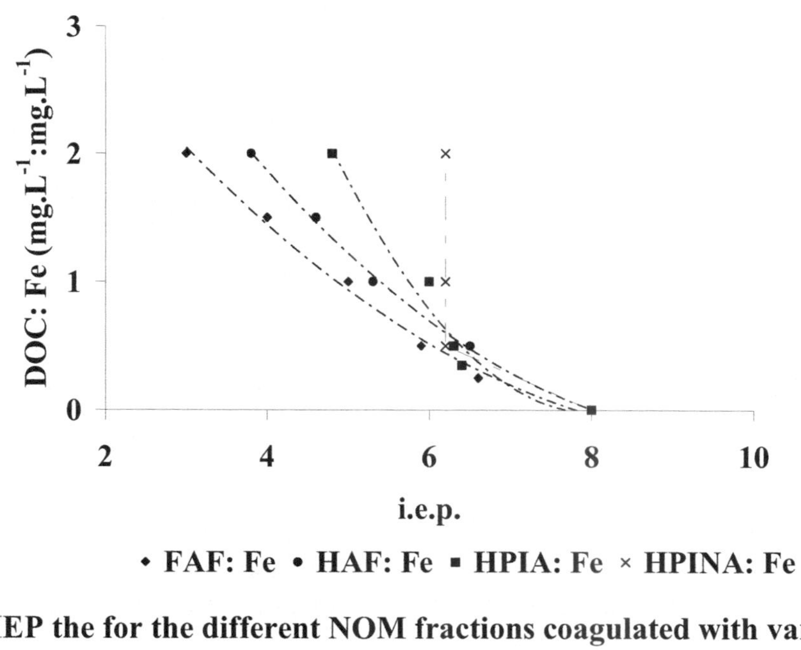

Figure 4.7 IEP the for the different NOM fractions coagulated with varying doses of iron.

Reports in the literature suggest competitive interaction between coagulants and NOM are favored towards larger MW, more hydrophobic, more highly charged components (Gu *et al.*, 1995; Huang and Shiu, 1996) due to the combination of charge neutralization, complexation/precipitation and ligand exchange mechanisms which are thought to dominate NOM removal (McKnight *et al.*, 1992). The current work agrees with that as the hydrophobic rather than hydrophilic material dominates the interactions between NOM and the metal coagulant. This is confirmed by studies that have examined DOC removal across different fractions where only hydrophobic fractions are significantly removed and hydrophilic non acids are almost completely not removed during coagulation (Fearing *et al.*, 2004a). Consequently, the hydrophilic concentration of a water acts as a good indicator of the likely residual concentration achievable by coagulation.

The comparison between humic and fulvic acids is less clear as they are both hydrophobic and have MW distributions that overlap considerably. To illustrate, in the case of the water reported in the current study, HPSEC chromatograms of the HAF and FAF had a distribution between 5.27 and 11.2 minutes and 6.50 and 11.2 minutes respectively, showing an overlap across the 0.5-5 kDa range (Goslan, 2003). Reports in the literature discuss the issue in relation to the principle removal pathways for the two components. HAF is thought to be removed through a combination of charge neutralization (for colloids), complexation/precipitation and ligand exchange adsorption (Huang and Shiu, 1996). Whereas FAF is thought to be mainly removed through an adsorption pathway (McKnight *et al.*, 1992). Evidence for this has been shown by recording pH changes due to the exchange of OH^- groups as the adsorption process proceeds (Gu *et al.*, 1995). The situation is further complicated as evidence suggests adsorption follows a two stage sequence of fast followed by slow adsorption during which competitive displacement occurs where high absorption affinity compounds replace low affinity ones (Avnea and Koopal, 1999). Replacement is thought to be aided as residual charge on the adsorbed molecules repel each other and reduce the affinity of large MW compounds and enabling smaller MW material to adsorb shifting the balance of molecules bound to the floc.

However, it should be noted that most of the reported work has involved synthetic solutions of known chemistry or preformed adsorptions sites. Consequently interpretation to the current situation of poly dispersed NOM solutions and its interaction to added coagulants is unclear and ultimately the exact hierarchy is unknown. Whilst the exact explanation of the observed interactions is unclear it does however provide a useful guide to understand differences in observed removal patterns and the properties of the resultant flocs. Given that optimum removal occurs over a range of zeta potentials around the zero value (Parsons *et al.*, 2004) the concentration and character of the FAF fraction is likely to control the dose requirements to maintain the system within the appropriate range. Whereas the balance of hydrophobic to hydrophilic compounds will control the optimized residual concentration. Therefore, optimizing the coagulation process based on charge is likely to result in improved hydrophobic NOM removal, although the same level of success may not be observed with the hydrophilic fractions.

Coagulation Trials

The aim of the following bench scale experiments reported here was to identify a zeta potential range that can be robustly used to optimize coagulation of high DOC, low alkalinity waters. This is in addition to changing (i) NOM fraction make-up and characteristics;

(ii) coagulant dose (DOC: coagulant ratio); (iii) season and (iv) water source. In order to investigate this, raw water samples for both Albert and Bamford WTWs were coagulated during the period of September to December 2003. As discussed previously, Bamford and Albert raw water sources both originate from moorland peat catchment areas and possess similar raw water characteristics.

Analysis of the performance of the coagulation process, for both Albert and Bamford September 2003 samples, revealed the existence of an operational window either side of the IEP. An operational envelope can be described where removal is high and effluent quality is low and stable. Turbidity residuals of below 1 NTU were still achieveable at a zeta potential of -20 mV. Whereas for DOC removal the operational window was narrower with a stable residual of 1.0 ± 0.2 mg.L^{-1} only achieved between zeta potential values of -10 mV and +3.5 mV (Figure 4.8). This is similar to the range reported by Gregory and Carlson (2003), who considered optimum alum doses for TOC removal to be in the zeta potential range 4.0 ± 3 mV. Ratnaweera *et al.* (1999), suggest that colloidal destabilization occurs before complete charge neutralization of surface charge which is why good turbidity residual is still possible although the zeta potential may not necessarily be zero. Therefore, it is the organic concentration which influences the coagulation conditions and as a result optimized DOC removal lead to good turbidity removal (Dennett *et al.*, 1995; Edzwald, 1993).

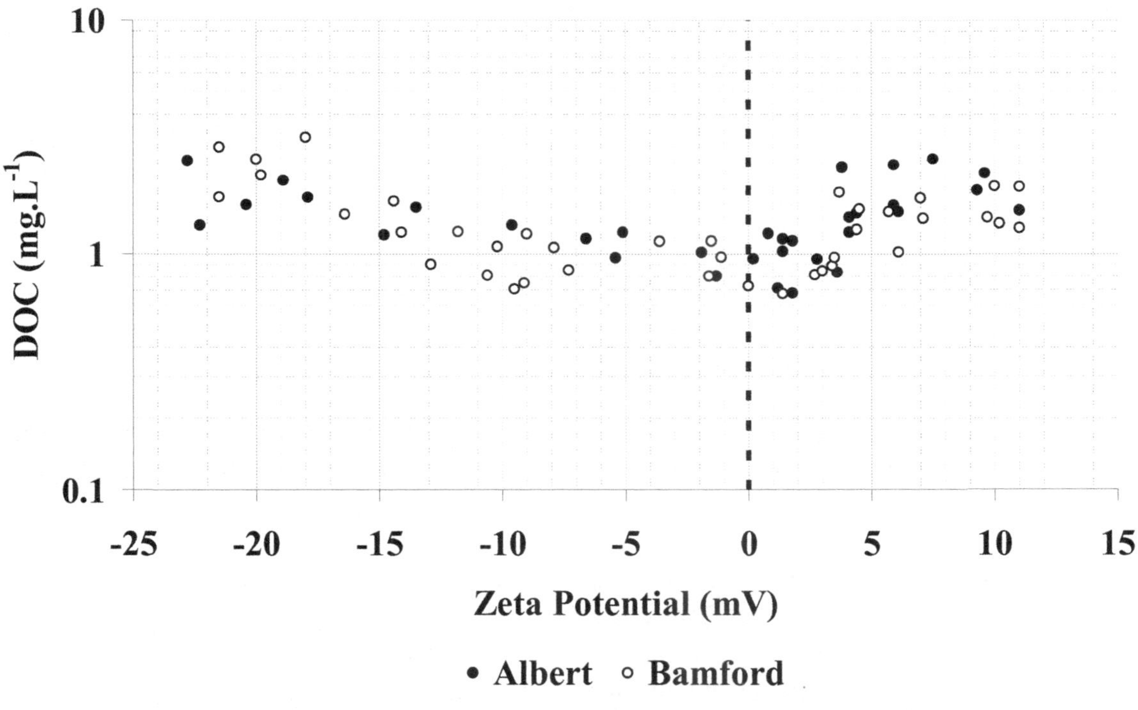

Figure 4.8 Zeta potential vs. Removal performance in terms of DOC residuals for Albert and Bamford September raw water samples coagulated with 5-15 mg.L^{-1} as Fe.

DOC:Fe ratio

The interactions between NOM and ferric sulfate were monitored through changes in surface charge with varying pH in order to establish the effect on the IEP of the system. For instance, a sample of Bamford raw water from September 2003 was coagulated with 5 mg.L^{-1}

ferric sulfate as Fe at varying coagulation pHs. The IEP occurred at a pH of approximately 4.5 (Figure 4.9). Increasing the dose, to 15 mg.L^{-1} as Fe, only required a pH of 5.7 in order to keep the zeta potential constant at approximately 0 mV. Hence, the charge of the NOM-coagulant complex is dependent on both the mass ratio of the NOM to metal ion, and the pH of the system (Jiang and Graham, 1998). Tseng *et al.* (2000), looked at treating snowmelt runoff water and concluded that 2.1 ± 0.1 mg.L^{-1} more alum was required for each additional mg of DOC. However, the relationship may not always be stoichiometric, for example, coagulating a sample of Albert raw water from September 2003, also with 5 mg.L^{-1} ferric sulfate as Fe, required a pH of 5 in order to achieve the IEP. This was despite the fact that Albert raw water contained 2.1 mg.L^{-1} more DOC than Bamford in September.

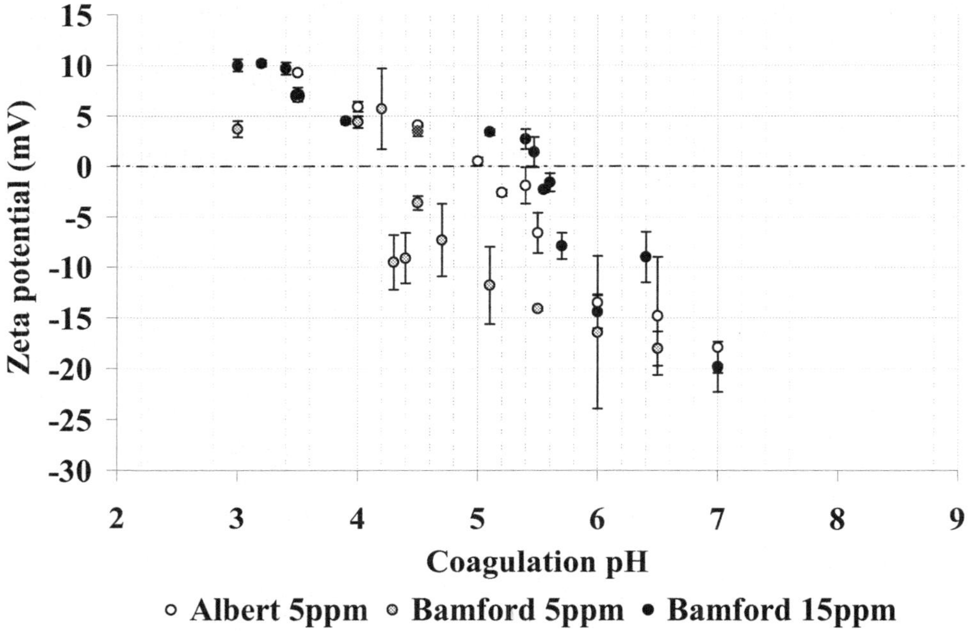

Figure 4.9 Coagulation pH vs. zeta potential for September 2003 raw water samples from Bamford, coagulated with 5 and 15mg.L^{-1} as Fe, and from Albert, coagulated with 5mg.L^{-1} as Fe.

To assess the effect DOC:Fe ratio has on the zeta potential window, Albert raw water collected in September 2003 three different DOC: Fe ratios (mg:mg) of 1.3:1, 0.7:1 and 0.4:1 were used. The results show that despite the varying coagulant dose and DOC:Fe ratio, the lowest residual DOC was achieved when the zeta potential was minimized (Figure 4.10). This was subsequently confirmed with HPSEC chromatograms of coagulated Albert water (Figure 4.11), which demonstrated little difference in the residual organics at a zeta potential of 0 mV and a coagulant dose of either 5 or 10 mg.L^{-1} (as Fe).

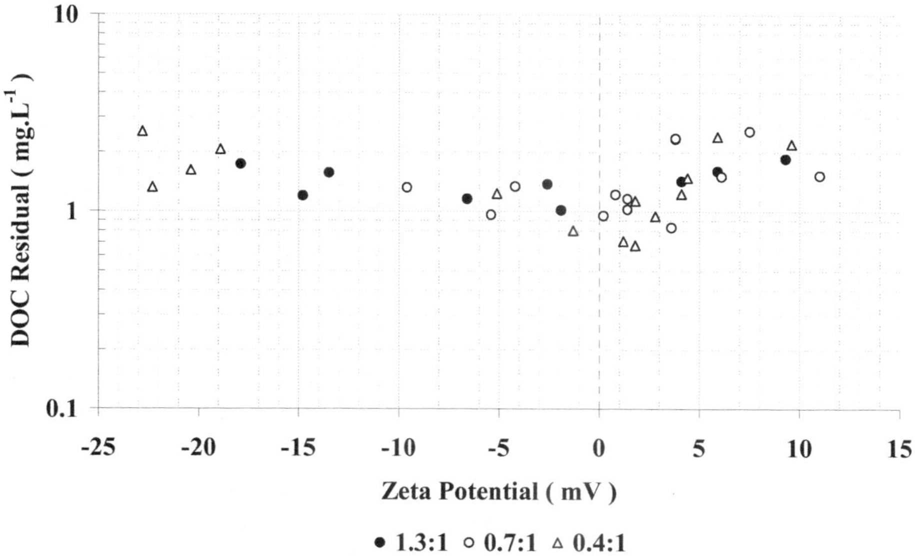

Figure 4.10 Zeta potential vs. DOC residual for Albert raw water from September, coagulated with ferric sulfate to a varying DOC:Fe (mg:mg) ratio at pH's between 3 and 7.

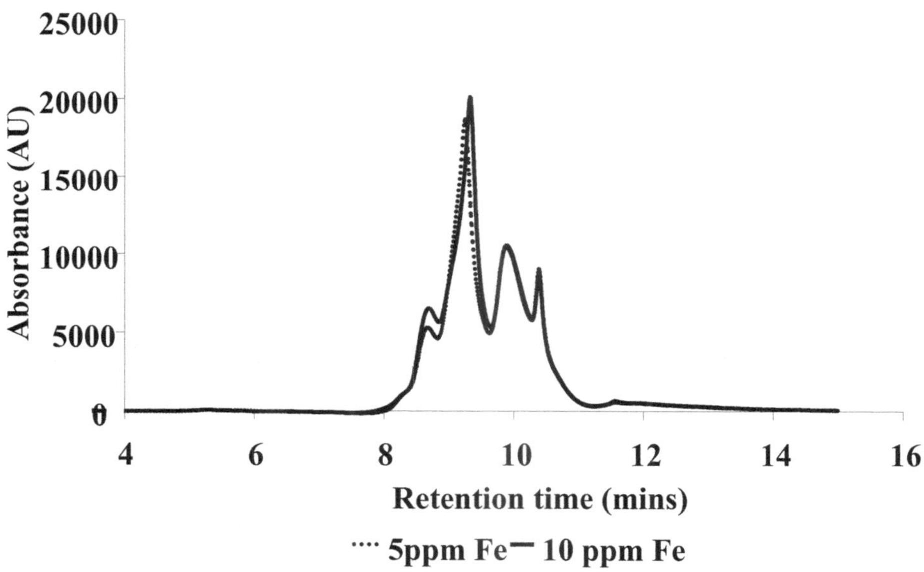

Figure 4.11 HPSEC chromatograms of Albert raw water from September, coagulated with 5 and 10 mg.L^{-1} ferric sulfate as Fe and a resulting zeta potential of zero.

The HPSEC data confirms that a proportion of the NOM fraction which cannot be removed by conventional coagulation with metal salts. The higher MW, more hydrophobic NOM fraction is generally more sorbable and easily removed during coagulation, whereas there is also a non sorbable fraction, with more hydrophilic tendencies, which cannot be removed (Fearing *et al.*, 2004a; Edwards, 1997; White *et al.*, 1997; Krasner *et al.*, 1994; Randtke, 1988; Owen *et al.*, 1993). Therefore, increasing the coagulant dose further is unlikely to result in a lower residual. To illustrate, Krasner *et al.* (1994), obtained similar results when they blended two waters with varying concentration of low MW DOC, but found the residual was 1.3-1.4 mg.L^{-1} DOC, regardless of the blend.

Seasonal Variation

Fractionation data revealed seasonal changes in NOM fraction-make up, especially during the autumn and winter months following periods of heavy rainfall. Variations in NOM fraction make-up, especially hidden changes when the DOC concentrations appear stable, have significant implications for the coagulation process. Hydrophobic fractions contribute the largest proportion of colloidal charge and their increased presence will exert a greater coagulant demand required to minimize the surface charge (Edzwald, 1993; Collins *et al.*, 1986). However, if the increase in organic concentrations coincides with the hydrophobic content of the water, then optimizing the coagulation process should result in a more robust performance.

In order to investigate this further, raw water samples collected from Albert WTW from September to December 2003, where the DOC levels increased from 6.6 to 9.7mg.L^{-1}, were coagulated with an increasing iron dose (16.5-24.3 mg.L^{-1} of ferric sulfate as Fe) in order to maintain the same DOC:Fe ratio (0.4mg :1mg). However, despite the increasing organic concentration in the raw water, the shape of the plot of DOC residual versus zeta potential remain unchanged (Figure 4.12). Again if the increased coagulant demand is satisfied, in order to achieve a constant zeta potential of approximately 0 mV, then the performance is not affected by the increased organic levels.

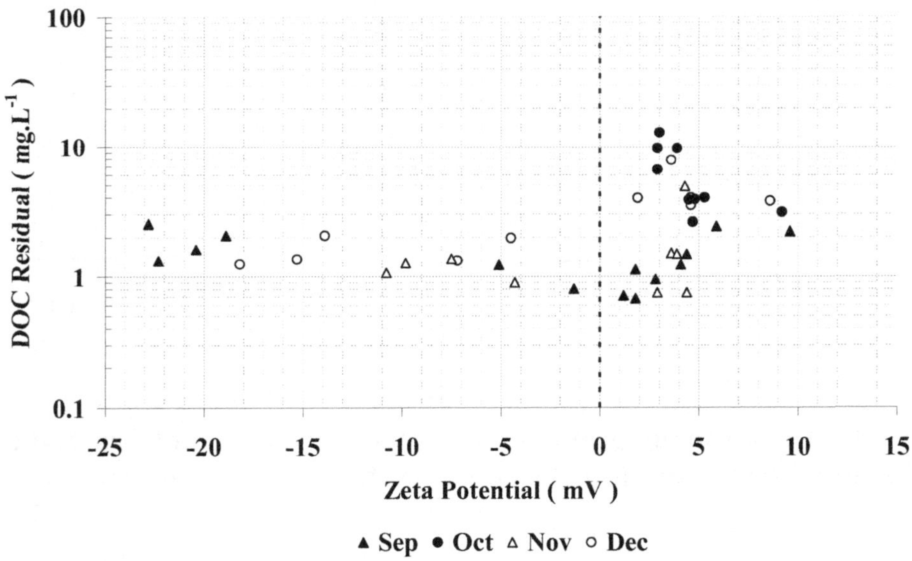

Figure 4.12 Zeta potential vs. removal performance in terms of DOC residual for Albert raw water coagulated with ferric sulfate to a 0.4:1 DOC:Fe (mg:mg) ratio at pH's between 3 and 7.

Samples were also collected from Bamford WTW during the same period, September-December, 2003 and coagulated with ferric sulfate (11.3-25.5 mg.L^{-1} as Fe) in order to maintain the same DOC:Fe ratio of 0.4:1. DOC concentrations exhibit slightly more variation than Albert, ranging from 4.5 to 10.2 mg.L^{-1}. However, combining coagulation results from both Albert and

Bamford raw waters, for the period September to December 2003, revealed a similar shaped plot for DOC residuals against zeta potential, regardless of water source (Figure 4.13).

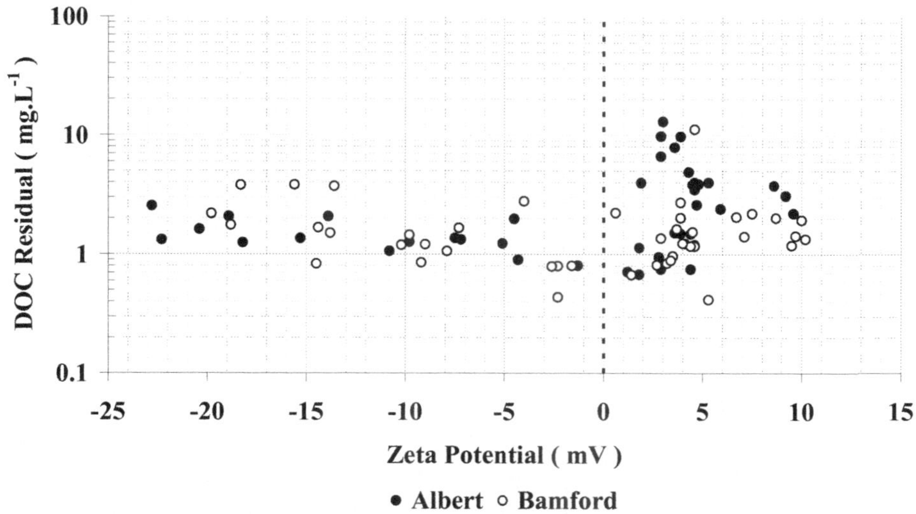

Figure 4.13 Zeta potential vs. removal performance in terms of DOC residuals for Albert and Bamford raw water from September to December, coagulated with ferric sulfate to a 0.4:1 and 0.5:1 DOC:Fe (mg:mg) ratio respectively, at pH's between 3 and 7.

Results presented here have shown the relationship between zeta potential and residual concentration, which appears relatively to be insensitive to season, source and coagulation conditions. This indicates that by using this approach it should be possible to identify the zeta potential range for which optimum removal is achievable. However, profiles still demonstrate a lack of symmetry. For instance, with increasing negative zeta potential values a gradual decline in performance was observed, corresponding to a negative gradient of 0.06, whereas for positive zeta potential values there was a much steeper reduction in performance at operational boundary. These observations suggest the importance of patchwork agglomeration where only partial neutralisation of the charge is required (Ratnaweera *et al*, 1999), whereas the large negative values that still provide optimal treatment suggest the potential importance of polymer bridging from the large MW material.

Pilot plant trials

The laboratory scale tests reported above identified a target zeta potential range required for optimum removal during the coagulation process. The tests were repeated using three pilot plants of varying scale to investigate the effect of scale on the identified operational envelope. The pilot plants incorporated a number of different treatment options used during this project (Table 4.4). Two pilot plants were operated using raw water from Albert reservoir, the DA20 and Cranfield pilot plants, during various periods throughout the year. The third pilot plant, at FCWTF, was operated using raw Poudre river water during the spring snowmelt runoff period.

Table 4.4
Pilot plant operations

Plant	Scale	Processes	Coagulant dose	Operation
Cranfield (Albert)	450 $L.h^{-1}$	Coagulation Flocculation	Fe (5-15 $mg.L^{-1}$)	Jan-04
DA20 (Albert)	15 $m^3.h^{-1}$	Coagulation Flocculation DAF	Fe (5-15 $mg.L^{-1}$) Al (10 $mg.L^{-1}$)	Mar-04 Aug-Sept-04
FCWTF (Poudre)	1.81 $m^3.h^{-1}$	Coagulation Flocculation Lamella plate settling Filtration	Fe (10-17 $mg.L^{-1}$) Al (1.3-4 $mg.L^{-1}$)	Apr-June-04

Cranfield University Pilot Plant.

The Cranfield University pilot plant (Figure 4.14) consisted of coagulation and flocculation processes and was operated in January 2004, using raw water tankered directly from Albert WTW. The plant provided dosing of coagulant with rapid mixing (200 rpm for 2 minutes) and two stage slow speed flocculation (30 rpm for 10 minutes per stage). The pH was controlled by the addition of 0.1M NaOH (Fisher Scientific, UK*)* into the rapid mix tank. Ferric sulfate (Ferripol XL, EA West, UK) was also added directly into the rapid mix tank, with doses varying between 5 and 15 $mg.L^{-1}$ as Fe. For all the pilot plant trials, coagulant dose and pH were varied in order to manipulate the zeta potential and, for each set of experimental conditions, flocculated water samples were taken after conditions had stabilized for one hour, and were analyzed for DOC, UV_{254}, HPSEC, pH and zeta potential.

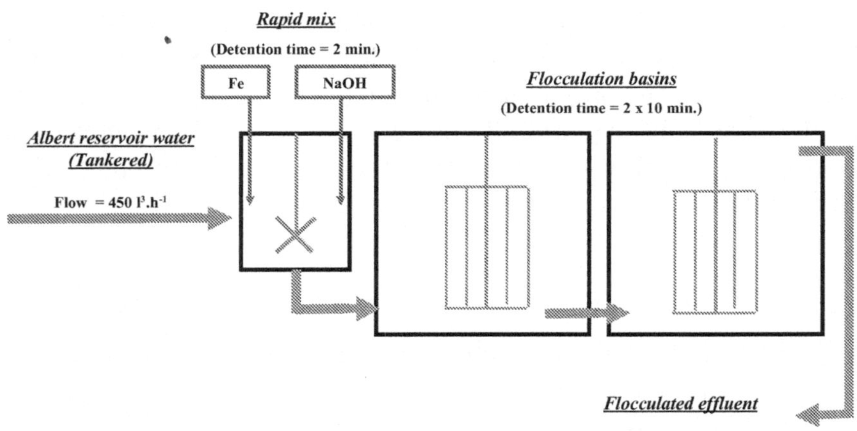

Figure 4.14 Cranfield University pilot plant schematic

DA20 Pilot plant

The DA20 pilot plant (Figures 4.15) was located at the Albert reservoir and was operated during two periods, March and August/September 2004. The full flow (15.$m^3.h^{-1}$) was passed through an inline static mixer (Series 100, Statiflow, UK), where the aluminum or ferric based coagulants and a 12% lime suspension (Kalic supplied by Buxton Lime, UK*)* were introduced consecutively. This was followed by three stage plug flow flocculation (5 rpm for 20 minutes) and dissolved air flotation (6.3 $m.h^{-1}$, 8% recycle and 5 bar saturator pressure). Ferric sulfate (Ferripol XL, EA West, UK) was the primary coagulant used, with doses varying between 5 and 15 $mg.L^{-1}$ as Fe. However, during the August and September 2004 period the DA20 was also operated using aluminum sulfate (Alum, EA West, UK), at a dose of 10 $mg.L^{-1}$ as Al. For all pilot trials, coagulant dose and pH were varied in order to manipulate the zeta potential. Samples were taken after conditions had stabilized for 1½-2 hours and analyzed for DOC, UV_{254}, HPSEC, pH, turbidity and zeta potential.

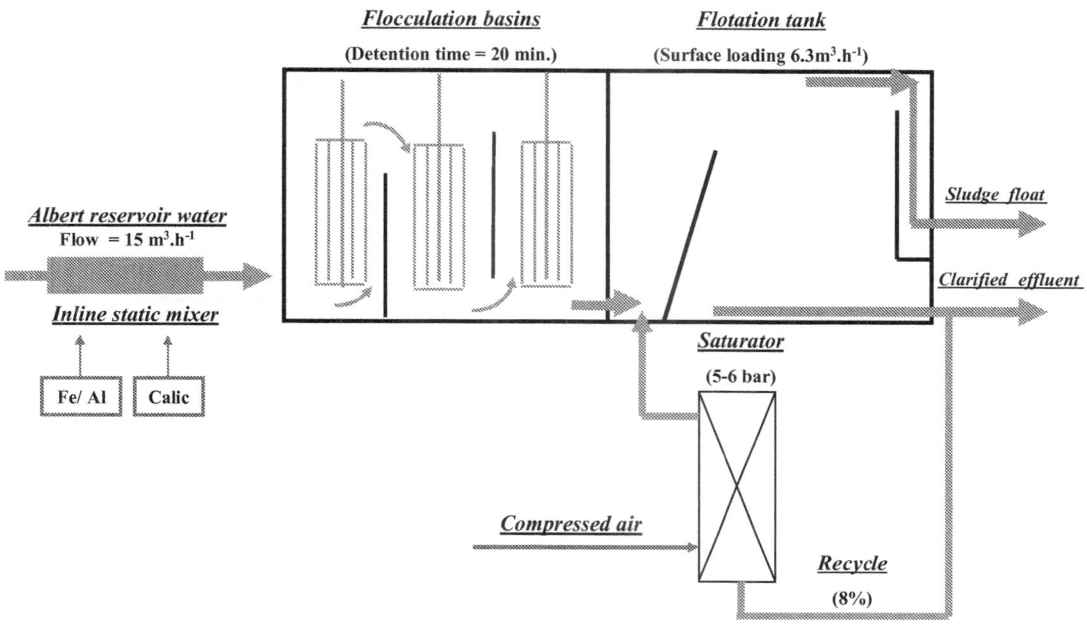

Figure 4.15 DA20 pilot plant schematic.

FCWTF pilot plant

The FCWTF pilot plant (Figure 4.16) included two parallel trains designed to reflect the operational characteristics of full-scale treatment including rapid mix (1000 rpm), 3 stage plug flow flocculation (11, 8 and 3.5 rpm) and lamella plate settling (0.83 $m^3.m^{-2}.h^{-1}$). This pilot plant has also been used in previously published experimental work at FCWTF (Gregory and Carlson, 2003; Billica and Gertig, 2000). For this current research, raw Poudre River water was pumped into the pilot plant at a flow of 1.8 $m^3.h^{-1}$. A 30% lime slurry (Mississippi Lime Company, MO, USA) and carbon dioxide (Fisher Scientific, USA, 10-752-1G regulator and Omega, USA, model FMA1916 gas mass flow controller) were added to the raw water before it was split between the two trains, in order to achieve an alkalinity of 25-35 $mg.L^{-1}$ as $CaCO_3$ to adjust the coagulation pH. Trials were conducted using both aluminum sulfate (General Chemical, USA) and ferric sulfate (Eaglebrook Inc, USA) at doses of 1.3-4 $mg.L^{-1}$ as Al (14-44mg/l alum) and 3-5 $mg.L^{-1}$ as Fe (10.4-17.4 $mg.L^{-1}$ ferric sulfate) respectively. For each set of experimental conditions, samples were taken over a period of 15 minutes after conditions had stabilized for 2-2½ hours and analyzed for TOC, DOC, turbidity, zeta potential, HPSEC, UV_{254}. Genesis® for Windows

(V3.5), SCADA industrial software program (Iconics Inc, Foxborough, MA) was used to interface with the pilot facility and provide a means of online data acquisition. Parameters such as flow, pH, and settled turbidities were also continuously data logged at 3 minute intervals. Operation of the FCWTF pilot plant filters will be discussed further on in the chapter.

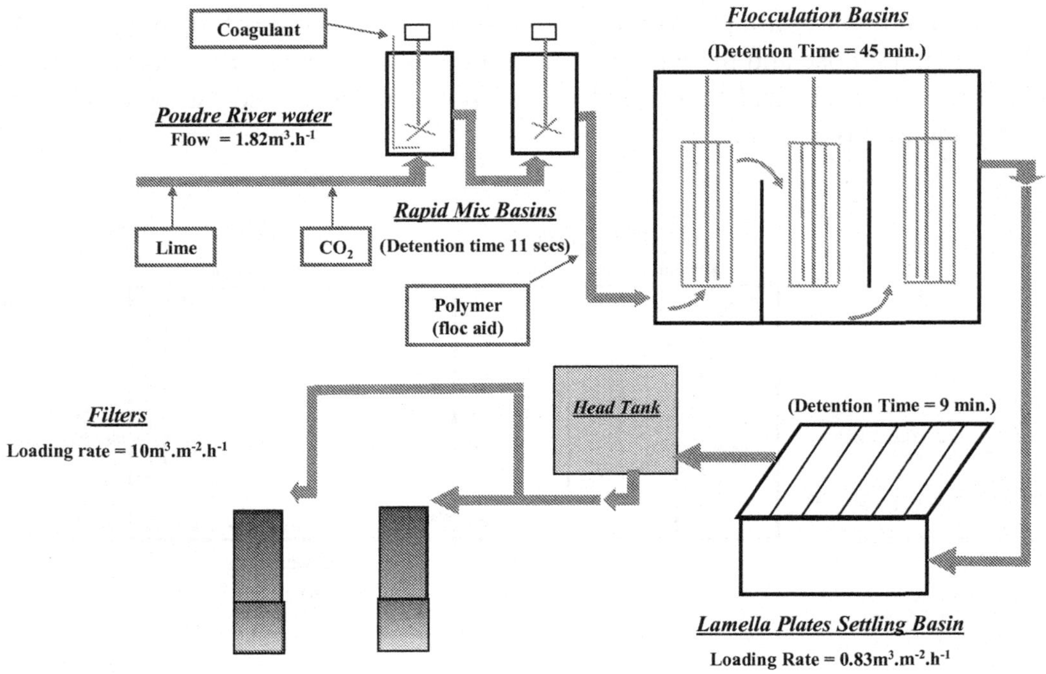

Figure 4.16 FCWTF Pilot plant schematic.

Operational window

As with the bench-scale trials, the operational window for DOC removal also occurs over a narrower zeta potential than for turbidity as shown by normalized plots taken from the DA20 plant during trials in March 2004 (Figure 4.17). Turbidity residuals remained stable as the zeta potential increased from 0 mV to –15.5 mV, whereas both the DOC and UV_{254} residuals increased gradually over the same zeta potential range. Therefore, with regards to this research, results will be discussed in terms of the conditions required for optimum DOC removal.

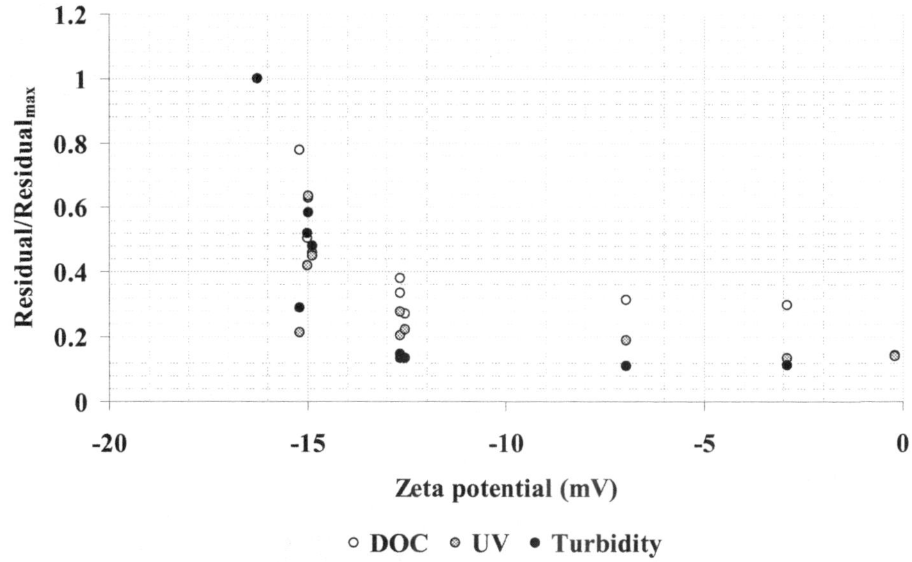

Figure 4.17 Normalized plot of DOC, UV_{254} absorbance and turbidity residual against zeta potential for the DA20 pilot plant after clarification.

The DA20 pilot plant was operated with Albert raw water during March 2004 at at a ferric dose of 12 mg.L^{-1} (as Fe), and a pH between 5.2 and 7.5, in order to manipulate the zeta potential (-15.3 to +2.8). This resulted in varying degrees of removal, with DOC residuals between 2.6 and 0.8 mg.L^{-1}. HPSEC chromatograms demonstrate that as the zeta potential is reduced, more organic material is removed. (Figure 4.18).

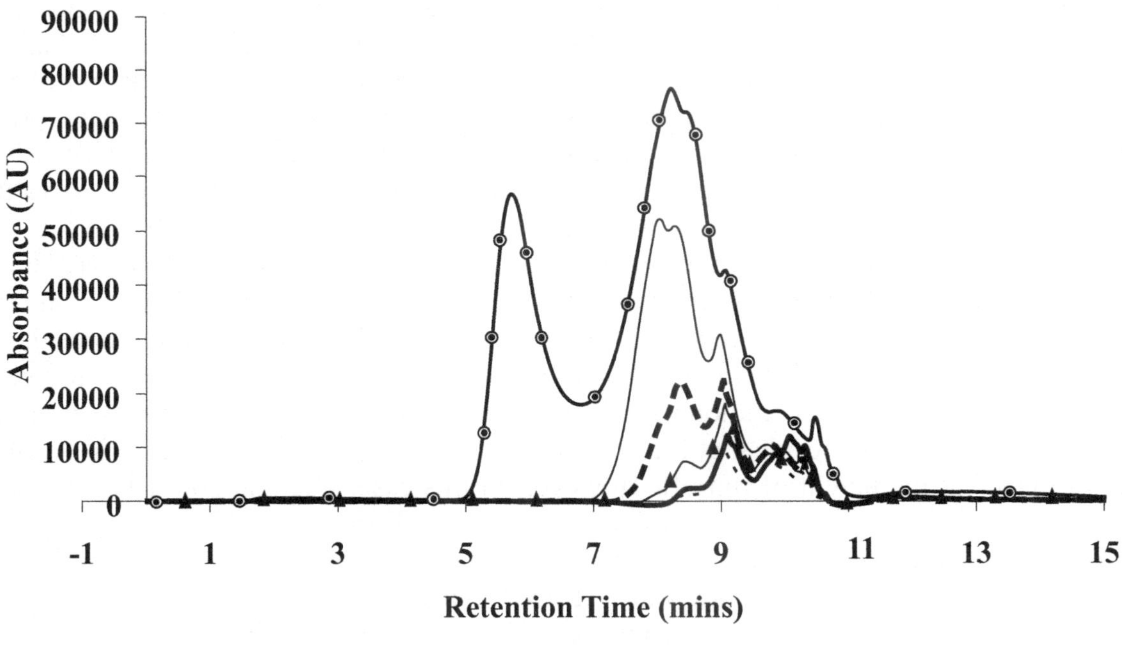

Figure 4.18 HPSEC chromatograms for Albert WTW (DA20) March 2004 raw water, coagulated within 12 mg.L^{-1} as Fe, at varying zeta potentials (-15.3 to +2.8 mV)

Similar results were observed at FCWTF. The pH was varied between 5.3 and 8.1, at an alum dose of 32 mg.L^{-1}, resulting in a zeta potential range of -15.3 to +4. Again this resulted in varying degrees of organics removal (Figure 4.19). This can be seen clearly in the reduction of the second peak (8.7 minutes retention time) of the HPSEC chromatogram. The removal of the lower MW material is less influenced by the change in zeta potential.

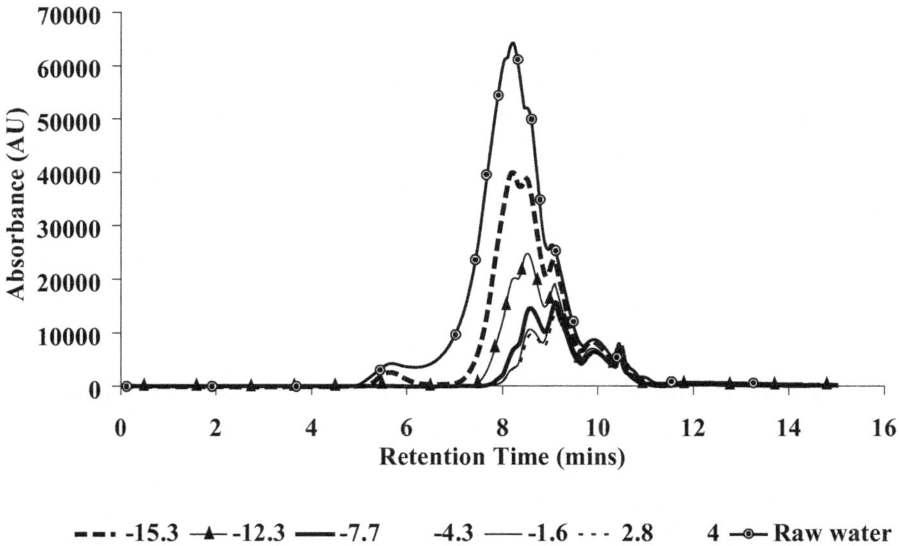

Figure 4.19 HPSEC chromatograms for FCWTF raw water, coagulated within 32 mg.L^{-1} alum, at varying zeta potentials (-15.3 to +4 mV)

Direct comparison of the residual DOC concentration to the zeta potential reveals that residual concentrations are minimized at low zeta potentials. At FCWTF, residual DOC remained at 1.5 ± 0.2 mg.L^{-1} between zeta potentials of -10 to +3 mV. Comparison of results from the different water samples present similar profiles with respect to zeta potential (Figure 4.20). Across all the sites and sources tested a value of -10 mV was observed to be within all the operational windows and potentially represents a safe target threshold value for coagulation monitoring and control.

However, the residual within the operational window has been shown to differ. The DOC residual at Albert WTW increased from 1.2 mg.L^{-1} in January 2004, to 2.0 mg.L^{-1} in August 2004. This corresponded to a 1.3 mg.L^{-1} increase in initial HPINA concentration over the same period. The DOC residual achieved at FCWTF exhibited less variation over the period studied. Although the total DOC concentration rose from 4.4 mg.L^{-1} to 7.4 mg.L^{-1}, and removal increased from 40 to 73%, the DOC residual at the optimum was unchanged, at 2.16 mg.L^{-1} in April 2004 and 2.03 mg.L^{-1} at the height of run-off. However, in this case, fractionation data revealed that the hydrophilic content of the raw water only increased from 1.6 to 1.9 mg.L^{-1}. Further investigation into the differences in the achieveable residual DOC concentration from each trial uncovered a relationship with the raw water HPINA concentration (Figure 4.20b). The gradient of the regression line (R^2 = 0.91) represents a correlated ratio of 90%, indicating the majority of the residual is contained within the HPINA fraction. Previous research has concluded that the higher MW, more hydrophobic NOM fraction is generally more sorbable and easily removed during coagulation, and there is also a non sorbable fraction, with more hydrophilic tendencies, which is not (Edwards, 1997; White *et al.,* 1997; Krasner *et al.,* 1994; Randtke, 1988;

Owen *et al.,* 1993). However, the amount of DOC removed was still greater than the total hydrophobic content, indicating that it is possible to remove some hydrophilic material although not all, as also confirmed by White *et al.* (1997).

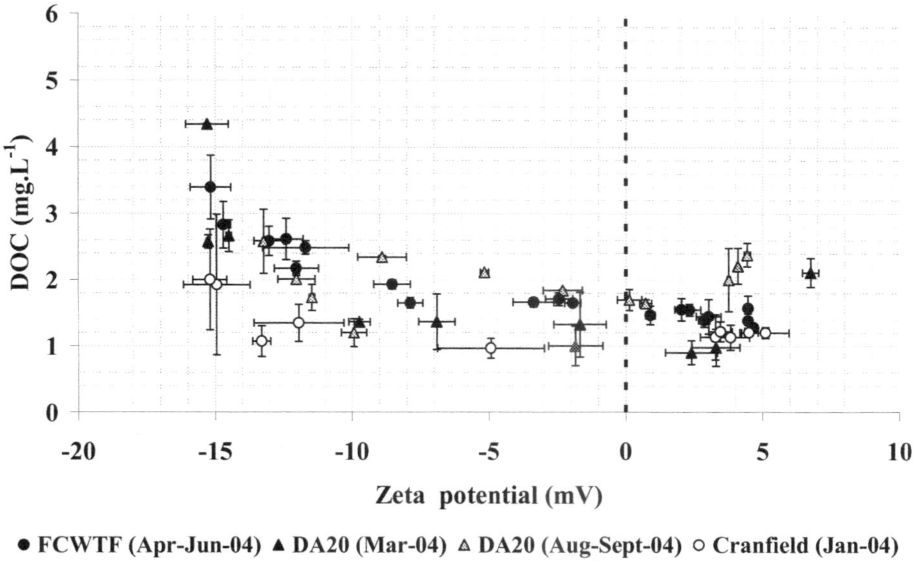

Figure 4.20 DOC residual vs. zeta potential for all water samples coagulated with ferric.

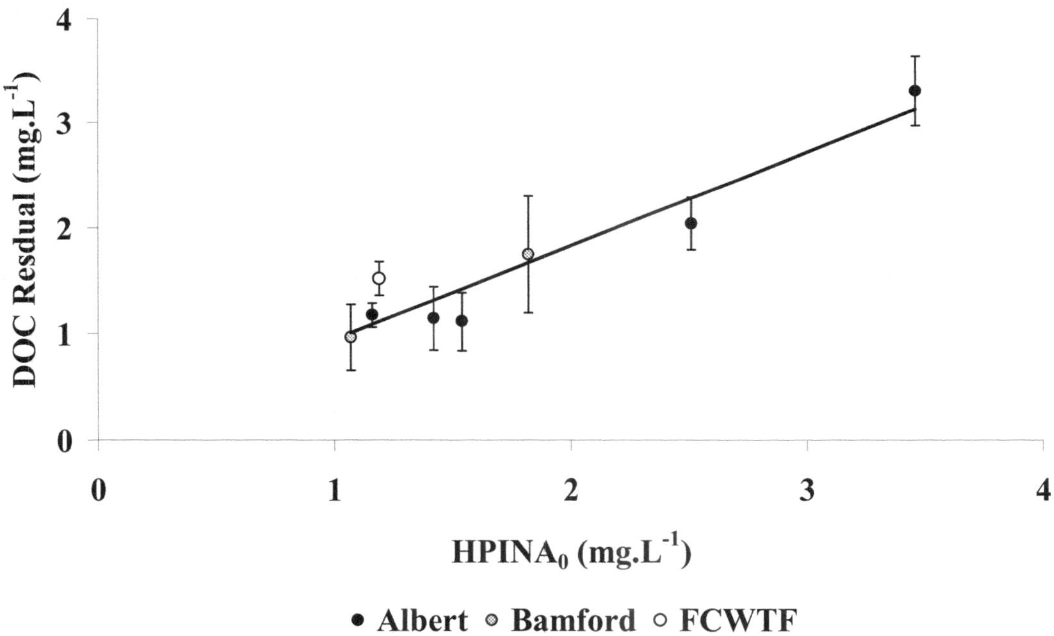

Figure 4.20b Residual DOC vs. initial HPINA concentration

Aluminum vs. iron.

Trials were undertaken simultaneously on the two identical trains of the FCWTF pilot plan in order to compare the performance of aluminum sulfate (alum) and ferric coagulants. Results from both coagulants display similar profiles for DOC residuals with respect to zeta potential, with optimum removal achieved at a zeta potential of approximately zero (-4 to +4 mV). However, a lower DOC residual was possible with ferric, removing 0.5 mg.L^{-1} more DOC than alum (Figure 4.21). Bell-Ajy *et al.* (2000), who looked at 16 different sites with initial TOC concentrations ranging from 1.5 to 16.3 mg.L^{-1}, also found it was possible to remove on average 0.86 mg.L^{-1} more DOC with ferric chloride than alum. In fact, a number of authors have achieved higher DOC removal with ferric salts compared to aluminum salts (Hwang *et al.,* 2001; Edzwald and Tobiason, 1999). It has been suggested that this is due to the fact that alum based coagulants are more selective and ferric can remove more DOC at higher does, especially when treating elevated levels of organics (Budd *et al.,* 2004; Kastl *et al.*, 2004). HPSEC chromatograms of optimum performance with both ferric and alum (Figure 4.22), show that ferric removed more of the UV_{254} absorbing compounds.

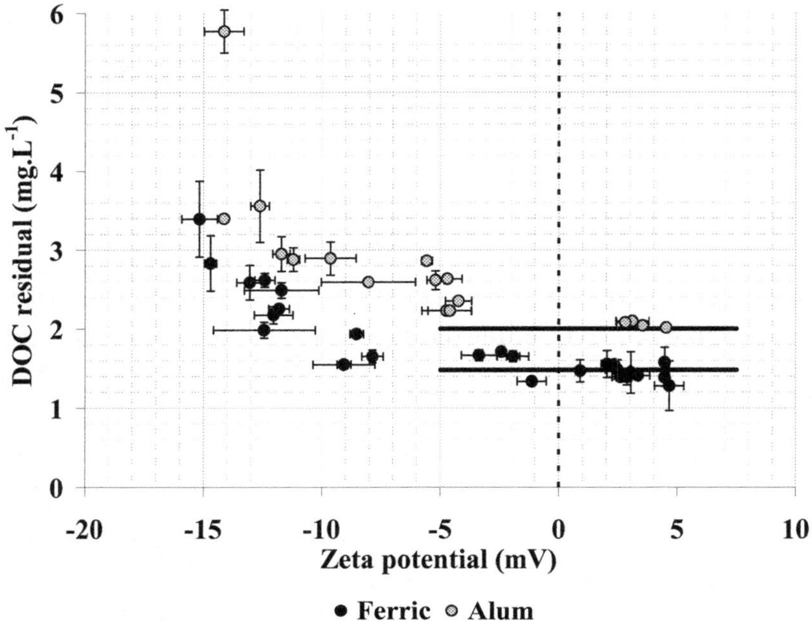

Figure 4.21 DOC residual vs. zeta potential for FCWTF, coagulated with either ferric or alum.

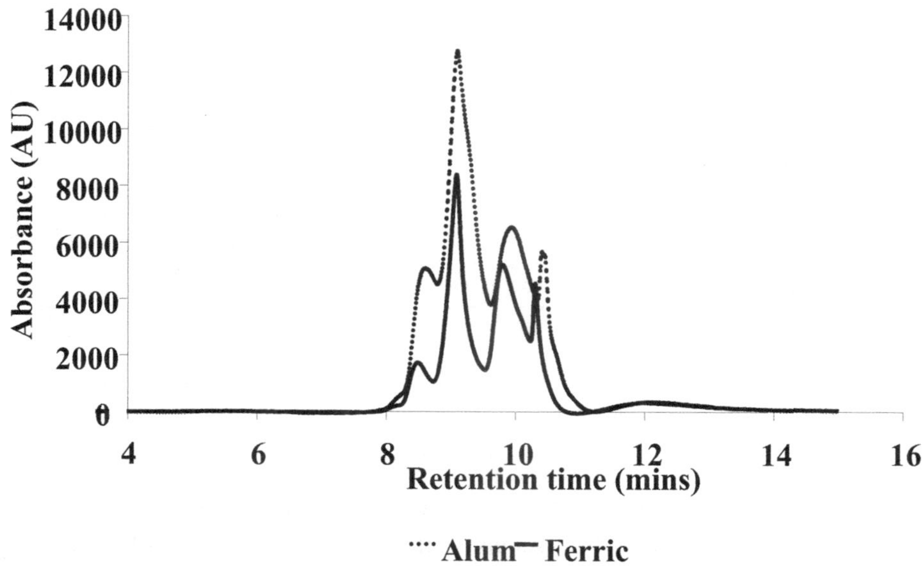

Figure 4.22 HPSEC chromatograms for treated water, coagulated within optimum range, with either ferric or alum at FCWTF.

Ferric and alum doses were increased and the coagulation pH was varied in order to maintain a constant zeta potential, +4 mV for alum and +3 mV for ferric, (Figure 4.23). The DOC residual remained constant at approximately 2.16 ± 0.20 $mg.L^{-1}$ for a pH range 4.9-6.5 and an alum dose 32-44 $mg.L^{-1}$(2.9-4.0 $mg.L^{-1}$ as Al). Similarly with iron, no additional DOC removal was observed when the dose was increased from 5 to 15 $mg.L^{-1}$ (as Fe), with the DOC residual remaining constant at 1.4 ± 0.1 $mg.L^{-1}$. The same was true when operating at a negative zeta potential (Figure 4.24). At FCWTF the alum dose was increased from 20 to 40 $mg.L^{-1}$ (0.9-3.6 $mg.L^{-1}$ as Al) coinciding with a pH increase from 6.1 to 7 in order to maintain the zeta potential to between -4 and -5 mV. However, the DOC residual remained relatively constant at 2.4 ± 0.20 $mg.L^{-1}$. Krasner *et al.* (1994) carried out jar tests on California State Project Water at incremented alum doses from 22 to 111 $mg.L^{-1}$. They also found that after a certain dose of 47 $mg.L^{-1}$, DOC removal reached a plateau. Gregory and Carlson (2003) found improved performance when the pH was increased and the zeta potential kept constant and according to Amal *et al.* (1992), the improved removal may be due to a more even distribution of charge and an increased number of collisions. Results from the current study do not provide any evidence to support this and improved removals were not observed at the higher pH's. Hence these findings promote the theory that attaining the optimum zeta potential is more important than both coagulation pH and coagulant dose in achieving optimal removal.

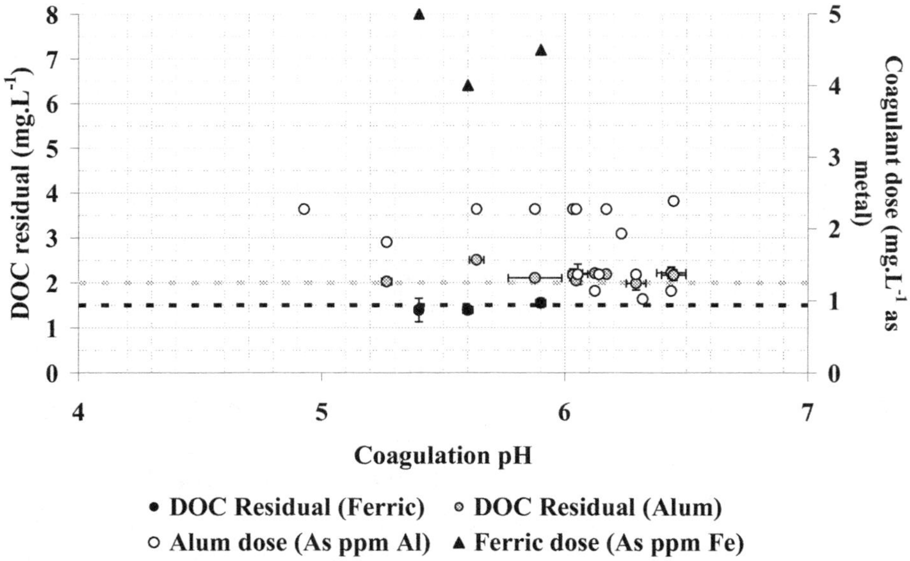

Figure 4.23 Coagulation pH vs. DOC residual and coagulant dose for FCWTF samples operating within optimum zeta potential range +4 mV (alum) and +3 mV (ferric).

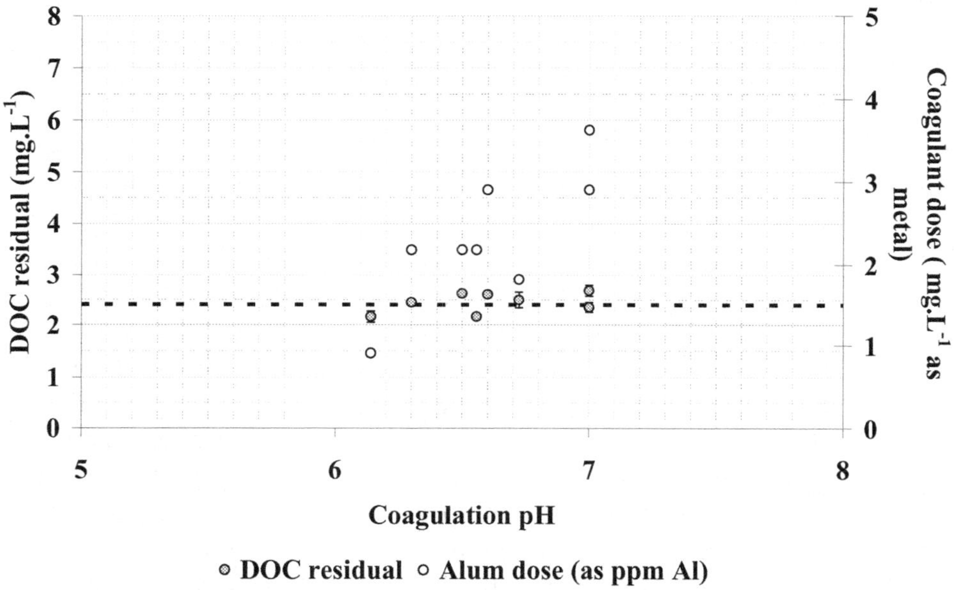

Figure 4.24 Coagulation pH vs. DOC residual and coagulant dose for FCWTF samples operating at a zeta potential range -4 to -5 mV.

Iron and aluminum coagulants were also trialed using the DA20 during August and September 2004 at Albert reservoir and compared with the results obtained from FCWTF (Figure 4.25). In general, a lower DOC residual was possible with the ferric coagulant, removing at 0.5-0.8 mg.L^{-1} more DOC than alum. Overlaying data from clarified spot samples obtained from a number of upland reservoirs and river based water works across the UK, also follows a similar trend, depending on whether iron or aluminum is used in the coagulation process.

Coagulation at these sites had not been optimized using zeta potential and therefore at the time of sampling and may not have been operating at optimum.

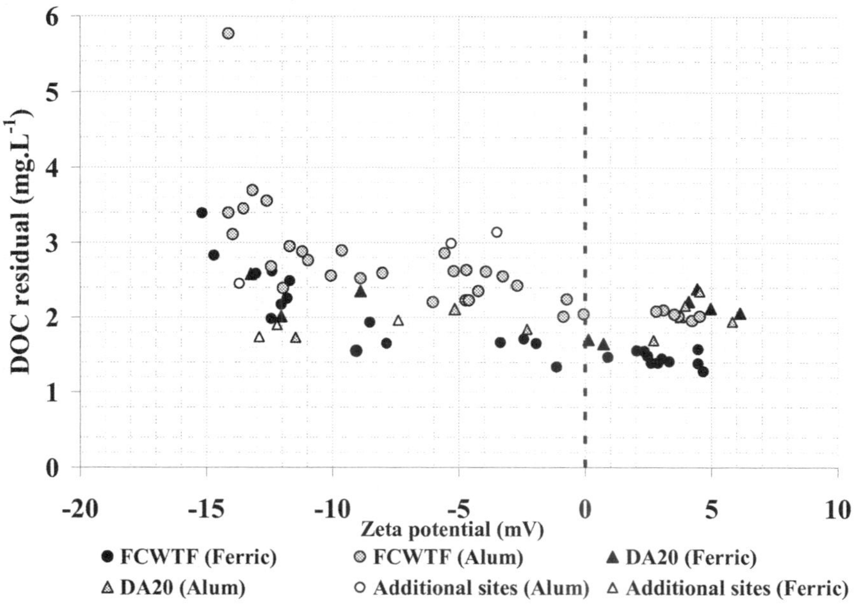

Figure 4.25 DOC residual vs. zeta potential for: FCWTF DA20 pilot plant trials coagulated with both ferric and alum overlaid with additional clarified water spot samples from a range of water treatment works across the UK.

Summary

Results from this investigation indicate the existence of operational windows of zeta potential within which treatment efficiencies are high and effluent concentrations are low. This operational window appears unaffected by the changing organic concentrations and characteristics. However, the increased hydrophobic content of the water, and associated increase in raw water charge density, does result in an increased coagulant demand required to minimize the surface charge of the organics. The variation in NOM fraction make-up and the coagulant used also influences the achievable residual DOC concentration at optimum zeta potential, especially when there is a portion of the hydrophilic NOM which cannot be removed through conventional coagulation processes, with either aluminum or iron based coagulants. The variations in NOM composition cannot be detected by conventional DOC monitoring but zeta potential measurement has been demonstrated to be a useful tool for optimizing coagulant dose.

IMPACT ON FILTRATION

Water treatment is a connected series of process units and as such the operation and performance of upstream units impacts on the downstream processes. A key example of this in water treatment is the use of depth filtration downstream of chemical pretreatment and clarification in the form of either DAF or sedimentation. Mechanistically, performance of all the processes is related to the size, density and charge of the particles to be removed and as such zeta potential is very important. For example, in the case of filtration zeta potential monitoring provides a good indicator of the level of retardation to attachment that will be experienced within the bed (McCarthy and Zachara, 1989). Consequently, previous studies have demonstrated the importance of optimisation based on charge measurements rather than dose and pH to provide a more robust operational approach to filter operation (Gregory and Carlson, 2003). This section of report describes results at bench, pilot and full scale in terms of operational zeta potential ranges for good filtration for comparison to the coagulation results reported above.

Method

Bench-scale

A kaolin suspension (100 $mg.L^{-1}$) was filtered through a bench-scale filtration rig at a loading rate of 10 $m^3.m^{-2}.h^{-1}$. The filter column was 290 mm in diameter and contained a bed of 30 cm British standard sieve (BSS) 16/30 sand on top of a 10 cm gravel layer. The zeta potential of the kaolin suspension was measured before filtration and the effluent turbidity recorded. The pH of the kaolin suspension was adjusted to varying pH's between 2 and 10, using NaOH (0.1-1 Mol) and the process repeated. This was to determine if the magnitude of the negative zeta potential value had an effect on the filtration performance.

Pilot-scale

Filtration experiments were also undertaken with the FCWTF pilot plant. The settled water from each train, obtained from coagulation trials detailed previously, was fed into two of the four dual media filters (15.24 cm diameter) containing 76.2 cm of anthracite and 30.48 cm of sand. The filters were operated in declining rate mode at an initial loading rate of 10 $m^3.m^{-2}.h^{-1}$. The zeta potential of the settled water was recorded, along with the filter effluent turbidity. The filters were backwashed with water before the beginning of each run, and no air scouring was used. The procedure consisted of a flow of 0.019 $m^3.h^{-1}$ for 10 minutes, with a five minute ramp up and ramp down, giving a total backwashing time of 15 minutes. Filter columns were tapped to ensure initial conditions, such as media packing and depth, were the same for each run.

Full-scale

Full-scale filtration data was obtained from a trial at FCWTF, during the period June 2nd to June 4th 2004. The plant has a design capacity of 325 MLD and uses the conventional treatment processes of coagulation with alum, flocculation, sedimentation, dual media filtration and disinfection. A polymer is added to aid flocculation and produce acceptable settled turbidities. Settled turbidities and filter effluent turbidities were monitored using SCADA system. Settled water samples were also taken in order to obtain spot zeta potential measurements during the period.

Results and discussion

Filtered turbidity remained stable at 0.1 NTU for settled water and between 0.1 and 0.2 NTU for dispersed kaolin when the zeta potential was low. A clear zeta potential boundary exists for both systems beyond which effluent turbidity rapidly increased and filtration was compromised (Figure 4.26). The range was system specific such that good filtration was observed up to zeta potentials of -15 mV for the settled water and -10 mV for the dispersed kaolin. As the filtration media was consistent in both studies the observed differences reflect the impact of particle character on filtration probably as a result of the organic layers on the particles within the settled water. The observed zeta potential range and achievable residual turbidity both concur with previous AwwaRF research on filter maintenance and operation (Logsdon *et al*, 2002) and suggest that a more robust operational practice is likely to maintained providing the magnitude of the negative zeta potential does not exceed -10mV.

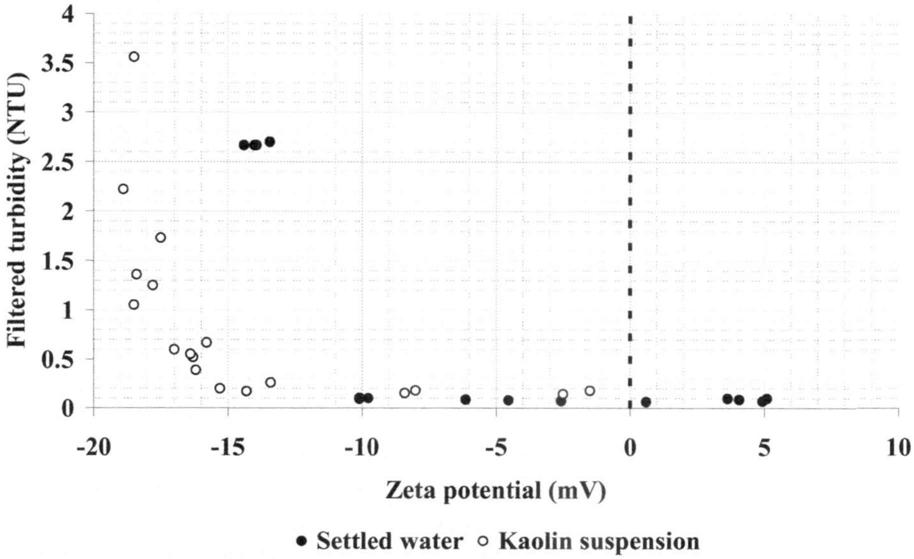

Figure 4.26 Filtrate turbidity against zeta potential for settled water and kaolin suspension.

The results from the bench scale were supported by observation of the full scale plant at FCWTF (Figure 4.27). To illustrate, for the period between June 2nd and June 4th 2004 the settled turbidity increased from a base line 1-2 NTU to a maximum of 8.6 NTU whilst the zeta potential was maintained within the range -1.8 to +0.2 mV. Filtered effluent turbidity remained stable at below 0.1 NTU throughout the run. The implication is that poor chemical pretreatment, generating high zeta potentials, cannot be corrected for during filtration whereas poor operation of upstream clarification units as seen by increased turbidly onto the filter can, to a degree, be successfully managed by good filtration operation. The operational cost of poor upstream management is reduced filter runs times as the bed becomes laden with material more quickly.

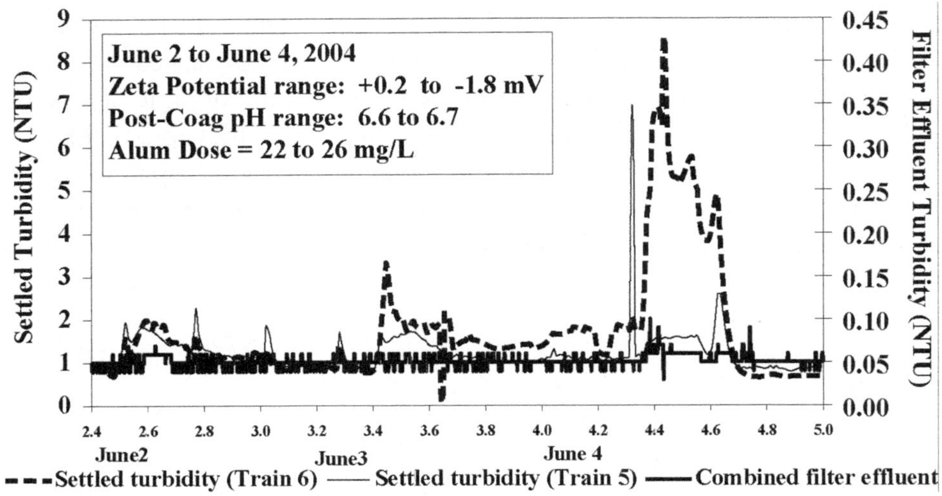

Figure 4.27 SCADA data from the full-scale FCWTF for settled turbidities (Trains 5 and 6) and combined filter effluent.

Detailed analysis of the filter runs on the pilot plant facility indicated that the filter effluent turbidity remained between 0.06 and 0.12 NTU for all alum doses, with and without polymer, as the negative zeta potential did not exceed -10 mV and concurs with previous trials at the same facility (Gregory and Carlson, 2003). Consequently the addition of polymer as a filter aid has no appreciable benefit in terms of the achievable turbidity from the filter. Instead the impact of polymer addition can be seen in terms of the filtration run time. To illustrate, in the case of a settled turbidity of between 1.1 and 3.3 NTU and an operating zeta potential of -3±1 mV the observed run time increased from less than 2 hours without polymer to more than 16 hour with polymer (Figure 4.28). In both case the settled water turbidity was very similar indicating that the polymer affected the bed loading profile enabling greater capacity of material to be captured before excessive head loss was developed.

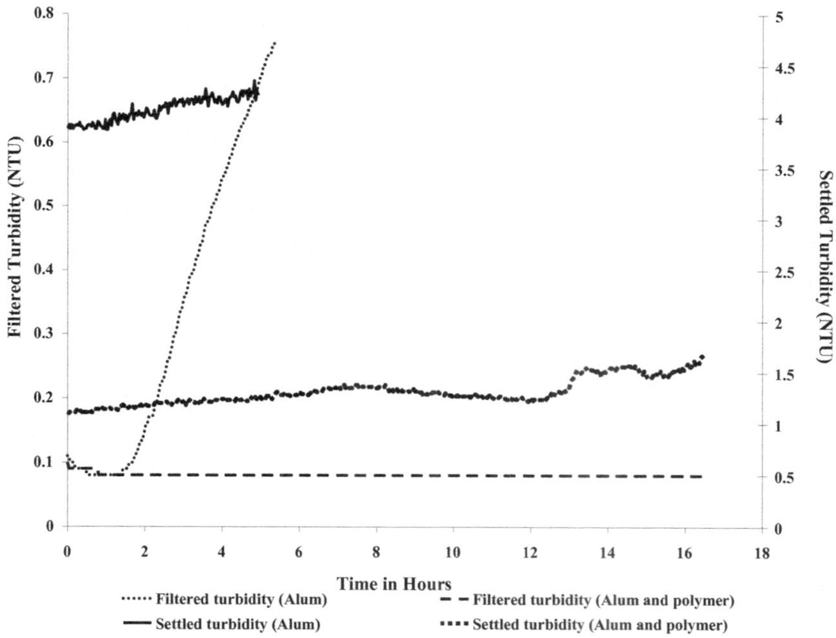

Figure 4.28 Filter effluent turbidity and run times after alum coagulation with and without the addition of polymer.

TREATMENT USING MIEX®

Process

Ion exchange is by no means a new process for the treatment of NOM. "Organic scavengers" have been in use since the 1940s as pretreatment for ion exchange demineralization plants treating upland waters for power station boiler feed where NOM has long been associated with the fouling of anion exchangers in this type of application, and fixed bed units using highly macroporous Type 1 or Type 2 strongly basic anion exchangers operating in the chloride cycle were developed for this "scavenging" duty. Organic molecules with carboxylic groups are readily sorbed by the resin but may become sterically trapped in the pores. Chloride cycle operation maximizes the swelling of the resin beads, expanding the resin matrix and allowing these large molecules to be desorbed. Regeneration with brine is normal but, depending on the nature of the organics, the addition of caustic soda to the brine regenerant is sometimes beneficial (Tilsley, 1979).

Fixed bed systems are large because of the large resin inventory associated with the slow kinetics of organic removal – hydraulic loadings of 15BV/h are normal for this duty – and the highly crosslinked resins are very expensive as are the pressure vessels required for the process, consequently very few ion exchange plants are in operation for the treatment of humic rich waters for potable supply. Against this background Orica Watercare, South Australian Water Corporation and the Commonwealth Scientific and Industrial Research Organisation (CSIRO) developed the magnetic ion exchange (MIEX®) process specifically for the removal of DOC from drinking water. Currently there are few published studies on the use of MIEX® and all the published data is based on waters in the US and Australia. Here a study on a range of waters from the UK sources including removal of fractions, MW ranges and process optimization is presented.

The MIEX® process is described in detail in Chapter 2 and involves adsorbing the DOC on to the MIEX® resin in a stirred contactor that disperses the resin beads to ensure maximum surface area availability. The magnetic core of the resin beads means that they agglomerate rapidly into fast settling particles that can be separated and regenerated by alkaline brine. The recovery rate is generally greater than 99.9% (Smith *et al.*, 2001) and any resin that is carried over is removed in downstream clarification processes.

The factors that may affect the performance of MIEX® are dose concentration, contact time, raw water quality, temperature, pH and mixing speed. Here the optimization of dose and contact time were carried out and the performance of the resin evaluated by trialing it on raw waters from differing geographical locations and catchments with widely variable characteristics in terms of DOC, UV_{254}, THMFP and MW. Also its effectiveness on the isolated NOM fractions from Albert WTW was assessed.

Bench Scale Jar Testing

Bench scale testing on solutions of raw water and water containing isolated fractions from the raw water collected in November 2000 were prepared to simulate the bulk water in terms of FAF, HAF, HPIA and HPINA concentration. This was achieved by dosing a predetermined amount of the isolated fraction into DI water whilst stirring on a magnetic stirrer. This was equivalent to the amount of fraction obtained from the water source. The DOC concentration of the fractions containing less than 3 mg/L was increased in order that the results of the coagulation could be measured accurately in terms of DOC and UV-Abs. The pH of the stock solutions was then measured and the pH adjusted to approximately 6 with NaOH (0.1M) and hydrochloric acid (HCl, 0.1M). To the stock solutions $NaHCO_3$ (20 mg L^{-1}) was added to achieve approximately 10 mg L^{-1} of alkalinity as $CaCO_3$, which again was approximately equal to that found in the raw water, and the pH measured again.

The coagulation and flocculation experiments were carried out using a Phipps & Bird PB-900 six-paddle jar tester. Aliquots of the solution (1 L) prepared previously were taken from the feed solution. Whilst stirring at 200 rpm for one minute, ferric sulfate (Ferripol XL, EA West) coagulant was dosed to the jars leaving one as a blank and the requisite amount of alkali added to adjust the pH to the chosen value; once again the pH was recorded. The jars were then stirred for 15 minutes at 30 rpm before settling for an additional 15 min before sampling. Samples of each of the jars were taken by filtering through glass fiber filter paper to remove any solids and each sample was analyzed for DOC, ultra-violet absorbance at λ=254 nm (UV-Abs) and hence SUVA.

MIEX® alone

MIEX® resin was prepared by measuring equal doses of resin to water in measuring cylinders and allowed to settle for approximately two hours, any necessary adjustments were made using a plastic pipette before an additional settling period of approximately one hour. Samples of water to be tested (1 L) were prepared and placed on the jar tester. The jar tester was set to 150 rpm and allowed to stabilize. The resin was shaken in the measuring cylinder and added to the water sample with any residual being rinsed into the jar with DI water. At the end of the allotted experiment time the treated water was immediately filtered to approximately 0.45 μm using glass fiber filter papers. If sampling was required during the process, five minute staggering of the resin additions is needed to allow for filtering.

MIEX® + Coagulation

MIEX® resin (200 mL) was prepared as in section 2.4 and added to the water to be treated (10 L) giving a MIEX® concentration of 20 mL L^{-1}. This was stirred at ~150 rpm for 5 minutes before the resin was removed via decantation and filtering. The resultant water was then subjected to the bench scale jar testing using a range of Ferripol doses as described previously in Chapter 5. The treated water was then tested for UV-Abs, DOC, SUVA and HPSEC.

Analytical Techniques

The samples were analyzed using the following methods. DOC (mg/L) was measured using a Shimadzu TOC-5000A, UV-Abs (m^{-1}) was measured using a Jenway 6505 UV/Vis spectrophotometer, SUVA ($m^{-1}.L.mg^{-1}$) was calculated as a ratio of the UV-Abs to DOC and THMFP was carried out using a method adapted from procedure 5710 in 'Standard Methods for the Treatment and Examination of Water and Wastewater' (American Public Health Association, 1992). The method involved buffering samples at pH 7, chlorinating samples with excess free chlorine and storing the sample at 20°C for 7 days to allow the reaction to approach completion. The total THM (chloroform, dichlorobromoform, dibromochloroform and bromoform) concentration was measured using a SRI 9300A gas chromatograph. HPSEC was carried out using an HPLC (Shimadzu VP Series, Shimadzu, Milton Keynes, UK) with UV detection set to 254 nm. The mobile phase was 0.1 M sodium acetate at a flow rate of 1 ml min^{-1}. The column was a TSK - gel G3000SW 7.5 mm (ID) × 30 cm and the guard column was TSK gel 7.5 mm (ID) × 7.5 cm (Tosoh Biosep GmbH, Stuttgart, Germany).

MIEX® optimization

A series of experiments to determine optimum dose and contact time for the MIEX® on Albert raw water shows that the best DOC and UV_{254} removal (90%) for both occurs at maximum dose (30 mL L^{-1}) and contact time (60 minutes). However the use of a relatively high MIEX® dose (20 mL L^{-1}) for a shorter contact time (30 minutes) shows that approximately 80% DOC and UV_{254} removal is achievable compared to 60% and 40% DOC and UV_{254} removal for a low MIEX® dose of 2 mL L^{-1} for the same contact time (Figures 4.29 and 4.30).

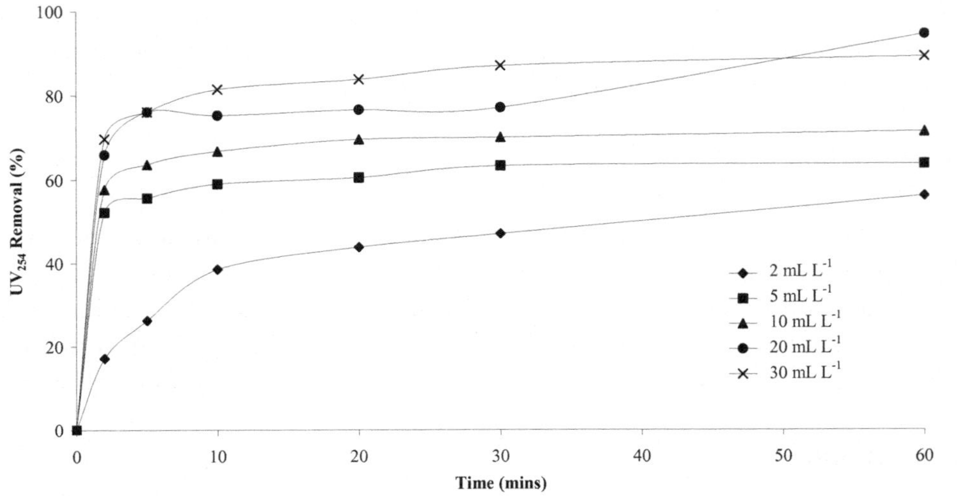

Figure 4.29 UV$_{254}$ optimization of MIEX® for the treatment of Albert raw water

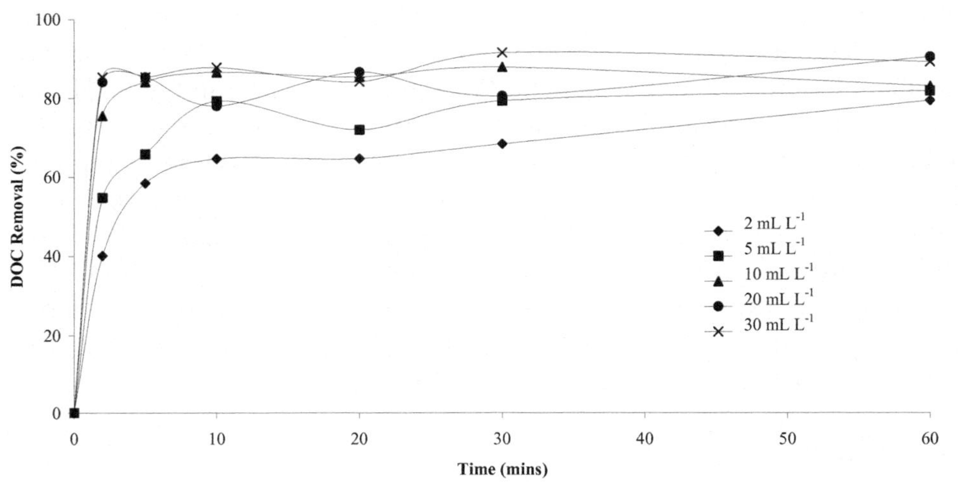

Figure 4.30 DOC optimization of MIEX® for the treatment of Albert raw water

The HPSEC data for raw water treated with 20 mL L^{-1} MIEX® for 30 minutes shows that peak 1 at 5.7 minutes (>5000 Da) is relatively unchanged by the addition of MIEX®. Peaks 2, 3, 4, 5 and 6 at 8.7, 8.9, 9.3, 9.7 and 10.3 minutes respectively (<5000 Da) are reduced significantly. However peak 7 at 10.7 minutes is less well removed (Figure 4.31).

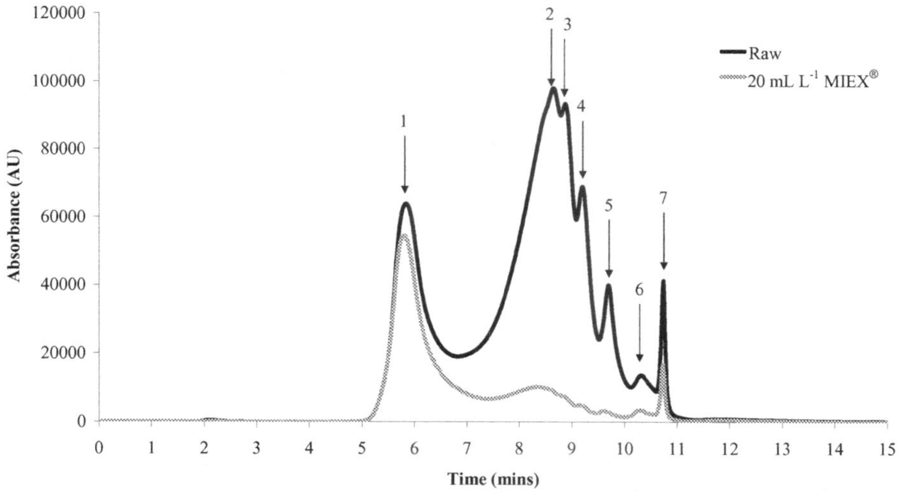

Figure 4.31 Effect of MIEX® treatment on the MW profile of Albert raw water

From this data, an optimum MIEX® dose and contact time of 20 mL L^{-1} and 30 minutes respectively was chosen for further studies. Finally, another basis for using a higher MIEX® dose (>20 mL L^{-1}) was that at low concentrations there was high dispersion of the resin beads and less agglomeration leading to slower settling rates.

MIEX® + coagulant optimization

The addition of a ferric based coagulant to Albert raw water showed that the maximum removal efficiencies were seen for a coagulant dose of approximately 15 mg L^{-1} at a pH of 4.5. The removal efficiencies achieved under these conditions were 94% and 95% for DOC and UV_{254}, respectively. Coagulating water previously treated with the optimized dose of MIEX® (20 mL L^{-1}, 30 minutes) showed that for a coagulant dose of 4 mg L^{-1} as Fe at a pH of 4.5 the maximum removal efficiencies for DOC and UV_{254} were 97% and 88% respectively (Figure 4.32). These results were comparable to waters of similar SUVA value from other sources.

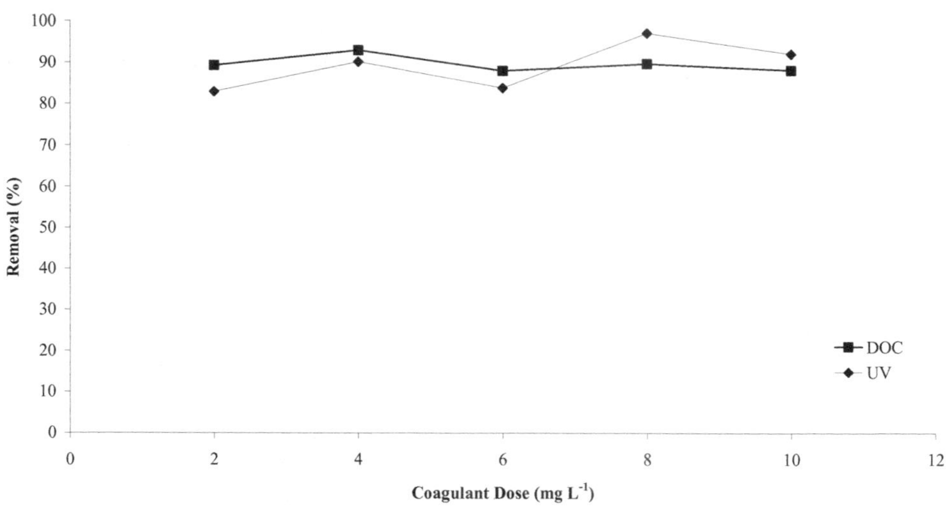

Figure 4.32 Optimization of MIEX® plus coagulation treatment of Albert raw water by UV_{254} and DOC removal

The HPSEC plot of MIEX® treated water coagulated with varying doses of coagulant show that peaks 1-4 (>2000 Da) are further removed whilst only a limited extra removal for peaks 5-7 (<2000 Da) is achieved by coagulation (Figure 4.33).

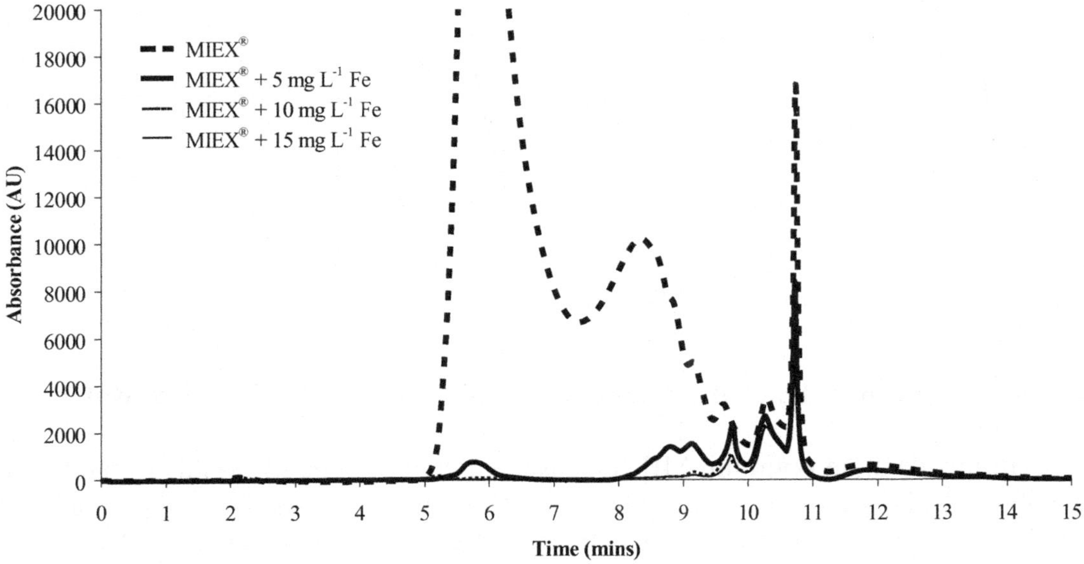

Figure 4.33 Optimization of MIEX® and coagulation treatment of Albert raw water UV_{254} MW profile

The results of the HPSEC show only the effect on the UV absorbing fraction of the water, and as previously stated the UV absorbing fractions (HAF and FAF) are regarded as the easiest to remove. Many of the recalcitrant components of the water do not adsorb UV at 254 nm or have a low MW that may be the cause of high THM levels in the treated water observed at many WTWs (Matilainen *et al.*, 2002). With this in mind, it is necessary to study the removal of the DOC as well. When MIEX® is employed on its own approximately 60% of the DOC is removed and 70% of the UV absorbing material. When an optimum ferric dose is used, the DOC and UV_{254} removal is improved up to 75% and 90%, respectively. A combination of MIEX® and a lower dose of ferric coagulant (5 mg L^{-1}) achieved a similar DOC removal and almost complete removal of the UV_{254} (Figure 4.34).

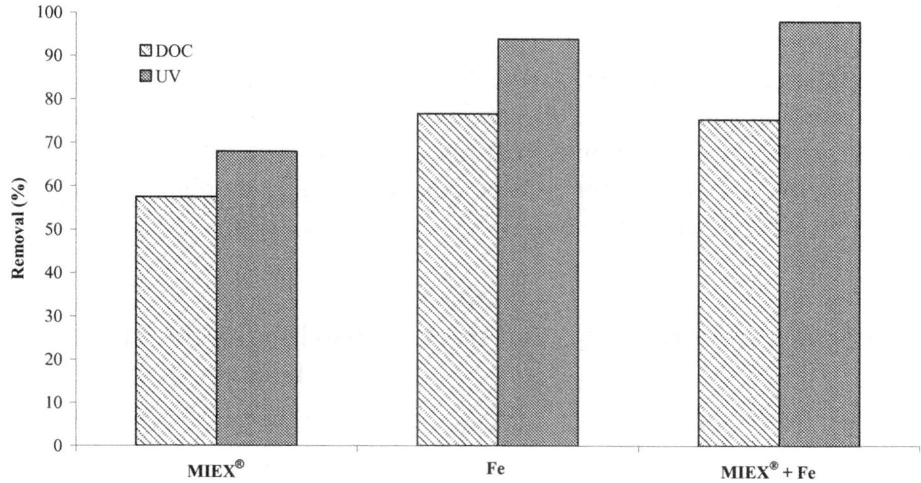

Figure 4.34 Comparison of conventional coagulation, MIEX® treatment and a combination of MIEX® and coagulation for the removal of UV_{254} and DOC

Similar results have been published where MIEX® in addition to a reduced coagulant dose of 60 mg L^{-1} ferric sulfate gave an average DOC level of 3.9 mg L^{-1} compared to an average DOC level of 5.5 mg L^{-1} when a ferric sulfate dose of 240 mg L^{-1} was used (Bourke *et al.*, 2002).

Studies were also carried out 'on site' to determine the optimum conditions for removal of THMFP precursors, and it was shown that as before a lower contact time but high dose of MIEX® was equally as effective for the removal of THMs compared to low dose, long contact time (Figure 4.35). This was combined with ferric coagulation, and the amount of coagulant required to achieve equivalent removals was considerably lower than needed when conventional coagulation was employed.

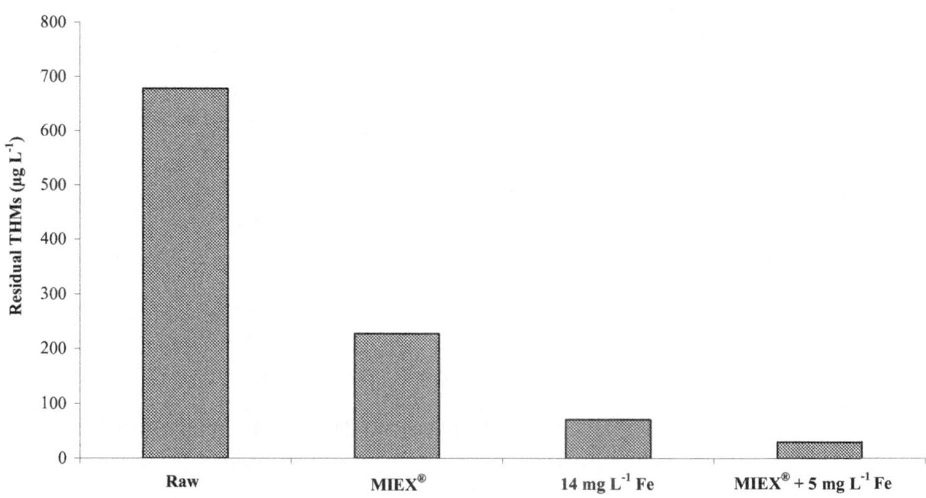

Figure 4.35 Residual THMs of raw, MIEX® treated, conventional coagulation and a combination of MIEX® and coagulation

Seasonal Effect on the Use of MIEX®

Previous research has focused on the treatment of waters at one time during the year (Singer and Bilyk, 2002; Drikas *et al.*, 2003 and Drikas *et al.*, 2003b). Here the effect of MIEX® on the treatment of water was tested on Albert raw water at three different times during the year. The nature and quality of the water varies considerably throughout the year as does the distribution of the NOM fractions (Table 4.5).

Table 4.5
Seasonal changes in DOC, UV_{254} and fraction distribution for Albert raw water

	pH	DOC (mg L^{-1})	UV_{254} (m^{-1})	SUVA (L $mg^{-1}m^{-1}$)	Fraction Distribution FAF (%)	HAF (%)	HPIA (%)	HPINA (%)
Apr 02	5.9	7.5	38.1	5.08	52	20	12	16
Oct 02	6.9	11.7	52.3	4.47	52	23	9	16
Mar 03	6.6	11.0	58.3	5.30	44	27	19	10

Consequently, the maximum removals achievable using MIEX® (10 mL L^{-1}, 60 minute contact time) for both DOC and UV_{254} varied markedly from 66% to 82% for DOC and 66% to 87% for UV_{254} (Table 4.6).

These removals generally relate to the fact that the highest removal (April 2002) is seen when both the DOC and UV_{254} are at their lowest, and likewise when the DOC and UV_{254} are at their highest (October 2002) the lowest removals are seen.

Table 4.6
Removal efficiencies for the MIEX® treatment of seasonally different waters

	DOC			UV_{254}		
	Initial (mg L^{-1})	Final (mg L^{-1})	Removal (%)	Initial (m^{-1})	Final (m^{-1})	Removal (%)
Apr '02	7.8	1.4	82.1	32.5	4.4	86.5
Oct '02	11.2	3.8	66.1	50.7	17.5	65.5
Mar '03	11.0	3.2	70.9	51.3	11.5	77.7

Fraction removal

The effect of MIEX® on the individual isolated fractions was examined in this study to assess the removal efficiency and relate this to the assumptions regarding preferential removal of lower MW material. It also enabled the MIEX® to be directly compared to the optimum removal achievable using coagulation. The results confirm that MIEX® removes a large amount of the FAF (90%) with respect to UV_{254} and achieves similar removals to ferric for both the HAF and HPIA (90% and 75%, respectively). It was also noted that MIEX® performs better at removing the HPINA fraction which has been shown in the past to be the most recalcitrant fraction when coagulation is employed (Bolto *et al.*, 2001). In terms of DOC removal the FAF and HAF fractions are readily removed by both the MIEX® and conventional coagulation (60%-70% respectively). This is slightly unexpected as the MIEX® previously appeared to be less adept at removing high MW compounds that would be expected to be found in HAF. A disadvantage of MIEX® compared to coagulation is its reduced removal efficiency of DOC for the HPIA fraction (Figures 4.36 and 4.37).

Although there are no published data on the effect of MIEX® on isolated NOM fractions, results published by Singer and Bilyk (2002) show that treating a high SUVA water (4.5 $m^{-1}.L.mg^{-1}$) with MIEX® prior to conventional coagulation increases the DOC removal from 50% to 87%. It can also be seen that using MIEX® for the treatment of low SUVA water also increases the DOC removal from 23% to 46%.

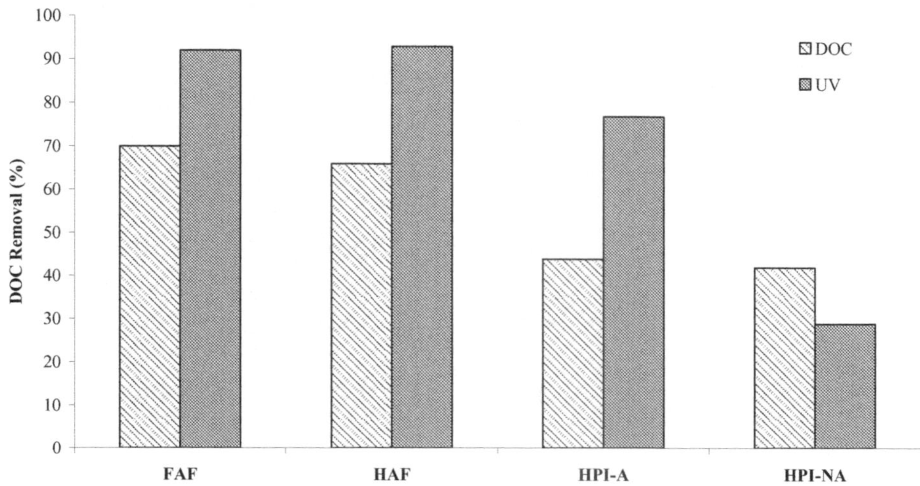

Figure 4.36 Maximum DOC and UV_{254} removal for the isolated fractions using MIEX® treatment only (10 mL L^{-1}, 60 minutes)

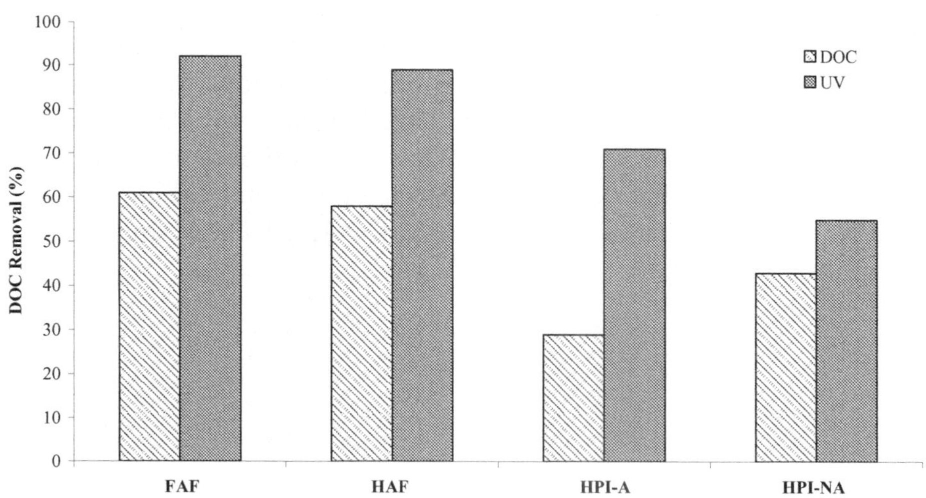

Figure 4.37 Maximum DOC and UV_{254} removal for the isolated fractions using ferric treatment only (each fraction optimized for dose and pH)

Evaluation of other UK water sources (2003-2004 survey)

Water from five different WTWs was treated with MIEX® to assess the effect of differing water quality and catchments (moor land reservoir, low land river and upland impounded) have on performance. The moor land water sources are from Albert (Yorkshire Water) and Bamford (Severn Trent), the low land river samples are from Draycote (Severn Trent) and Loftsome Bridge (Yorkshire Water). The upland impounded water is from Amlaird (Scottish Water). A summary of the main water chemistry characteristics are shown in Table 4.7.

Table 4.7
Raw water chemistry characteristics of trialed water sources

Water Source and Sampling Date	Albert (10.06.2003)	Bamford (3.09.2003)	Draycote (10.11.2003)	Loftsome Bridge (22.2.2004)	Amlaird (17.12.2003)
pH	6.0	5.7	8.2	7.8	6.0
DOC (mg L^{-1})	8.18	8.06	15.63	4.30	20.85
UV_{254} (m^{-1})	47.3	34.7	34.0	19.5	86.3
Turbidity (NTU)	5.02	1.35	0.75	14.6	3.08
SUVA (L $mg^{-1}m^{-1}$)	5.78	4.31	2.18	4.53	4.14

The HPSEC data for the raw waters studied are shown in Figure 4.38. The waters from Albert and Bamford show similar MW ranges although peak 1 (>5000Da) for Bamford is approximately 50% of the intensity compared to that of Albert. The water from Amlaird has significantly higher absorbance intensity than the other waters due to its high color and UV_{254} absorbance shown in Table 4.7. The water collected from Draycote is mainly composed of compounds that have a mid to low MW range and has no absorbance in the high MW region (5-7 minutes). Finally the MW profile of Loftsome Bridge raw water exhibits a very low intensity throughout the MW range and therefore HPSEC with UV_{254} detection would be of little use as an indicator for removal efficiency.

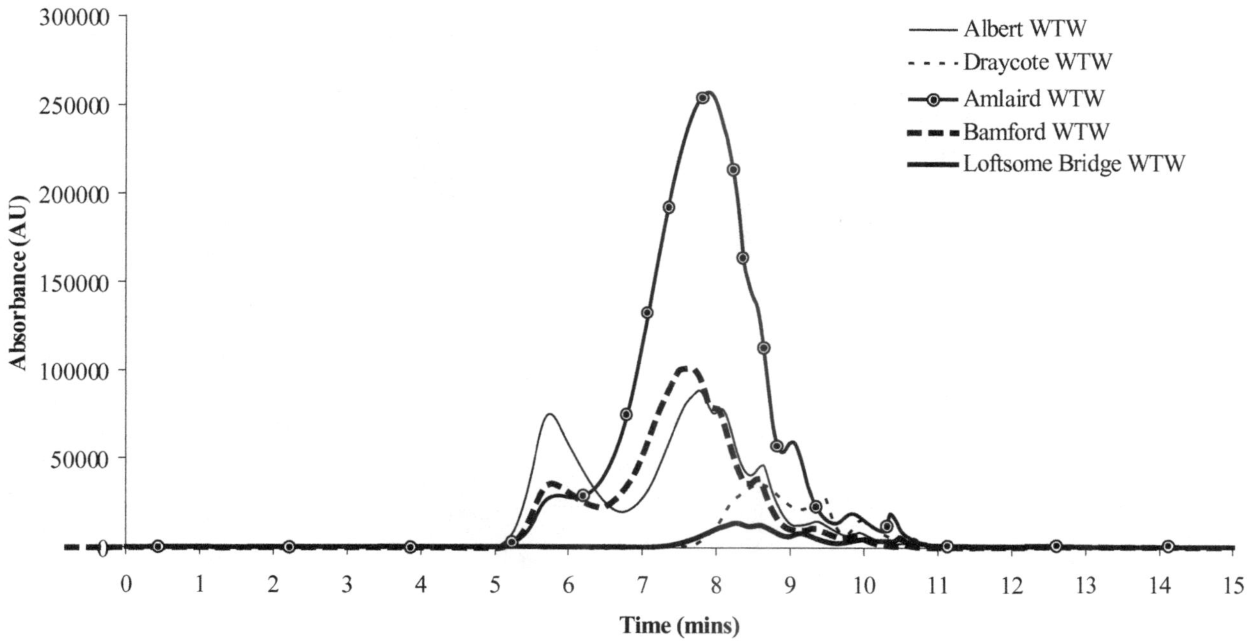

Figure 4.38 Comparison of MW profile for the trialed raw waters

Initially the MIEX® optimization of the trialed waters was carried out to determine the contact time and dose required to achieve optimum removal. The optimum was not necessarily the highest removal efficiency but rather the lowest dose and contact time that achieved comparable removals (Figure 4.39). The results for Loftsome Bridge WTW are not shown here due to the low UV_{254} absorbance.

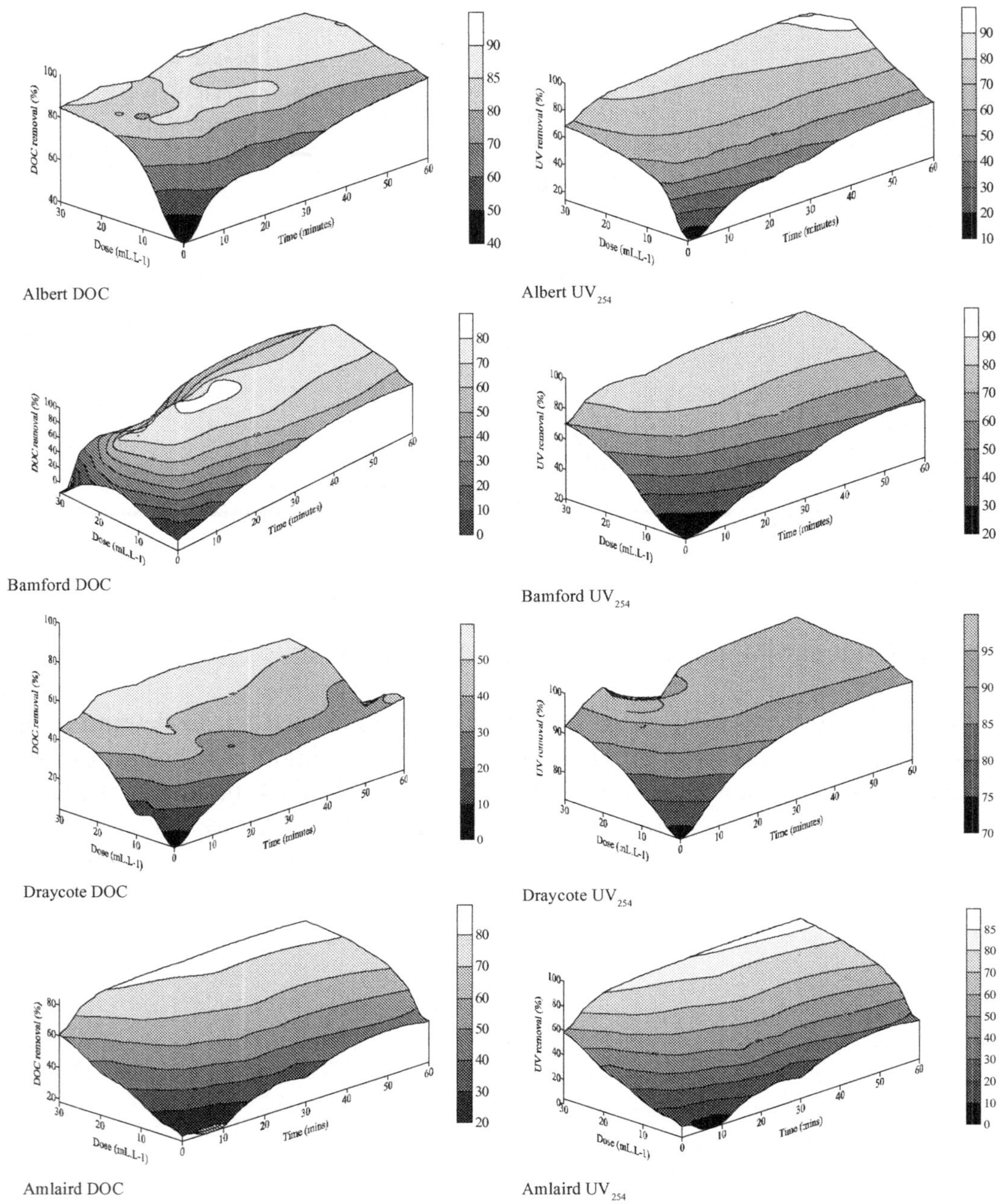

Figure 4.39 MIEX® optimization of waters studied during evaluation

Albert WTW

The results obtained by coagulation and MIEX® plus coagulation for the removal of DOC on water collected in June 2003 are similar with over 90% removal. Similarly the optimum UV_{254} removals for coagulation and MIEX® plus coagulation is approximately 96%. These removals are higher than experienced previously with conventional coagulation during other seasonal periods. This may be attributed to a higher proportion of hydrophobic material in the water during this period which as this leads to higher observed removal efficiencies.

This data along with further HPSEC data for raw water treated with MIEX® and ferric individually shows that MIEX® is better suited to removing mid and lower MW parts of the raw water and less effective for the high MW portion. Ferric coagulation at optimum dose and pH (14 mg L^{-1} as Fe at pH 4.5) shows excellent removals of the high and mid MW material in the water. Although comparable removals of the lower MW compounds is also observed, a decrease in efficiency would be seen if the dose was reduced in order that coagulant costs and sludge are to be minimized or there was an increase in DOC levels (i.e. coagulant to DOC ratio reduced). A combination of both MIEX® and a considerably lower dose of ferric (5 mg L^{-1} as Fe) show an improvement in UV_{254} removal in terms of HPSEC analysis (Figure 4.40).

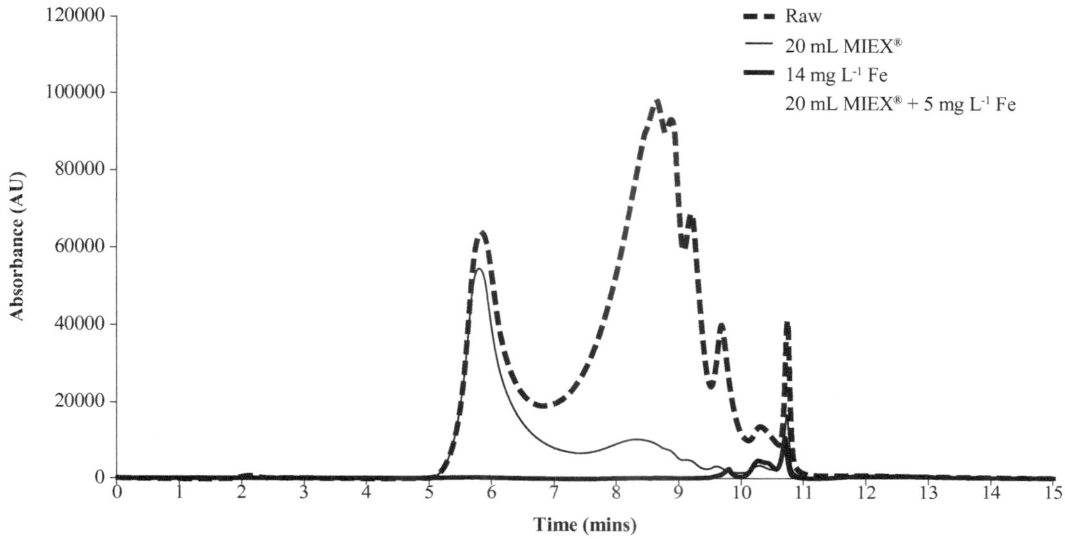

Figure 4.40 Comparison of MIEX®, ferric and a combination of MIEX® plus ferric on the removal of UV_{254} found within raw reservoir water (Albert).

Bamford WTW

From the raw water characteristic data shown in Table 4.2 and the HPSEC of the raw water in Figure 4.37 it can be seen that Bamford raw water is similar to that of Albert with the exception of SUVA which is lower (4.31 L mg^{-1} m^{-1} compared to 5.78 L mg^{-1} m^{-1} for Albert). The DOC removals achieved by both coagulation and MIEX® plus coagulation show that a lower dose of coagulant (6 mg L^{-1}) when combined with MIEX® achieve similar removals (~80%) to that of conventional coagulation (15 mg L^{-1} as Fe at pH 4.5). Similarly the UV_{254} data shows that the use of MIEX® prior to coagulation does not significantly increase removals (87% and 84% for MIEX® treated and conventional coagulation respectively).

Draycote WTW

Draycote raw water has considerably different characteristics to the other waters reported previously. The UV_{254} is relatively low when compared to the DOC (SUVA = 2.18 L mg^{-1} m^{-1}), suggesting that the water contains more low MW hydrophilic material. As a result, coagulation is not as effective for the removal of the DOC. Approximately 46% DOC removal is observed for a coagulant dose of 10 mg L^{-1} as Fe at pH 4.5. A lower coagulant dose (2 mg L^{-1}) in combination with treatment with MIEX® treatment (20 mL L^{-1} for 30 minutes) achieves a DOC removal of approximately 76%. Comparable results are observed for the UV_{254} removal, where 71% and 97% removals are observed for coagulation (10 mg L^{-1}) and MIEX® plus coagulation (20 mL L^{-1}, 30 minutes plus 2 mg L^{-1} Fe), respectively.

The HPSEC data for coagulation, MIEX® treatment and combination of both confirms that coagulation alone achieves the lowest UV_{254} removal over the full MW range. This is to be expected as coagulation is usually less effective for low SUVA waters. The absence of the high MW material usually seen at 5.7 minutes is most effectively removed by coagulation (Figure 4.41).

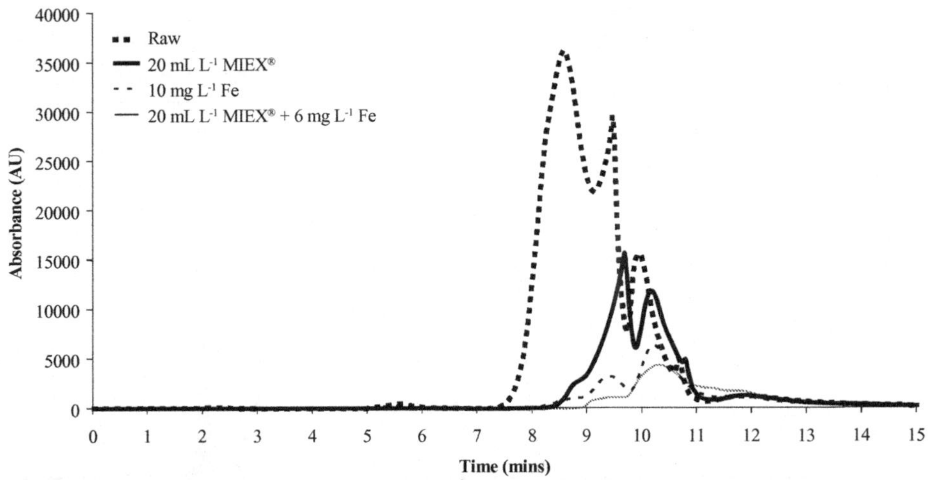

Figure 4.41 Comparison of MIEX® and ferric on the removal of UV_{254} found within raw reservoir water (Draycote)

Loftsome Bridge WTW

Loftsome Bridge water is low in both UV_{254} absorbing compounds and DOC but has a relatively high SUVA value (4.53 L mg^{-1} m^{-1}) suggesting that it should be suitable for treatment by coagulation. Surprisingly, only 30% removal of DOC is achievable by coagulating with 10 mg L^{-1} Fe at pH 4.5. Slightly higher removals are achievable by coagulation following MIEX® treatment (33%), with a reduced coagulant dose of 1 mg L^{-1}. Loftsome Bridge water has a high alkalinity content (~250 mg L^{-1} as $CaCO_3$) and exhibited some peculiar results when the DOC concentrations were analyzed (Table 4.8).

Table 4.8
Organic carbon analysis by standard TOC method (Loftsome Bridge)

	Fe	MIEX®	TC	IC	DOC	TC Rem	IC Rem	DOC Rem
	Mg L^{-1}	mL L^{-1}	mg L^{-1}	mg L^{-1}	mg L^{-1}	%	%	%
1	0	0	34.16	29.86	4.30			
2	10	0	9.83	6.85	2.99	71.1	77.1	30.5
3	0	20	30.80	29.69	1.11	9.8	10.6	74.1
4	2	20	6.011	3.132	2.88	82.4	89.5	33.0

Sample 3 (MIEX® 20 mL L^{-1}) shows better DOC removal than sample 4 (MIEX® + Fe). The spurious results can be attributed to the accuracy of the TOC analysis. The manufacturer's specification states a 5% error on both the total carbon (TC) and inorganic carbon (IC). Due to the high values for both TC and IC, the combination of these errors gives a range of -1.5 to 5.1 mg L^{-1} for the MIEX® treated water. The use of non purgeable organic carbon (NPOC) as an indicator for the removal efficiency was assessed. With this method, the inorganic carbon is purged out acid digestion and nitrogen stripping prior to detection by and should lead to a reduction in the instrument error. The results showed that a 60% reduction in the NPOC was observed for a coagulant dose of 10 mg L^{-1} at pH 4.5. When a coagulant dose of 2 mg L^{-1} is used following MIEX® treatment (20 mL L^{-1}, 30 minutes) a 65% removal efficiency is measured (Table 4.9). Also, a SUVA value of 3.46 L $mg^{-1}m^{-1}$ is determined when NPOC is used to measure the organic content, indicating that the water would be less suitable for coagulation than was previously thought.

The use of NPOC as a method for determining the effectiveness of a treatment process on high alkalinity waters was shown to have less error associated with its measurement in terms of organic carbon content. Similarly high UV_{254} removals of 87% were achieved for both coagulation (10 mg L^{-1} as Fe) and MIEX® followed by coagulation (1.5 mg L^{-1} as Fe). The HPSEC data for Loftsome Bridge was inconclusive due to its low initial absorbance mentioned previously.

Table 4.9
Organic carbon analysis by NPOC method (Loftsome Bridge)

Fe mg L^{-1}	MIEX® mL L^{-1}	NPOC mg L^{-1}	NPOC Removal %
0	0	5.63	
10	0	1.91	66.1
0	20	2.62	53.5
1.5	20	2.08	63.0

Amlaird WTW

Amlaird raw water had the highest DOC of any of the waters tested (20.85 mg L^{-1}) and the highest concentration of high to medium MW species (Figure 4.37). Consequently a high dose of coagulant (>15 mg L^{-1}) was required to achieve a 85% DOC removal. The effect of treating the water with MIEX® prior to coagulation was that for a coagulant dose of 0.5 mg L^{-1}, a DOC removal efficiency of 93% was achievable. Due to the exceptionally high UV_{254} in the raw water, 99% removal was achievable by both conventional coagulation and coagulation of MIEX® treated water.

However, when looking at the HPSEC data for the conventional and MIEX® coagulation it can be seen that coagulation alone does not remove all of the material eluting between 8 and 9.5 minutes (~3000 Da) whereas the MIEX® treated coagulation appears to remove all of this material leaving only a low concentration of low MW material eluting at approximately 10 minutes (2000 Da) (Figure 4.42).

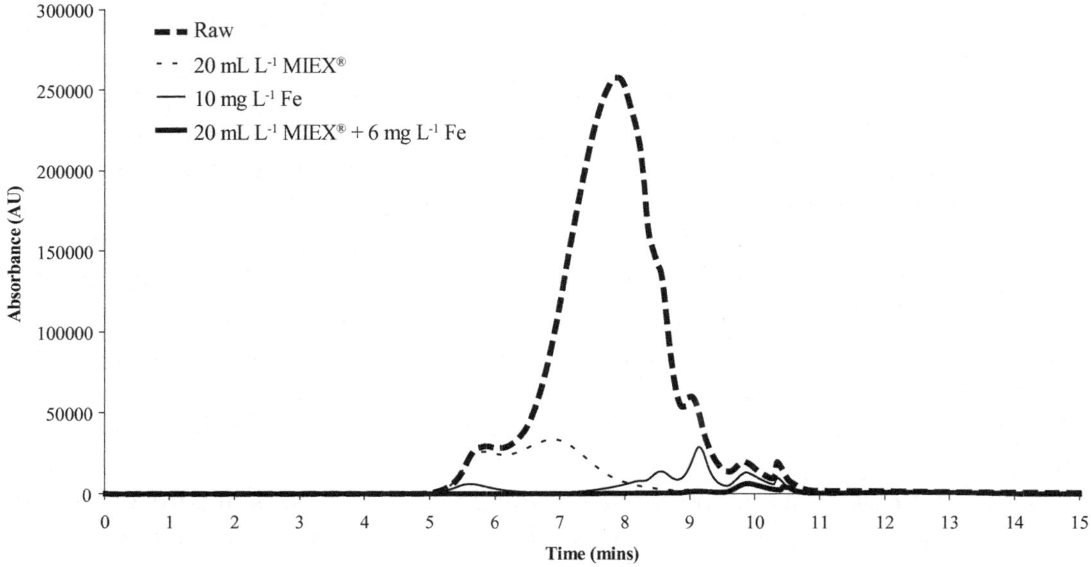

Figure 4.42 Comparison of MIEX® and ferric on the removal of UV_{254} found within raw reservoir water (Amlaird)

Summary

A summary of the removal efficiencies for all of the waters studied shows that the combination of MIEX® and a reduced coagulant dose improves the DOC removals in all cases. The removal of the UV_{254} is not as effective for MIEX® treated water. This is because the material that absorbs UV at 254 nm is an indicator for conjugated species and not all DOC absorbs UV at 254 nm. As a result of a lower dose of coagulant being used following MIEX®, this UV absorbing material is not as well removed although the differences in removal are not significant. All MIEX® doses and contact times used in the trials were 20 mL L^{-1} for 30 minutes (Table 4.10).

Table 4.10
Comparison of the DOC and UV_{254} removals for the studied water sources with conventional and MIEX® pre treated coagulation

Water Source	SUVA (L $mg^{-1}m^{-1}$)	Treatment	Coagulant Dose (mg L^{-1})	Residual DOC (mg L^{-1})	DOC Removal (%)	Residual UV_{254} (m^{-1})	UV_{254} Removal (%)
Albert	5.78	Fe	14	0.5	93.9	1.9	96.0
		MIEX®+ Fe	5	0.6	92.9	1.4	97.0
Bamford	4.31	Fe	15	1.3	83.8	0.3	99.2
		MIEX®+ Fe	6	1.1	86.9	0.6	98.7
Draycote	2.18	Fe	10	5.2	45.7	9.9	70.9
		MIEX®+ Fe	2	2.3	75.9	1.2	96.5
Loftsome Bridge	3.46	Fe	10	1.9*	66.0	2.5	87.2
		MIEX®+ Fe	1.5	2.1*	63.0	2.3	88.2
Amlaird	4.14	Fe	15	3.4	83.8	0.2	100.0
		MIEX®+ Fe	0.5	1.5	92.8	1.4	98.4

*-Measured as NPOC

MIEX® summary

Comparing the results of conventional coagulation with MIEX® shows that conventional coagulation achieves better removals for raw waters with high SUVA values (>4 L mg^{-1} m^{-1}) and high DOC concentration (>8 mg L^{-1}). Draycote has a low SUVA (2.18 L mg^{-1} m^{-1}) and is consequently more difficult to treat with coagulation alone. This may be explained by coagulation acting as a charge neutralization/complexation process at low pHs and hydrophilic waters are usually less charged than hydrophobic waters making this process limited. The pre treatment with MIEX® almost doubles the DOC removal efficiency.

Figure 4.43 is an attempt to correlate optimum DOC removal achieved for the different waters against their raw water SUVA values. The figure suggests that treating water with ferric is more efficient if the raw water SUVA is greater than about 5 L mg^{-1} m^{-1}, whilst MIEX® followed by ferric coagulation performs better at lower SUVA values. This suggests that MIEX® achieves a greater reduction in the hydrophilic fractions than does ferric coagulation.

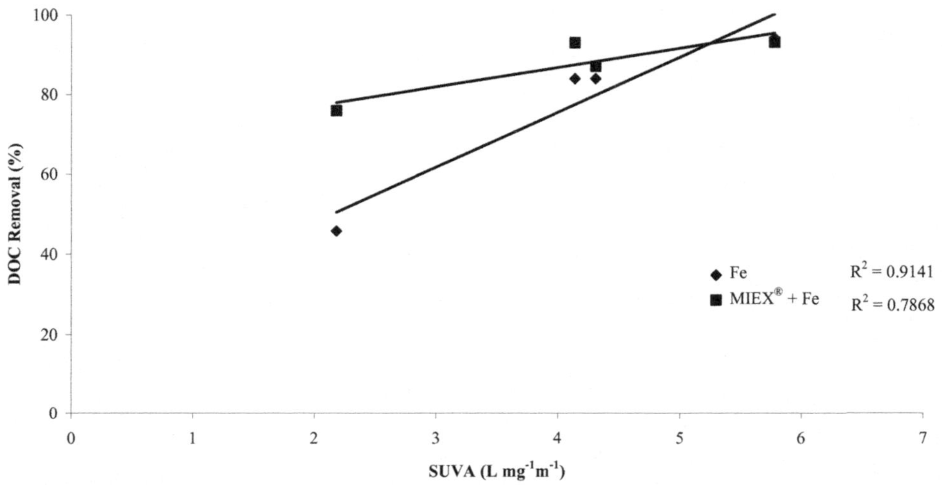

Figure 4.43 DOC Removal vs. SUVA for waters treated by conventional and MIEX® pre treated coagulation

By contrast, the combination of MIEX® and alum gives consistently higher DOC removal than alum alone across the range of SUVA values. (Figure 4.44).

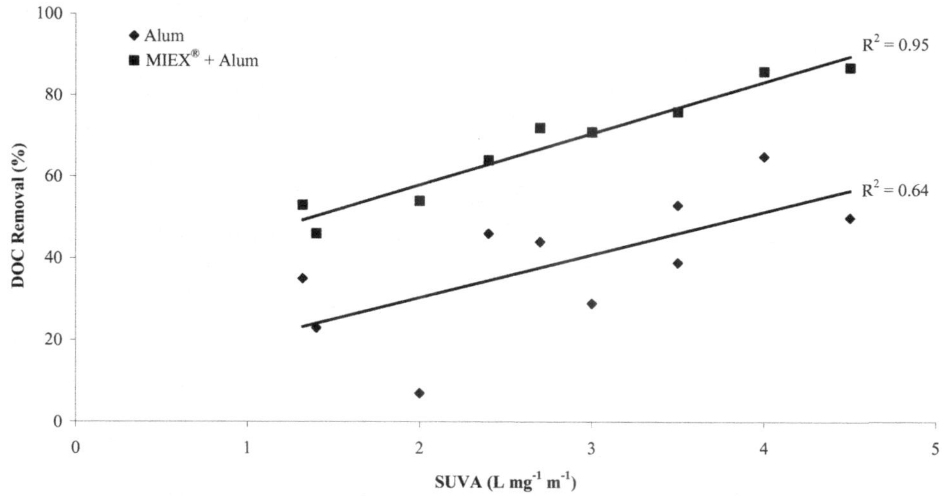

Figure 4.44 DOC vs. SUVA correlation for waters treated by conventional and MIEX® pre treated coagulation (Singer and Bilyk, 2002; Drikas *et al.*, 2003; Drikas *et al.*, 2003b).

The use of MIEX® as a pre treatment prior to coagulation appears to offer a very robust option for the treatment of humic rich waters. The advantages of its use are the reduction in costs for both coagulant purchase and sludge disposal. Also the variations in the water quality from seasonal changes and geographical location appear to have little effect in the efficacy of the process.

PROCESS COMPARISON

The literature review identified three options for treating organic rich waters (i) coagulation (ii) ion exchange (MIEX®) and (iii) pDADMAC. The sections have focused on optimizing the use of coagulation and MIEX. Here a series of jar tests were undertaken to evaluate the three options on a single water sample from Albert WTW. The water had an initial DOC of the raw water was 8.5 (± 2.5) mg L^{-1} and a UV_{254} of 50.6 (± 2.5) m^{-1}.

Table 4.11 shows a significant variation in the optimum pH depending on the choice of coagulant, and typical relationships between coagulation pH and both residual turbidity and UV_{254} are shown in Figure 4.45. In the case of the ferric based systems optimum DOC removals of 83, 87 and 87 % were achieved at pH 4.5 for the ferric only, ferric + pDADMAC and MIEX® + ferric systems respectively. MIEX® pre-treatment showed good DOC removal of 75 % and the subsequent ferric dose could be reduced to 4 mg L^{-1} to achieve similar removals to the higher ferric dose. In comparison, an optimized DOC removal of 72 % was observed when coagulating alum at pH 6.5 and 50% when coagulating with pDADMAC at pH 6.2. Settled turbidities for all of the optimized systems were between 1-2 NTU.

Settled turbidities for all of the optimized systems were in the range 1-2 NTU, and comparison of the different treatments using HPSEC showed that the different options removed different MW portions of the NOM (Figure 4.46). High MW molecules pass through the HPSEC column at a faster rate than low MW species, therefore the lower peak numbers of the raw water profile are indicative of higher MW relative to the higher peak numbers. Through calibration of the column, each of the peaks can be roughly equated to a specific MW range with peaks 1-2 being >5000 Daltons (Da), peaks 3-4 between 2000-5000 Da and peaks 5-6 <2000 Da. For each of the treatment options shown, preferential removal of the high MW organics was seen with poor removal of the low MW compounds. This was especially observed for MW of <2000 Da. Ferric and ferric plus polymer showed better removal of the 2000-5000 Da organics when compared to aluminum, whilst the polymer showed poor removal at MW <5000 Da.

Table 4.11
Summary of optimum coagulation conditions and removal characteristics.

System	Optimized coagulant dose ($mg.L^{-1}$)	Optimized pH	DOC removal (%)	UV_{254} removal (%)
NOM + Fe	8 (as Fe)	4.5	83	93
NOM + Al	10 (as Al)	6.5	72	81
NOM + poly	10 (as poly)	6.2	50	71
NOM + Fe + poly	8 (as Fe) + 1 as (poly)	4.5	87	90
NOM + MIEX	20*	5.5	76	77
NOM + MIEX + Fe	20* + 4 (as Fe)	4.5	87	98
Kaolin + Fe	8 (as Fe)	6	N/A	N/A

*($mL.L^{-1}$)

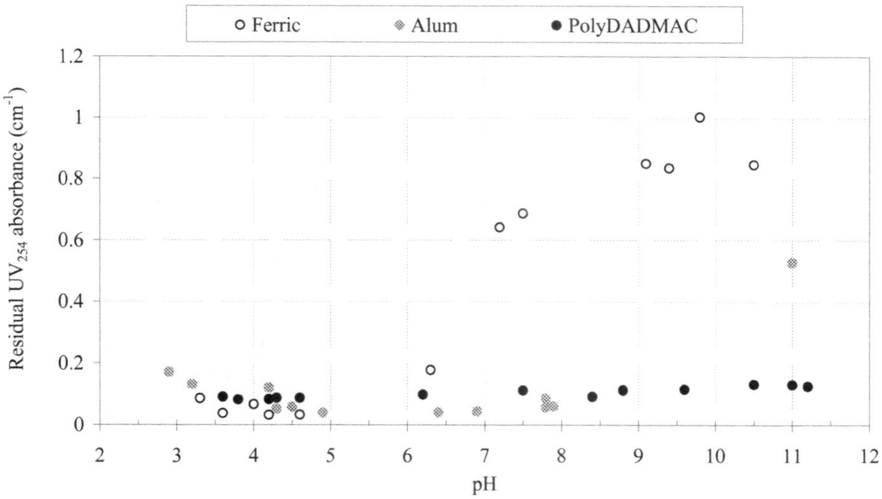

a) UV_{254} residual for the treatment options.

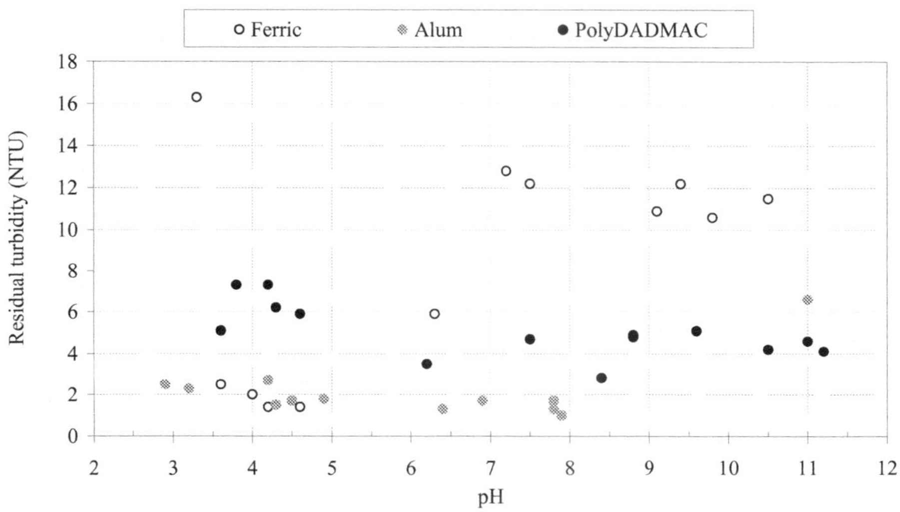

b) Residual turbidity for the treatment options.

Figure 4.45 The coagulant optimization experiments for the different treatment options

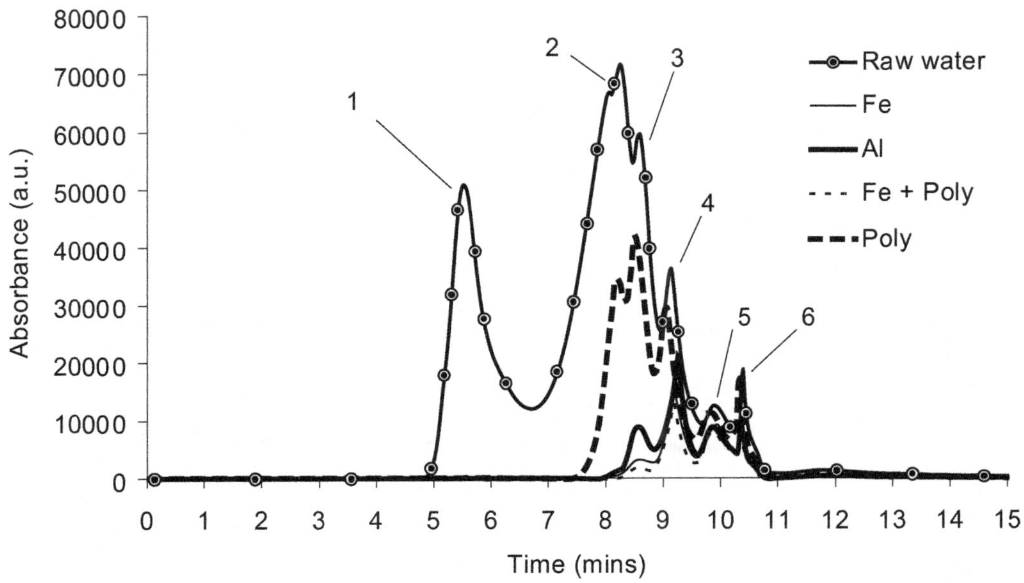

Figure 4.46 HPSEC chromatogram profiles of raw reservoir water and the treated water using different coagulants.

Comparison of the ferric and MIEX® systems showed a significant change in the MW removals (Figure 4.47). The coagulant and ion exchange resin preferentially remove different portions of the NOM. The ion exchange resin remov the medium to low MW whilst barely impacting on the high MW material (peak 1). When the ferric coagulant was added after MIEX®, the high MW compounds were then subsequently removed. The combined system giving rise to improved NOM removal across all MW.

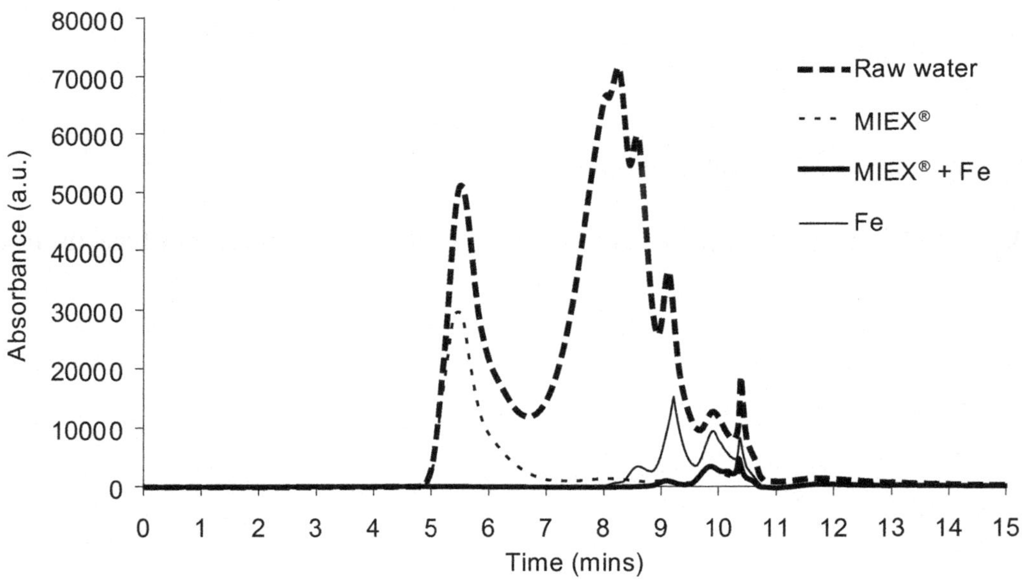

Figure 4.47 HPSEC chromatogram profiles of MIEX® treated waters compared with the raw water and iron treatments.

CHAPTER 5
REVIEW OF FLOC STRENGTH, BREAKAGE AND MEASUREMENT

FLOC STRENGTH

Floc size is an important parameter in many solid-liquid separation processes and, if flocs are grown to optimum size, it is essential that they are not subsequently broken into smaller fragments. Small particles are generally less efficiently removed (Boller and Blaser, 1998). Smaller particles settle more slowly than larger particles of similar density, and flocs formed for removal in dissolved air flotation that subsequently break up into many smaller parts may be captured less efficiently by air bubbles. In addition, flocs that are removed using membrane filtration will foul membranes if small pieces of floc break off and plug membrane pores.

Although most water treatment unit operations are designed to minimize shear forces, inevitably flocs are subjected to shear at certain points, such as the impeller zone of flocculating tanks, pumps, in processes like dissolved air flotation (DAF) and in transfer over weirs and ledges (McCurdy *et al.*, 2004). If flocs are to remain intact through these high shear regions they must have inherent strength.

Floc strength is dependent upon the strength and number of individual interparticle bonds between the components of the aggregate (Parker *et al.*, 1972; Bache *et al.*, 1997), so a floc will break if the stress applied at its surface is larger than the bonding strength within the its structure (Boller and Blaser, 1998). Increased floc compaction is considered to increase floc strength due to an increase in the number of bonds holding the aggregate together. Leentvaar and Rebhun (1983) also list the size and shape of floc microparticles as being an important consideration for floc strength.

Whilst the importance of floc strength has long been recognized, the development of a satisfactory technique to quantify it has proved to be difficult. This is partly due to the inherent complexity, fragility and variation in floc size, shape and composition of flocs and also due to a generally accepted view that there are two modes of floc rupture (Parker *et al.*, 1972; Francois, 1987; Yeung and Pelton, 1996; Mikkelsen and Keiding, 2002). These have been classified as surface erosion and large-scale fragmentation. Surface erosion is the removal of small particles from the floc surface resulting in an increase in the small particle size ranges. Large scale fragmentation is the cleavage of flocs into pieces of a similar size without an increase in primary particle concentration. The problem of describing strength arises from the fact that these two rupture modes are thought to be caused by different stresses (Yeung and Pelton, 1996). Fragmentation is thought to occur from tensile stress acting normally across the whole floc, whilst erosion is due to the shearing stress acting tangentially to the floc surface (Figure 5.1). In addition there are complex interpretations of floc strength data arising from relative eddy size which will be discussed in later sections.

A review of the literature shows there to be no established standardized floc strength test, although a number of techniques have been evaluated. Floc strength may be broadly considered in terms of the energy required to break flocs under tension, compression or shear (Zhang *et al.*, 1999). However, finding ways of quantifying the energy input for floc breakage has not been easy. There is therefore a need for a more thorough understanding of how floc strength can be measured and what information can be found from floc strength tests.

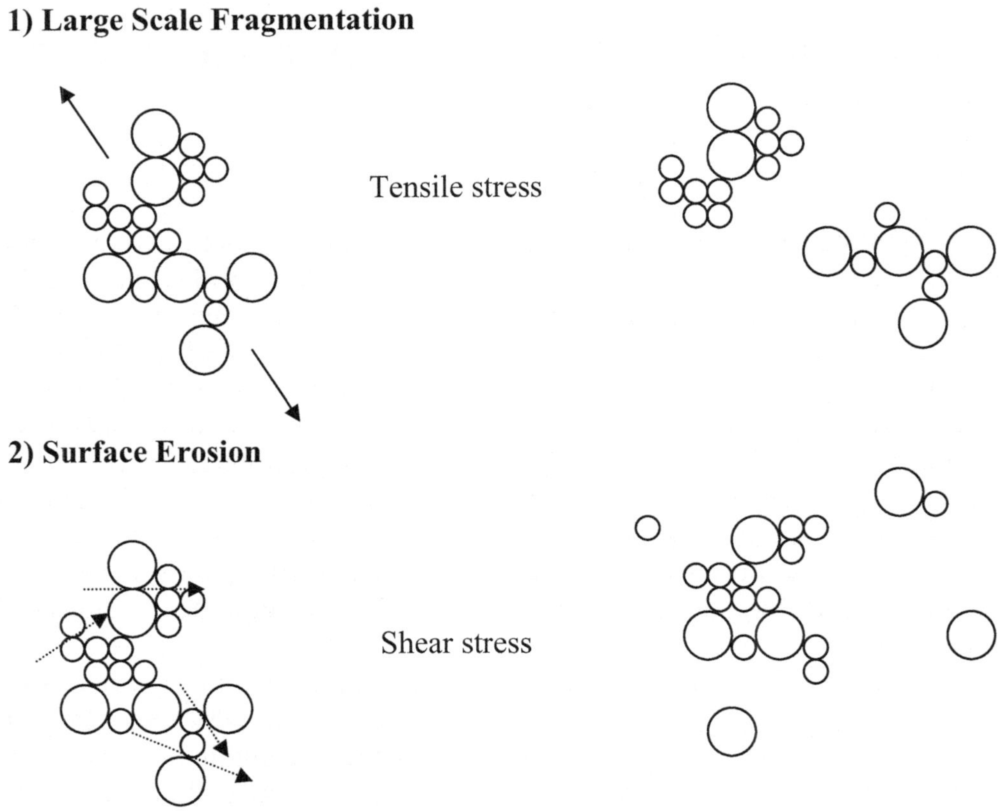

Figure 5.1 Two proposed mechanisms for the breakage of flocs under different shear conditions.

FLOC FORMATION AND BREAKAGE

Floc strength is directly related to floc structure and is therefore highly dependent upon the floc formation process. The combined processes of coagulation and flocculation aim to increase particle size for increased removal efficiency of very small particles, colloids and micro-pollutants. Coagulation is the process of chemically changing colloids so that they are able to form bigger particles by coming close to one another. This may be achieved by particle destabilization by double layer compression or physical enmeshment of colloids within coagulant precipitates or chemical reaction or chemical sorption (Cornwell and Bishop, 1983). Flocculation is the process of transferring coagulated colloids into contact with each other to form larger aggregates (Klimpel and Hogg, 1991; Gregor *et al.*, 1997). The exact process of particle destabilization and the subsequent colloid aggregation is complex. It is generally considered to be a two stage process of particle transport and particle attachment (Armirtharajah and O'Melia, 1990). Agglomerating particles must firstly collide with each other and secondly must adhere upon collision. Gregory (1988) treats these steps as being independent and separate from one another. However, flocs do not continue growing and reach a steady state size for a given shear condition. It has been generally accepted that floc growth is held in check by floc breakage so that the rate of aggregation is considered a balance between floc formation and floc breakage (Parker *et al.*, 1972; Francois, 1987; Spicer and Pratsinis, 1996; Ducoste and Clark, 1998; Biggs and Lant, 2000). The stability of flocs in suspension is therefore dependent upon how easy they are to break, with aggregate breakage being directly related to the strength and number of the bonds holding the floc together. During the rapid initial formation of microflocs, aggregation dominates over floc breakage, however the importance of breakage increases as floc size increases until a steady state floc size distribution is reached. Thus the steady state floc size is governed by the prevailing shear stress conditions within the containing vessel. The rate of particle collision and the nature of the particle interactions are fundamental to the rate of floc growth (R_{floc}) and may be summarized as in Equation 5.1 as the difference between the rate of aggregation and the rate of floc breakage (R_{br}). The former can be written as the rate of particle collision, R_{col} and a collision efficiency factor, α. The factor α. is the fraction of collisions which result in attachment. Thus the overall rate of floc growth may be written as:

$$R_{floc} = \alpha R_{col} - R_{br} \tag{5.1}$$

When the two terms on the right hand side of Equation 5.1 are equal, the net rate of floc growth is zero and the floc size attains a limiting value. The collision efficiency is not constant, but depends on the effective shear rate and particle size. For given shear conditions α. decreases as the floc size increases, which is another reason why flocs reach a limiting size (Brakalov, 1987). Another important consideration is that floc breakage may be irreversible to some extent, so that broken fragments do not readily re-form (Francois, 1987; Spicer *et al.*, 1998; Gregory and Dupont, 2001). For irreversible breakage the collision efficiency will be reduced and, for completely irreversible breakage α. = 0.

Gregory and Dupont (2001) have compared a number of different coagulants and monitored the effect of shear on grown floc suspensions using a photometric dispersion analyser (PDA) – a technique that determines aggregation rate based upon transmitted light developed by Gregory and Nelson (1986). Suspensions were coagulated for 11 minutes at a stirring rate of 30 rpm which was then increased to 300 rpm for 50 seconds and then reduced back to 30 rpm for 5 minutes. For the coagulation of kaolin using alum, the flocculation index (FI) - a combined measure of both particle size and number - was 3.0 at the end of the slow stir period. This was reduced to an FI of 0.5 after the increased shear. After 5 minutes at the original 30 rpm, the FI had increased and reached a plateau at an FI of 1.0 suggesting the original floc size was unattainable. The same effect was observed for three polyaluminium chloride (PACl) commercial polymer products. Similar work has shown the same behaviour for the breakage and re-growth of a coagulated natural water using alum, PACl and a combination of the two (McCurdy *et al.*, 2004). Alum and PACl had poor floc re-growth, whilst using both coagulants together showed a higher degree of recovery. Mono-disperse polystyrene beads coagulated with alum have been shown to have 60-80% recovery depending upon the intensity of the breakage shear (Spicer *et al.*, 1998). A number of mechanisms have been proposed to explain the phenomenon of irreversible floc breakage. Spicer *et al.* (1998) suggest that when inter-particle bonds within a floc are broken they lose some of their 'stickiness' such that when subsequent collisions with other particles occur, the attachment efficiency is much reduced. Yeung *et al.* (1997) use a similar analogy of re-using adhesive tape on a dirty surface for the reduction in re-flocculation. This may be due to floc re-structuring during exposure to increased shear, such that bonds previously linking parts of a larger floc are broken and are then able to re-form or re-structure within the smaller aggregate producing a more compact floc. However, irreversible breakage may also be indicative of chemical bonds being broken that are then unable to reform. In the case of polymeric coagulants, bridging flocculation is the main binding mechanism holding the particles together. High shear is likely to break polymer chains or change the organisation of the polymer to prevent new bonds forming (Yukselen and Gregory, 2004).

Floc breakage is not always irreversible. A kaolin suspension coagulated with PDADMAC showed 80% recovery in floc size whilst a copolymer of acrylamide and cationic monomer showed complete recovery in floc size (Yukselen and Gregory, 2004). When latex suspensions are destabilised by adding an ionic salt, the original floc size can be reached after being broken by an increased shear and then subsequently returned to the original low shear environment (Spicer *et al.*, 1998). This may be due to physical attractive forces like van der Waals bonds, which can be broken and reformed without a reduction in the strength and number. This is supported by McCurdy *et al.* (2004), who found that flocs formed by a combination of PACl and alum showed better re-growth than those formed by either coagulant when used alone, probably due to an increase in the formation of strongly charged species for the promotion of coagulation by charge neutralisation. Chaignon *et al.* (2002) have shown that activated sludge flocs have similar reversible breakage behaviour for continuous breakage/re-growth cycles, suggesting the principal mechanism involved in bonding microbial flocs is charge neutralisation.

Most high molecular weight organic polymeric flocculants are thought to work through bridging mechanisms however, some high charge density polymers are thought to operate by an 'electrostatic patch' effect (Gregory, 1989). In these cases, patches of charged polymer form dense regions of charge on primary particle surfaces. Strong attractive forces can then occur between regions of opposite charge on the particle surface and on the adsorbed polymer. These patches are similar to charge neutralisation mechanisms, which may help to explain improved re-growth potential for some polymers.

To summarise floc formation and breakage, flocculating suspensions are governed by the prevailing shear conditions and will reach a steady state. When the shear increases above a critical level flocs will break until a new steady state is reached. In some cases, because of the irreversible nature of floc breakage, flocs are unable to re-grow if broken at a higher shear level. Measuring the shear energy required to break individual flocs or those in a suspension is therefore of high operational importance.

MEASURING FLOC STRENGTH

Two fundamental approaches have been taken in measuring floc strength; a macroscopic measure of the energy required in a system for floc breakage and a microscopic approach that measures the inter-particle forces within individual flocs. A brief summary of these techniques is shown in Table 5.1. Due to the fact that the applied shear determines floc size, most work on floc strength has been in the macroscopic field, using applied shear techniques.

Macroscopic floc strength tests

Macroscopic floc strength tests have arisen from the relationship between the applied hydrodynamic shear and the resulting floc size. Under low shear conditions, particles may be encouraged to aggregate, but, under increased shear, the aggregates begin to break up (Mikkelsen and Keiding 2002). Gregory (2003) states that when comparing different flocs, the size (or flocculation index) for a given shear level indicates floc strength. Whilst this is the case for the given shear condition under which the flocs were formed, it does not give an indication of how flocs will behave upon exposure to increased shear during transfer, pumping or high shear treatment processes such as DAF or high rate filtration. For this reason, floc strength is better measured by applying an increased shear or a normal stress to the formed aggregates and relating the energy dissipation or velocity gradient applied to the maximum or average floc size remaining (Yeung *et al.*, 1997, Boller and Blaser, 1998 and Lee and Liu, 2001).

Table 5.1
A review of the techniques used for determining floc strength.

Strength technique		Description	Strength calculation	
Macroscopic techniques	Impeller	Exposure of floc to single level of increased shear within a containing vessel and compare the ratio of the floc size before and after breakage (Francois, 1987; Fitzpatrick *et al*., 2003)	$Strength\ factor = \frac{d(2)}{d(1)} \times 100$ where d(1) is the average floc size of the plateau before breakage and d(2) is the floc size after the floc breakage period.	5.2
		Exposure of the floc to increased levels of shear at a controllable rate within a containing vessel and measure the energy input for floc breakage (Leentvaar and Rebhun, 1983; Francois, 1987).	$\log d_{\max} = \log C - \gamma \log G$ d_{max} is the maximum floc diameter (m); C is the floc strength co-efficient; G is the average velocity gradient (s^{-1}) and γ is the floc strength constant, an exponent dependent upon floc break-up mode and the size of eddies that causes the breakage. A plot of the maximum floc size remaining against the average velocity gradient gives a line with a characteristic slope indicative of floc strength and break-up mode.	5.3
	Ultrasonics	Application of a controllable ultrasonic field to a floc suspension and observe floc erosion (Wen and Lee, 1998; Chu *et al*., 2001).	$\frac{\delta}{\phi} = \frac{-0.78k^{0.5}\Delta\tau}{d_{fo}^{-D/3} D \frac{dy}{dj}\Big\vert_{j \to 0}}$ δ is the floc binding strength ($J\ m^{-2}$), ϕ is the power of the ultrasonic field per floc's volume and time ($W\ m^{-3}\ s$), $\Delta\tau$ is the ultrasonic time (s), d_{fo} is the floc size before sonification (m), d_f is the floc size after sonification (m), k is the proportionality constant (the ratio of the floc's cross-sectional area and $d_f^{2D/3}$) and j is the time of ultrasonification (s).	5.4
	Multigrid mixer	Flocs exposed to hydrodynamic stress from a controllable oscillatory mixer (Bache *et al*., 1999). Flocs placed in a vibrating column and subjected to varying amounts of oscillation. The vibration is converted into an energy input for the system.	$\sigma \approx \frac{\rho_w \varepsilon^{3/4} d^{1/3}}{v^{1/4}}$ σ is the floc strength ($N\ m^{-2}$), ρ_w is the density of water ($kg\ m^{-3}$), ε is the energy dissipation at height of floc rupture ($m^2\ s^{-3}$), d is the floc diameter (m).	5.5

(continued)

Table 5.1 (continued)

Strength technique		Description	Strength calculation
Microscopic techniques	Micro-mechanics	The breaking force required to pull apart a single floc in the tensile mode (Yeung and Pelton, 1996).	$\sigma = \frac{F}{\frac{\pi}{4}d^2}$ 5.6 σ is the floc strength (N m^{-2}) and d is the floc diameter (m), where F is the floc rupture force: $F = C_s D$ 5.7 F is the floc rupture force (N), Cs is the cantilever stiffness (N m^{-1}) and D is the cantilever deflection (m).
	Micro-manipulat.	The squeezing of a single aggregate in suspension between a glass slide and a fibre optic probe until floc breakage using a force transducer (Zhang *et al*., 1999).	$F = K(W_0 - W)$ 5.8 *F* is the floc breaking force (N), *K* is the sensitivity of the force transducer (N V^{-1}), W_0 is the baseline voltage of the force transducer

The majority of macroscopic floc strength tests have used impeller-based systems whereby a known shear is applied to a grown floc suspension within a vessel of 1-4 L volume. However the geometry of the vessel and impeller type varies between the techniques (Table 5.2). The breakage behaviour of flocs is monitored by following changes in floc size over a range of shear values and the range of shear stresses investigated varies from study to study. The method of floc size determination is crucial, and this also varies between techniques. It is important to ensure that the technique used to measure floc size does not act to damage the aggregates due to their highly fragile nature. Most of the techniques listed are non-destructive *in situ* measurements, although some rely on taking flocs from the containing vessel for observation using microscopy and image analysis (Wu *et al.*, 2003). It is clearly difficult to remove flocs in this manner without damage. The remaining methods have used commercial light scattering particle size instruments (Francois, 1987; Spicer *et al.*, 1998; Biggs and Lant, 2000) or analysis of video frames or photographs using image analysis (Leentvaar and Rebhun, 1983; Bache *et al.*, 1999; Bouyer *et al.*, 2001; Bache and Rasool, 2001).

Light scattering instruments measure particle size by passing a laser beam through a suspension of particles flowing through the instrument at constant rate during the measurement cycle (Farrow and Warren, 1993) and provide a non-destructive methodology for measuring floc size (Spicer *et al.*, 1998; Biggs and Lant, 2000; Lartiges *et al.*, 1995). Typical methods have a stirred vessel containing the aggregate suspension and are connected to the particle sizing device by plastic tubing. Inherent to this type of system is a requirement to pump the suspension through the optical unit of the size analyzer, and this can cause floc break-up. Spicer *et al.* (1998) compared 3 types of pumping techniques for delivery to the optical cell: a peristaltic pump, a syringe pump and a hand pipette. They concluded that a continuous recycle using a peristaltic pump on the return side of the measuring cell was the least damaging and allowed easy continuous monitoring of the suspension. Discharging the sample to waste after measurement overcomes the problem of returning boken flocs to the test vessel but results in a continuous decrease in the test volume (Francois, 1987). Consequently most researchers have worked with a continuous recycle of suspension through the instrument.

Similar dynamic systems have been employed using the PDA to give an indication of floc size (Burgess *et al.*, 2000; Gregory and Dupont, 2001; Yukselen and Gregory, 2004; McCurdy *et al.*, 2004). In this technique, a narrow light beam is passed through a flowing suspension. The transmitted light intensity (dc value) is measured along with the root mean square value of the fluctuating component (rms). A ratio of the rms:dc gives a very sensitive indication of particle aggregation and is known as the previously mentioned flocculation index. The PDA is reported to be a very good and easy to use comparative tool showing qualitative changes in floc aggregation (Gregory and Nelson, 1986). However, the instrument is unable to give an absolute particle size for comparison with other techniques. In addition, the FI is an indicator of both particle size and particle number and as such there is no way of knowing the precise contribution of each of these components in the final FI value (Gregory and Chung, 1995). However, aggregate size is probably the over-riding factor as previous work has shown that when flocs grow larger the FI value always increases.

A combination of photography/video and image analysis has also been widely used to monitor floc suspensions, such that a flocculated suspension can be observed by capturing images of a stirred suspension by focusing on a plane a short distance (0.3 – 1 cm) behind the wall of vessel containing the suspension (Leentvaar and Rebhun; 1983; Ducoste and Clark, 1998; Chakraborti *et al.*, 2000; Bache and Rasool, 2001). Calibration is achieved by focusing on a graticule suspended into the tank prior to flocculation experiments. The advent of powerful digital and CCD cameras and comprehensive image analysis software has allowed much quicker measurements of an almost inexhaustible number of different floc size measurements to be made from floc samples (Wang *et al.*, 2002).

For measuring floc size, no particle size method can be considered ideal. Aggregates are highly irregular and porous and so their scattering patterns are likely to be very different than for equivalent solid spheres of the same material in light scattering devices. Although the light scattering properties of aggregates have yet to be fully quantified, Farrow and Warren (1993) conclude that, similar to the PDA device, light/laser scattering and transmission techniques are good for showing qualitative (rather than absolute) changes in floc size for aggregation systems. In addition their capability for measuring a very wide range of floc sizes (20 nm to 2 mm) makes them very suitable for monitoring flocs and colloidal systems, although the expense of these instruments limits their wide-scale application. In the previously mentioned set-ups, the inclusion of the measuring instrument will undoubtedly change the shear profile in the containing vessel and as such may inadvertently affect the floc size. Photographic techniques are good in this respect, because the flocs are only exposed to the shear of the impeller in the containing vessel and do not have to pass through a pumped system. The limitations of photography are the time and care that must be taken in preparing complex background lighting arrangements to allow for suitable contrast between floc and background. Furthermore, very small flocs may be missed that are beyond the detection limits of the camera being used thus biasing towards the detection of large particles (Leentvaar and Rebhun, 1983). Bache *et al.* (1999) concluded that floc sizes below 30 μm could not be reliably measured using their camera system. This limitation may prevent the detection of floc erosion mechanisms as the small eroded particles may be missed by the technique. Masking effects can also occur in very concentrated floc suspensions. Given these constraints, care must be taken when comparing floc size data obtained from different methodologies.

Table 5.2

A review of the different methods employed in shear based techniques for determining floc strength.

Type of flocs	Impeller system	Method of determining floc size	Shear levels investigated	Reference
Tap water + ferric chloride Tap water + ferric chloride	2 L glass tank with a turbine type impeller	Photography + image analysis Non-invasive	G_{av} 160-500 s^{-1}	Leentvaar and Rebhun (1983)
Kaolin + alum	4 L stirred vessel (unknown impeller type)	Small angle light scattering (Malvern Mastersizer 2200) *in situ* measurement. Single pass to waste.	G_{av} 30-1000 s^{-1}	Francois (1987)
Polystyrene + alum	2.8 L baffled tank with Rushton impeller	Small angle light scattering (Malvern Mastersizer E) *in situ* measurement. Continuous recycled pump loop.	100-460 rpm G_{av} 50-500 s^{-1} G_{max} 597-5969	Spicer *et al.* (1998)
Activated sludge	1.2 L baffled mixing tank with six flat blade impeller	Small angle light scattering (Malvern Mastersizer E) *in situ* measurement. Continuous recycled pump loop.	100-700 rpm G_{av} 19.4-444s^{-1}	Biggs and Lant (2000)
Bentonite + alum	Jar tester (unknown volume and impeller type)	Video camera + image analysis Non-invasive measurement.	30-150 rpm G_{av} 9-33 s^{-1} G_{max} 35-360 s^{-1}	Bouyer *et al.* (2001)
Humic + alum	2 L square beaker with single flat blade impeller	Video camera + image analysis. Non-invasive measurement.	G_{av} 10-400 s^{-1}	Bache and Rasool (2001)
Sewage sludge + cationic polymer	Couette flocculator powered by variable speed motor	Image analysis with flocs removed *ex situ*	G_{av} 10-400 s^{-1}	Wu *et al.* (2003)
Kaolin + alum/ polyaluminium chloride	1 L jar test beaker with single flat blade impeller	Photometric dispersion analyser (PDA). Continuous re-cycled pump loop.	50-400 rpm G_{av} 23-520 s^{-1}	Gregory and Dupont (2001) & Fitzpatrick *et al.* (2003).

Floc strength factor

One of the simplest evaluations of floc strength is the calculation of a floc strength factor (Francois, 1987). This is effectively the ratio of floc size after and before breakage at a particular shear level and may be calculated as shown in Equation 5.2 in Table 5.1.

The higher the value of the strength factor, the less sensitive the flocs are to breakage as a result of increased shear and are therefore considered stronger. The strength factor is not a constant and changes depending upon the applied shear during rupture and so strength factors can only be compared for similar breakage conditions. Francois (1987) noted an increase in the strength factor from 23.9 to 29.3 for kaolin flocs formed with longer rapid mix times between 0-360 s and an increase in strength factor from 17.5 to 26.5 for flocs formed with increased slow stir shear levels between 21-54 s^{-1}. A breakage shear of 1396 s^{-1} for 1.5 minutes was applied in these experiments. Fitzpatrick *et al.* (2003) compared the strength factor of kaolin flocs and observed the impact of temperature (6-29° C) on floc strength using a PDA and a breakage shear of 520 s^{-1} for 10 seconds. Whilst the PDA does not give an absolute floc size, the FI value before and after breakage can be used as a surrogate for $d(1)$ and $d(2)$. Floc strength has been seen to decrease with increasing temperature. For example, the floc strength factor decreased from 21 to 14 for alum-kaolin flocs dosed at 3.4 mg L^{-1} Al over the experimental temperature range. A general trend was seen that showed floc strength decreased with increasing floc size, such that flocs formed at the original slow stir that had a high FI value had lower floc strength factors than those of lower FI value.

A comparison of kaolin floc strength factors for a range of different coagulants has shown that hydrolysing coagulants such as alum and polyaluminium chloride (PAX XL-9) give lower floc strength than cationic polyelectrolytes (pDADMAC and Zetag 64) (Yukselen and Gregory, 2004).

The floc strength factor allows a relatively quick and easy method of determining a floc strength value and indicates how small a floc will become relative to its previous size for a given increase in shear. However, as the breakage shear is invariably different between different studies it is hard to directly compare the results other than for general trends. There has been little work showing how the relationship between the floc strength factor changes with increasing breakage shear levels. More work needs to be done in this area to give a better comparison of floc strength using the strength factor and on a wider range of suspensions other than kaolin in order to compare strength values between different coagulated systems for a standardised breakage shear.

Shear based floc strength and breakage models

Shear based floc strength models have evolved from strong empirical evidence showing that the final floc size distribution is related to the applied shear (Parker *et al.*, 1972; Spicer *et al.*, 1998). Shear has usually been characterised by the average velocity gradient as shown in Equation 5.9 (Camp and Stein, 1943).

$$G = \sqrt{\frac{\varepsilon}{\nu}} \tag{5.9}$$

G is the average velocity gradient; ε is the rate of energy dissipation per unit mass of fluid and ν is the kinematic viscosity.

The energy dissipation term is given by Equation 5.10.

$$\varepsilon = \frac{P_0 N^3 D^5}{V} \tag{5.10}$$

ε is the energy dissipation per unit mass of fluid; P_0 is the impeller power number; N is the impeller speed; V is the stirred tank volume and D is the impeller diameter.

Due to the relationship between the velocity gradient in the flocculating vessel and aggregate size, Parker *et al.* (1972) suggested an empirical expression for the stable floc size (Equation 5.11).

$$d = CG^{-\gamma} \tag{5.11}$$

d is the floc diameter; C is the floc strength co-efficient; G is the average velocity gradient and γ is the stable floc size exponent.

Linearization of the equation allows values of γ and log C to be found from a log-log plot of floc size measurement against the average velocity gradient (Equation 5.12).

$$\log d = \log C - \gamma \log G \tag{5.12}$$

There is some argument in the literature as to whether the maximum floc size remaining in the system should be measured or the average floc size. The initial empirical observations relate the shear to the maximum particle size (d_{max}), as in Equation 5.3 (Table 5.1). Bache et al. (1999) used the 95 % floc diameter obtained using an external video camera. This was done to remove problems associated with resolution when measuring smaller floc sizes using this technique. However, both Leentvaar and Rebhun, (1983) and Francois (1987) concur that the same relationship is seen when using average floc diameters. When the values of γ have been compared for the mean, median and maximum floc size for the same coagulation conditions γ remained fairly constant – between 0.43 and 0.49 – for all three floc sizes (Leentvaar and Rebhun, 1983).

The value of Log C strongly depends upon the method used for particle size measurement and which characteristic value of d has been used. As there has been wide variation between different studies, Log C can only be used to compare floc strength within specific experimental systems. However, as the value of γ remains relatively constant whichever value of d has been used it is a useful value for comparing floc strength and break-up.

The steeper the slope γ, the greater the reduction in floc size is seen with increasing G. The curves shown in Figure 5.2 are three theoretical examples of different floc suspensions formed at a slow stir of G_{init} showing different resistance to floc break-up. One interpretation of the data is to consider the steepness of the slope as an indicator of floc strength. For example, considering line (a) the slope of the line is 0 and floc size is independent of the applied shear. The flocs do not break on exposure to shear and as such must be considered strong flocs. If the slope of the line is shallow as in line (b) then these flocs are better able to resist shear than the example in (c). Therefore flocs (b) should be considered stronger than flocs (c) as the average/maximum floc size does not decrease so rapidly. However, this is likely to be an over-simplification of the situation. Many workers ascribe the value of γ to the dominant mode of floc degradation from both theoretical and experimental analysis and increasing values of log C as an indicator of increased floc strength (Parker *et al.*, 1972; Leentvaar and Rebhun, 1983; Francois, 1987). Solving complex equations to describe floc break-up and turbulence patterns, Parker *et al.* (1972) were the first to theoretically describe the impact shear has on floc size. These types of models assume flocs are composed of mono-disperse primary particles and characterise eddy frequency and breakage capacity.

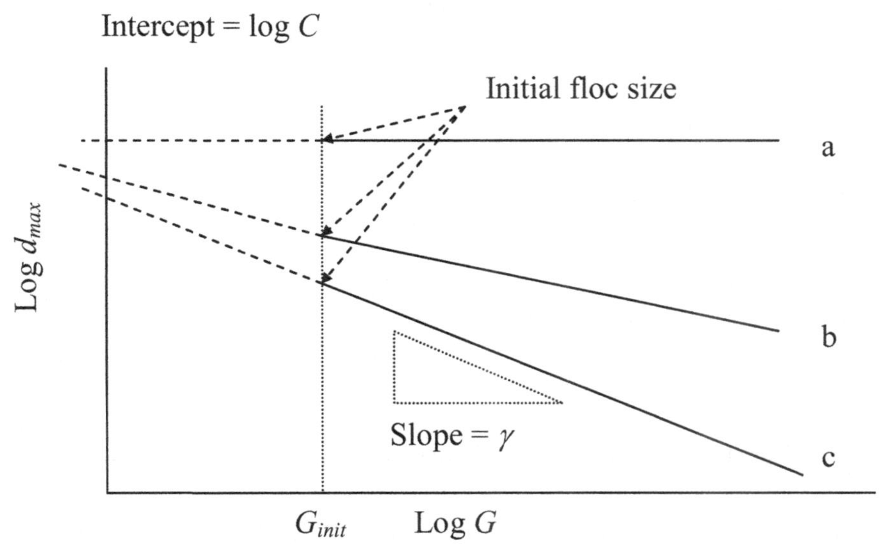

Figure 5.2 The relationship between the change in particle size and an increase in velocity gradient for 3 types of floc. Floc (a) is resistant to breakage, floc (b) is thought to break due to large scale fragmentation and floc (c) is thought to break by surface erosion.

The theoretical basis of the value γ may be explained by turbulent shear patterns relative to eddy size (Biggs and Lant, 2000; Bache 2004). In most flocculation processes, conditions are generally considered to be turbulent (Francois, 1987; Boller and Blaser, 1998). Eddy viscosity is the turbulent transfer of energy as a result of moving eddies, giving rise to tangential stresses. This is analogous to molecular viscosity in laminar flow where liquids resist flow as a result of intermolecular friction, therefore, when a liquid moves, energy is dissipated in a tangential direction to the moving viscous fluid. This energy is commonly referred to as viscous dissipation. Inertial convection is the release of energy in an outward direction, normal to the rotational eddy flow. Levich (1962) described turbulence using a scale whereby turbulent flow is described by the velocity and size of eddies. Very large eddies are responsible for the mixing of the system with little energy dissipation and therefore do not rupture or break flocs, whilst smaller eddies are responsible for most of the energy dissipation. The Kolmogorov microscale describes the length scale (η) of the energy-dissipating eddies (Equation 5.13).

$$\eta = \left(\frac{\mu^3}{\varepsilon}\right)^{\frac{1}{4}} \tag{5.13}$$

η is Kolmogoroff's microscale of turbulence, m; μ is viscosity and ε is the energy dissipation.

Inertial convection is responsible for energy dissipation of the larger eddies within the microscale range, whilst viscous energy is responsible for the energy dissipation of the smaller eddies in this range (Figure 5.3).

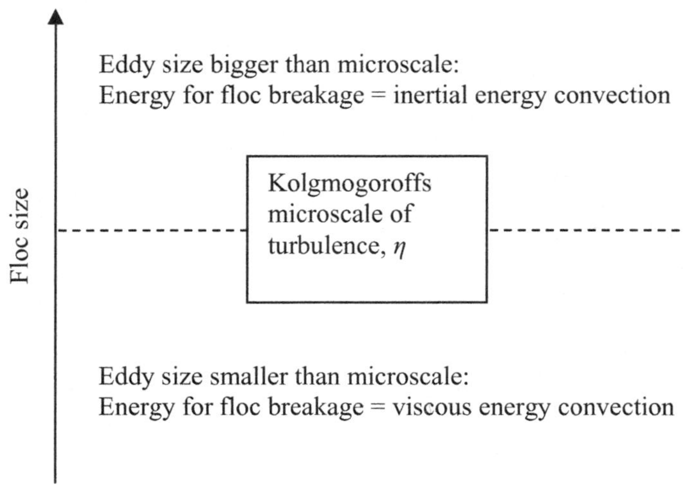

Figure 5.3 The importance of eddy size on floc break-up mechanism.

At high velocity gradients, the value of the micro-scale η is of a similar order of magnitude to the floc sizes whilst at low velocity gradients it is much larger, which helps to explain why flocs become more prone to breakage at high velocity gradients. Under normal flocculator conditions, viscous effects dominate (Boller and Blaser, 1998). Some authors suggest that when flocs are smaller than the micro-scale they become prone to breakage by surface erosion, whilst above the microscale flocs are thought to be more exposed to breakage by fracture (Thomas *et al.*, 1999). Other theoretical analysis suggests that when flocs are in the viscous energy region (i.e. below the critical microscale eddy size) values of γ of 0.5 have been calculated to indicate floc fragmentation, whilst erosion mechanisms are dominant if $\gamma = 1$. In the inertial convection zone a value of $\gamma = 0.5$ suggests large scale fragmentation events and $\gamma = 2$ suggests surface erosion (Parker *et al.*, 1972). The values of some of the previous experimental work for the calculation of floc strength co-efficient (Log C) and constant (γ) obtained for a variety of flocs are shown in Table 5.3. Good linearity is always seen between floc size and shear on a log-log scale for all of the experimental studies shown in the table. For example, both Bache et al. (1999) and Biggs and Lant (2000) had an R^2 correlation co-efficient in excess of 0.99.

Floc strength coefficient

For a fixed shear the larger the value of log C the stronger the floc (Bache, 2004). As has been explained, comparing values of log C between studies is not possible due to the different impeller/tank geometries and different floc sizing protocols employed, however a number of trends can be seen from within individual research. Wu *et al.* (2003) saw that an increase in the polymer dose lead to an increase in the floc strength coefficient for alum sludge flocs. A comparison of ferric hydroxide flocs in tap water and sewage sludge showed an increase in log C

from 1.5 to 1.9 with the authors stating that floc strength was higher in tap water than sewage (Leentvaar and Rebhun, 1983). For sewage flocs, it has been shown that adding polymeric flocculant doubled the floc strength co-efficient. Bache et al., (1999) and Bache and Rasool (2001) have investigated alumino-humic flocs. In the latter case, a useful comparison of flocs obtained from flocculators at 7 water treatment plants treating water of high colour (50-140° Hazen) has been made. The value of log C was highest for flocs where polymer was used and in the instance of water that was initially of high alkalinity.

Floc strength constant

Whilst the value of log C provides a good indication of how strong flocs are when they are formed at a given shear, it does not provide information on how flocs will respond to subsequent increases in shear. This information can be provided by the previously mentioned floc strength factor but also by the floc strength exponent (γ). The higher the value of γ the more prone the flocs are to breakage into smaller sizes with increasing shear.

For alumino-humic flocs under low alkalinity conditions, the value of the slope varied between 0.44 and 0.64 (Bache and Rasool, 2001). Looking at the data for comparable Al doses (2.4-2.7 mg L^{-1}), the humic floc degradation was reduced from 0.63 to 0.44 when polymer aid had been added. The value of γ was seen to increase when the water alkalinity was high suggesting that these flocs were less able to withstand increases in shear than the low alkalinity flocs. For sewage flocs, the addition of polymer has been shown to increase the degradation rate with increasing shear from 0.29 to 0.55 (Leentvaar and Rebhun, 1983). This does not concur with what has been seen for humic flocs. This is likely to be an indication of different polymer binding mechanisms occurring within different types of floc aggregate. Coagulant dose also has an impact on γ, for kaolin flocs, increasing coagulant dose from 4.02 to 5.02 mg L^{-1} Al increased the floc strength constant, whilst above this coagulant dose there was a decrease in γ (Francois, 1987). This suggests that there is an optimum coagulant dose in terms of floc strength.

For all types of floc the values of γ were around 0.5 (the exception being alumino-humic flocs formed in high alkalinity water were $\gamma = 0.81$) . Whilst these values of γ do not highlight major differences in degradation rate for different types of floc (e.g. sewage, kaolin and humic), the mode of breakage can be theoretically determined from the floc strength constant. If it is assumed that viscous effects are responsible for most of the energy dissipation in flocculating systems and the models of floc breakage are correct, then it appears that floc breakage is generally dominated by floc fragmentation as the value of γ remains around the theoretical value of 0.5 as derived by Parker *et al.* (1972). However, this strict theoretical interpretation is open to debate. For example fragmentation and erosion could occur simultaneously in a containing vessel. This may explain the result seen for the high alkalinity water, were the value of γ was half way between the theoretical values for fragmentation and surface erosion. In addition, large flocs in an aggregated system may be larger than the microscale (η) whilst the smaller flocs may be smaller than η. This has been shown experimentally for activated sludge flocs with the $d_{95} > \eta > d_{50}$, suggesting that the larger flocs in the system fragment and the smaller flocs erode (Biggs and Lant, 2000).

Table 5.3
The value of the floc strength constants and coefficients obtained from shear based techniques.

Type of floc	Coagulant type and dose	Floc strength coefficient value, log *C*	Floc strength constant, γ	Reference
Alum sludge	Cationic polymer (PL-320) 0 →30 mg L^{-1}	2.4 →5.9	N/A	Wu *et al.* (2003)
Ferric hydroxide precipitate in sewage effluent	Ferric chloride 10 mg L^{-1} as Fe	1.9	0.29	Leentvaar and Rebhun (1983)
	Ferric chloride + anionic polymer 10 mg L^{-1} as Fe + 3 mg L^{-1} polymer	4.1	0.55	
Ferric hydroxide precipitate in tap water	Ferric chloride 10 mg L^{-1} as Fe	2.5	0.51	
Alumino-humic (commercial humic solution)	Al based coagulant	3.1	0.44	Bache et al. (1999)
Alumino-humic in low alkalinity and high colour water	Al based coagulant			Bache and Rasool (2001)
	2.4 mg L^{-1} as Al	3.8	0.61	
	2.7 mg L^{-1} as Al	3.4	0.63	
	4.7 mg L^{-1} as Al	3.6	0.57	
	5.4 mg L^{-1} as Al	3.6	0.52	
	2.5 mg L^{-1} as Al + 0.1 mg L^{-1} polymer	3.8	0.44	
	3.7 mg L^{-1} as Al + 0.1 mg L^{-1} polymer	4.0	0.64	
Alumino-humic in high alkalinity and high colour water	Al based coagulant 5.0 mg L^{-1} as Al	4.1	0.81	
Alumino-kaolin in DI water	Aluminium sulfate	-		Francois (1987)
	4.02 mg L^{-1} as Al		0.44	
	4.52 mg L^{-1} as Al		0.48	
	5.02 mg L^{-1} as Al		0.61	
	5.52 mg L^{-1} as Al		0.50	
	6.02 mg L^{-1} as Al		0.43	
Activated sludge flocs	No coagulant dosed	2.9	0.35	Biggs and Lant (2000)

Velocity gradient term

Traditionally, the average hydraulic gradient (G_{av}) has been used as the descriptor for shear and as previously described, and many studies have reported a highly significant relationship between G_{av} and floc size (*d*). However, there is some debate as to the use of velocity gradient, and in particular the average velocity gradient, when considering floc strength and breakage. This is because G_{av} is unlikely to be able to accurately describe the complexities of turbulent flow in stirred vessels (Boller and Blaser, 1998). There is a large body of evidence suggesting that local *G* values are more important than the G_{av} in a stirred vessel (Francois, 1987; Essamanai and de Traversay, 2002). Maximum local *G* values are found close to the impeller region in a stirred vessel and are considered to be of most importance in determining floc size distributions in a tank (Francois, 1987: Bouyer *et al.* 2001). This is further supported by data showing that systems with different impellers or vessels may have the same average velocity gradient, but the resulting floc size distributions are not the same due to differences in G_{max} (Ducoste and Clark, 1998; Bouyer *et al.*, 2001).

The reported empirical relationship between G_{av} and *d* is valid for a given impeller and container geometry because they are directly proportional to one another. This can be shown by work by Bouyer *et al.* (2001) that has evaluated the hydrodynamic velocity profile in a jar tester using computational fluid dynamics (CFD). The relationship between the calculated G_{av} and G_{max} is very significant (Figure 5.4). The *G* value in the impeller region was shown to be four times the average velocity gradient at 30 rpm, whilst at 150 rpm the *G* value in the impeller region was more than 10 times the average velocity gradient. The maximum velocity gradient was seen in the area between the paddle and the vessel wall just as the paddle has passed through the region.

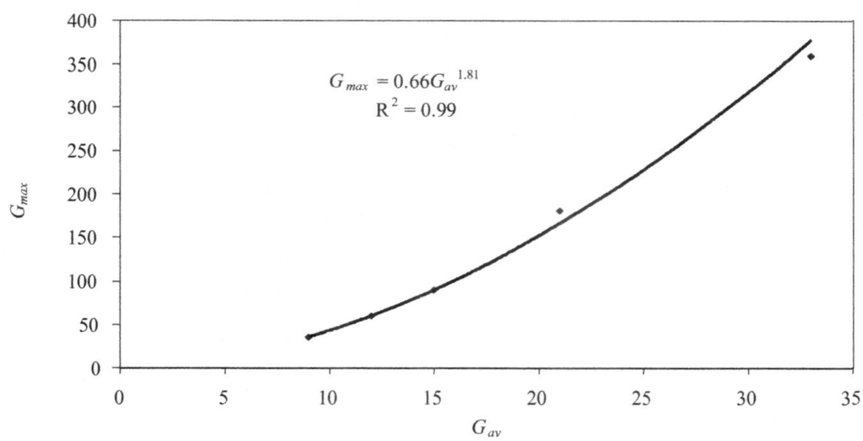

Figure 5.4 The relationship between G_{av} and G_{max} from data obtained using CFD (adapted from Bouyer et al. (2001).

Essamanai and de Traversay (2002) have used a combination of CFD modelling based on Navier-Stokes equations with experimental pilot work to produce a model of the different *G* distributions in a stirred flocculator. They have found that the maximum *G* value in the impeller region to be 38,000 s^{-1}, whilst the G_{av} in the impeller zone was 5549 s^{-1} and the G_{av} in the whole flocculator was 163 s^{-1}. These large orders of magnitude differences must be considered carefully as G_{max} was only a very small volume of the flocculator. The probability that a floc passes through this region must firstly be quantified because if the majority of flocs do not become exposed to this very high shear level, flocs cannot be broken by that shear level. Secondly, if a floc does enter the very high shear the time spent there must be long enough to break aggregate bonds. Further modelling work needs to be carried out in order to quantify critical G_{max} levels for floc breakage. In addition, although G_{max} may be the controlling factor of floc size it can also be attained in different ways. For example Spicer and Pratsinis (1998) calculated the G_{max} from Equation 5.9 based upon the impeller swept volume (compared to the G_{av} calculated from the vessel volume), whilst Bouyer *et al.* (2001) generated G_{max} using CFD.

Because there is still much work to be done in quantifying critical G_{max} the use of the G_{av} value should not be ignored and is probably acceptable for order of magnitude purposes to describe energy dissipation (Bache and Rasool, 2001). Furthermore, the concept of G_{av} is very well known, easy to calculate and universally used for the design of water and wastewater treatment facilities (Boller and Blaser, 1998).

Recent work by Ducoste and Clark (1998) has formulated an alternative empirical expression to that of Parker *et al.* (1972) that was satisfied regardless of variations in impeller geometry and tank size that removes the direct use of G_{av}:

$$d_{av} = \frac{h}{(P_0^{0.5} ND)^y} \tag{5.14}$$

d_{av} is the volume mean particle diameter, h is a constant, P_0 is the impeller power number, N is the impeller rpm and D is the impeller diameter.

The data suggest that the steady state floc size was controlled by the shear in the impeller zone, however more extensive quantification of this expression is required to confirm the observed relationship as the study only evaluated two impellers (a Rushton turbine and an A310 fluid foil impeller) and three square tank volumes of 5, 28 and 560 L.

Hydrodynamic shear based techniques using impellers have been widely used because of their similarity to operational flocculators and the likelihood that hydrodynamic shear stress is probably of great importance when considering floc breakage. In addition these techniques have highlighted differences in how a floc will respond to different increased levels of shear as determined by the value of γ. However, whilst the observation of an empirical relationship between shear and floc size is widespread, the problem has been in the interpretation and application of shear based models to the data. Whilst experimental work is fitted to these models there is little direct evidence to suggest either erosion or fragmentation is definitively taking place. Ideally, the particle size distribution of a 'fragmenting' floc system needs to be compared to an 'eroding' system to show the differences in particle sizes to confirm that the models are totally accurate. A final problem with these techniques has been an accurate description of the shear energy dissipated into impeller systems. G_{av} is probably not sufficient enough whilst critical threshold values of G_{max} have not been sufficiently well modelled to directly relate to the resulting floc size. This has lead to a number of workers tackling this problem by applying a reliable, accurate and controllable measure of the energy input and the following section will review a number of these techniques in turn.

Ultrasound

The use of ultrasound to condition sludge is a well known technique (Chu *et al.*, 2001). By applying a controlled ultrasonic field to a sludge suspension, flocs may be effectively 'exploded' into smaller parts to improve biodegradability. During ultrasonic treatment, pressure waves pass through a medium releasing large quantities of energy. This induces the formation and collapse of gas bubbles. The result is a release of energy in the form of temperature and turbulent eddies around the collapsing bubble. There has been some limited application of this in determining floc strength. Wen and Lee (1998) have developed a complex term for the calculation of a floc strength value from an applied ultrasonic field (see Table 5.1). Above a critical ultrasonic energy input it has been seen that kaolin flocs and activated sludge flocs break-up and decrease in size (Wen and Lee, 1998; Chu *et al.*, 2001). A direct measure of the ultrasonic energy per unit volume of sample can be made that reflects the flocs internal binding strength as the flocs break up. The resulting floc size can then be measured and compared to the original floc size using one of the previously mentioned techniques. For activated sludge thickened with alum, the effect of polymer addition on floc strength has been investigated (Lee and Liu, 2001). A non-ionic polymer was compared with a cationic polymer and it was seen that floc strength decreased with increasing polymer dose and that non-ionic polymer showed greater floc strength than cationic polymer. However, it was interesting to note that in all cases the sludge flocs without polymer were stronger than with polymer. The major problem associated with this technique is the impact of temperature. At an applied ultrasonic rate of 44 Watts ml^{-1} the temperature was seen to increase from 20 to 56° C in just over 1 minute. Therefore considerable temperature control needs to be applied in order to better stabilise the system conditions.

Wen and Lee (1998) investigated the strength of clay coagulated with cationic polymer for different pH and polymer dose. They found that at higher pH and polymer dose, the initial floc size was larger. The results were contradictory in that the large flocs formed at pH 7 and 10 were observed to break down at a faster rate than at pH 3 in terms of floc size. However, the binding strength based on Equation 5.4 was observed to increase with both polymer and pH. This is perhaps a reflection of there being two approaches of considering floc strength. The first is a measure of the maximum floc size attained at the end of the floc formation process such that larger flocs should be considered stronger than smaller flocs because they have reached a bigger size. The second considers the floc strength as a measure of the strength of floc bonds to withstand increases in energy once formed.

For activated sludge flocs the use of ultrasound is confused due to the effect of ultrasound on bacterial components of the floc. Ultrasound increases the formation of free radicals which can impact on bacterial metabolism and also cause cell lysis (Jorand *et al.*, 1995). Therefore, in these instances the applied energy may be causing other effects other than breaking apart primary particles within the floc. For water treatment flocs that are generally dominated by non-living chemical constituents, ultrasonic methods may have potential for determining the forces required to break flocs. From an operational sense it is difficult to relate the energy experienced by flocs from an ultrasonic field to the hydrodynamic shear conditions that are important in a flocculator. However, as inter-particle bonds are being broken in both techniques a comparison of the results between the different methods would be of great interest.

Oscillating multigrid mixer

Bache and Al-Ani (1989) developed a technique whereby flocs are exposed to stress in a vibrating settling column. An oscillating plunger was used to provide turbulent energy dissipation in the column. In this manner relatively uniform energy dissipation can be achieved in the horizontal plane with shear stress increasing as flocs pass down the column, therefore the height at which a floc ruptures in the column is proportional to a certain level of energy dissipation. A floc breaks when it passes through a region where the hydrodynamic forces are greater than the binding force at the point of rupture within the floc. The advantage of such a system is that the hydrodynamic shear stress at each point in the column may be more accurately modelled when compared to the more complex flow of impeller systems. In this technique floc size before and after breakage has been monitored using CCTV and image analysis. The underlying trend emerging from this technique is for large flocs to rupture higher in the column than smaller floc aggregates, showing the increased susceptibility of large flocs to increases in shear. This has been shown for kaolin-alum flocs (Bache and Al-Ani, 1989) and for humic-alum flocs (Bache et al., 1991). The calculation of a value for floc strength from these data is shown in Equation 5.5, (Table 5.1). This has been derived from a force balance between accelerations across the floc at the plane of rupture where turbulent energy acceleration have been defined as $\sqrt{3}\varepsilon^{3/4}v^{-1/4}$ from Levich (1962):

$$\frac{1}{4}\pi d^2\sigma = 2\frac{\pi}{6}\rho_w d^3\sqrt{3}\frac{\varepsilon^{3/4}}{v^{1/4}} \quad (5.15)$$

d is the size of floc sub-units, σ is the floc strength per unit area ρ_w is the density of water, v is the kinematic viscosity and ε the energy dissipation .

Floc strength calculations have shown that flocs sheared at high G_{av} values flocs were smaller but had higher strength. For example, flocs sheared at 50 s^{-1} had a d_{95} of 238 μm and a strength of 0.08 N m^{-2}, whilst flocs sheared at 230 s^{-1} had a d_{95} of 120 μm and a strength of 0.42 N m^{-2}. The relative floc strength of alumino-humic flocs was found to change with coagulant dose. For doses of between 2-7 mg L^{-1} Al, an optimum was seen at 4 mg L^{-1}.

Microscopic Floc Strength Tests

Whilst macroscopic techniques rely upon complex theory of turbulence, eddy size and floc breakage models, recent work has been carried out to gain a more direct measurement of floc strength by taking individual flocs and finding the force required to pull or compress a floc until breakage (Yeung and Pelton, 1996 and Zhang *et al.*, 1999). The principal advantage of such a technique being that an understanding of the mechanisms of how and where flocs break may be gleaned and a floc rupture force may be directly measured to give floc strength. This may eventually lead to a more thorough knowledge of the currently poorly-understood mechanisms of floc breakage. The following section briefly reviews two microscopic techniques.

Micromechanical approach

In this method, floc strength is measured by the tensile force required to break single flocs (Yeung and Pelton, 1996; Yeung *et al.*, 1997). A schematic overview of this technique can be seen in Figure 5.5. Calcium carbonate flocs coagulated in NaCl with polymeric floc aid were placed in a sample chamber consisting of two glass cover slips separated by 2 mm. The force required to rupture the flocs was determined by carefully attaching single flocs onto a glass rod cantilever beam and attaching and translating an opposing pipette horizontally until the floc broke. The deflection of the cantilever may be directly converted into a floc rupture force, provided the cantilever stiffness is known, using sensitive micromanipulators. Floc size before and after breakage was determined by averaging the major and minor axis of the aggregates of the original floc and the broken floc pieces. In their first study (Yeung and Pelton, 1996) floc strength was reported as simply the force required to break apart the floc. Initial floc sizes ranged from between 5-50 μm diameter. In this instance when this force is plotted against floc size no correlation was seen between floc strength and size. A change in floc strength was seen when two polymeric flocculants were compared with the first system having an average floc strength of 64 nN and the second was 110 nN. Boller and Blaser (1998) have transformed these data into a floc rupture pressure so that floc strength is found from the ratio of the rupture force to the cross sectional area of the smaller fractured aggregate (as shown in Equation 5.6 in Table 5.1) giving floc strengths in the region of 100-1000 N m^{-2}. In this way, the size dependency of floc strength was again seen, such that smaller flocs showed increased strength. The application of this technique has been taken further to show the effect of mixing speed on floc strength (Yeung *et al.*, 1997). A range of mixing speeds between 50-2500 rpm were investigated for paper pulp with calcium carbonate flocculated with polymeric flocculant. An optimum mixing speed was determined at 500 rpm that gave optimum floc strength of 1,500 N m^{-2}.

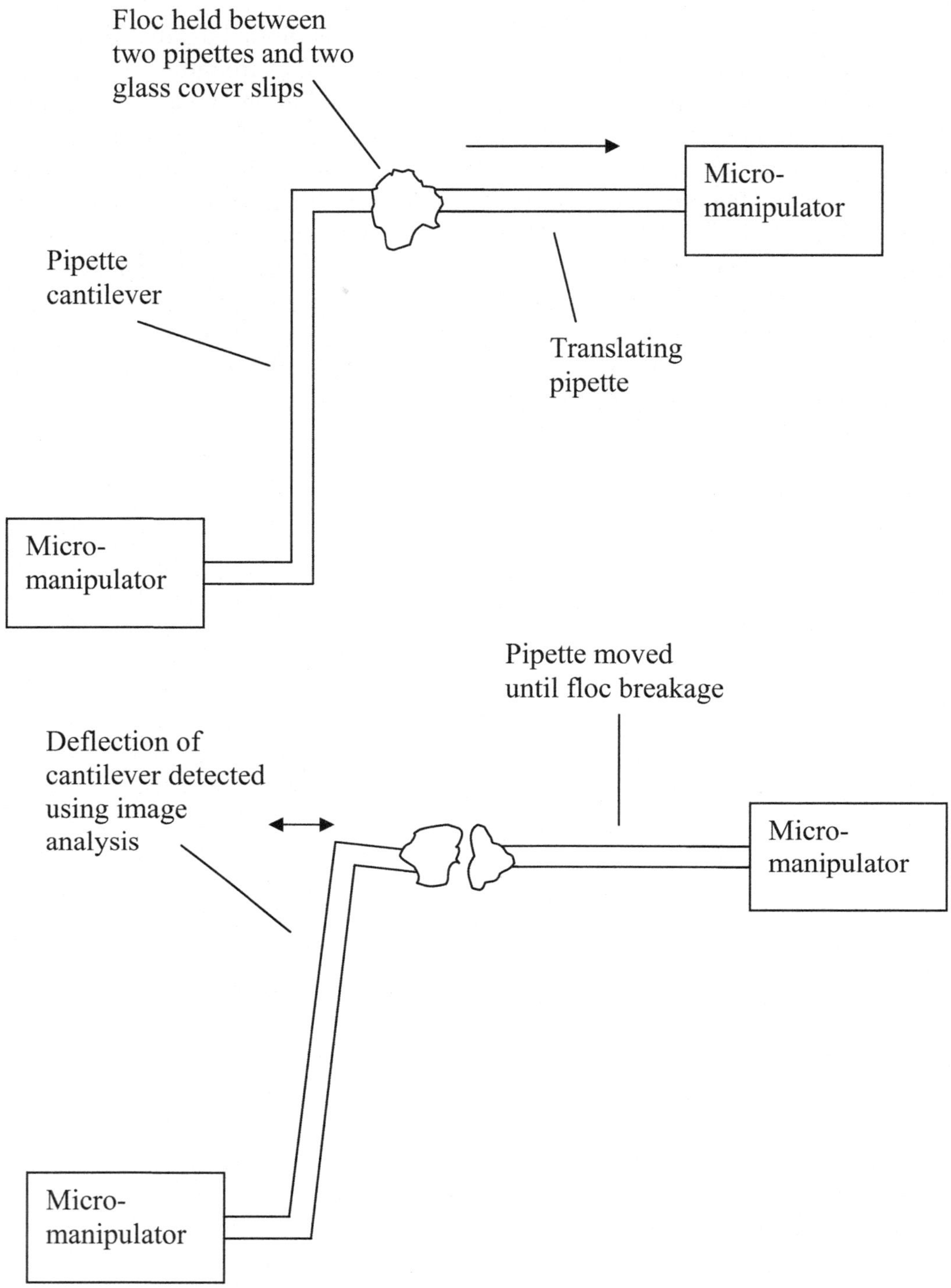

Figure 5.5 A schematic overview of the micromechanical floc strength technique (adapted from Yeung and Pelton 1996)

The technique provides an interesting insight into floc breakage in the tensile mode. It was observed that not all flocs breakages resulted in large scale fragmentation. This contradicts hydrodynamic shear models that suggest tensile stress is responsible for large scale fragmentation. It was hypothesised that flocs were breaking at their narrowest point because it is here that there are fewer attachment sites. The size ratio of floc fragments (d_1) after breakage with the initial floc (d_0) were compared with floc compaction (as determined by fractal dimension analysis). Values of d_1/d_0 approaching 0.5 indicate that the floc has been split into two similar sized pieces whilst near 0 and 1 a floc has been broken into a very small and large aggregate. The authors argue that the mode of floc breakage is therefore more to do with floc compaction than the differences in turbulence stresses and eddy size. Very compact flocs have fewer bonds in their margins, so surface erosion is more likely to occur. More loosely bound flocs are more likely to have weak points randomly located anywhere across their cross-section and are therefore more susceptible to large scale fragmentation. This questions one of the fundamental assumptions of the shear based floc strength characterisations based upon Equation 5.12 that flocs are homogenous structures composed of similar primary particle units.

The micromechanical technique has provided a useful addition to the field of floc strength quantification. However, as yet the application has been limited to the measurement of only a few hundred individual floc aggregates in total and across a relatively narrow range of floc types. More work is required in this field in order to assess its broad application to all types of floc. In addition, the importance of tensile stress as a mechanism for floc breakage in flocculators and other WTW processes has yet to have been fully quantified.

Micromanipulation

A further microscopic technique has measured the compression force required to break flocs between a glass slide and a fibre optic probe (Zhang *et al.*, 1999). In this technique, a sample of floc suspension was placed on a microscope slide, individual flocs could then be selected using a microscope (Figure 5.6). A 50 μm probe with one flattened end was mounted onto a force transducer and lowered just above a floc aggregate using a micromanipulator. The transducer then lowered the probe at a set speed until the aggregate broke from compression between the probe and slide. The applied force was then measured as the breaking force (Equation 5.7). The flocs under investigation were very small latex particles aggregated in a salt solution. The mean size of flocs aggregated by vortexing was 2.5 μm and the mean floc strength was 5.3 μN. However, the experimental procedure did not measure the sizes of the residual flocs once they had been broken. This meant the results could not be normalised to a floc rupture force over the floc cross sectional area and so compared to the other techniques. This method provides an alternative microscopic measure of floc size and again needs to be assessed over a range of floc types in order to rigorously evaluate the technique. However from an operational sense it is difficult to think of a situation where flocs would be exposed to such a compression stress under normal solid/liquid separation conditions and as such this value of floc strength may not be indicative of the likelihood of floc breakage in industrial unit processes.

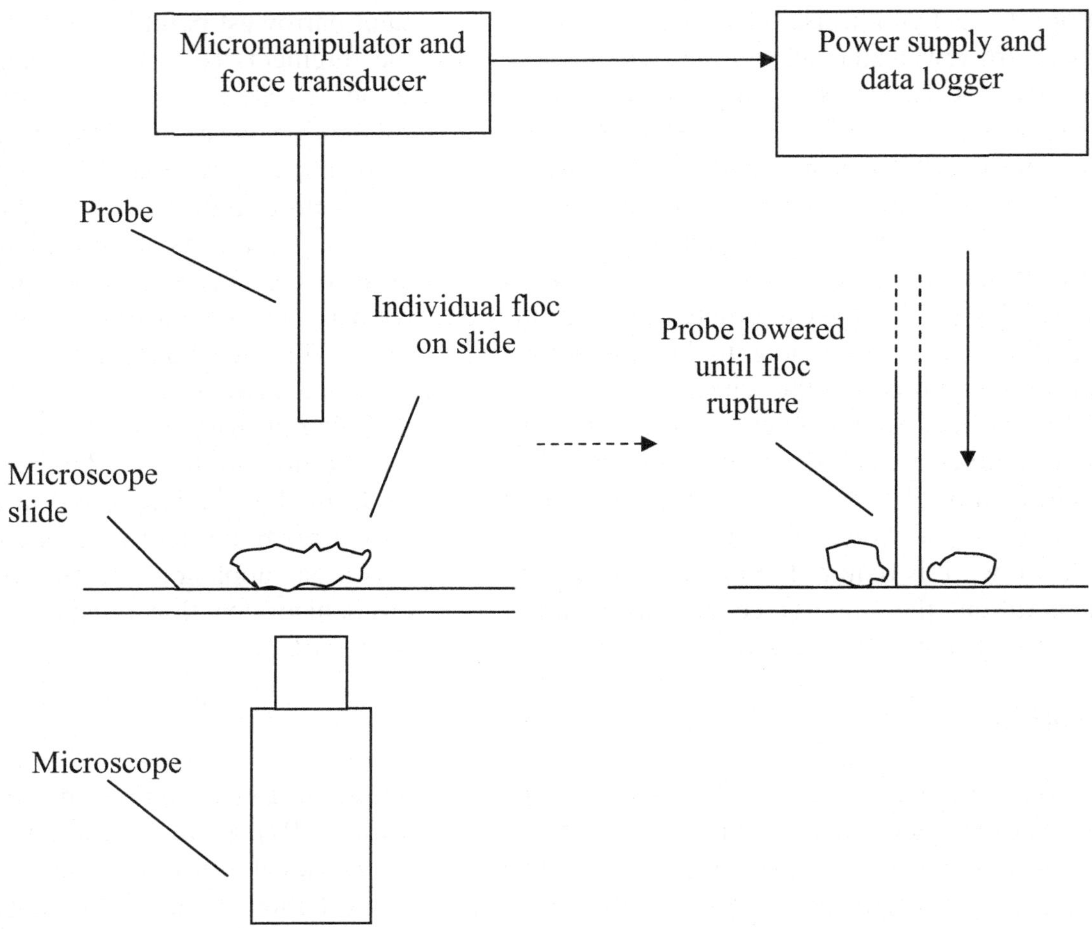

Figure 5.6 A schematic overview of the micromanipulation floc strength technique (adapted from Zhang et al. 1999)

TRENDS SEEN IN FLOC STRENGTH TECHNIQUES

Direct comparison of floc strength between the different techniques is made difficult because each technique measures floc strength in a different way giving either floc strength as a relative measure or an actual breakage force. However an attempt has been made in Table 5.4 to highlight the general trends that the floc strength measurements indicate.Table 5.5 emphasises some of these trends with a comparison of absolute values of floc strength from a number of techniques were the force per area of floc has been measured.

Table 5.4
The general trends observed in floc strength tests*

Floc characteristic	Trend observed
Floc size	Floc size increases → floc strength decreases
Coagulant dose	Optimum coagulant dose for floc strength
Polymer addition	Biological flocs: Addition of polymer → floc strength decreases Chemical flocs: Addition of polymer → floc strength increases
Type of floc	Bridging particle flocs > charge neutralised particle flocs > complexation flocs (e.g. NOM/coagulant)

* for this table, the symbol '>' indicates where flocs were stronger than another type of floc.

The most significant trend to emerge from these strength tests is the increase in floc strength with a decrease in floc size. This can be clearly seen from data in Table 5.5 for alumino-humic flocs with an increase floc strength of over five times with a halving of the initial floc size. This is further supported by polymer-calcium carbonate flocs. Polymer A produced flocs with an average diameter of 25 μm and an average floc strength of 100 N m^{-2}, whilst polymer B produced smaller flocs of 10 μm with an increase in average floc strength to 1000 N m^{-2}. A mechanistic explanation for this relationship has not yet been fully described. However, the reasons are likely to relate to floc compaction and the number of internal bonds. In much of the strength work, flocs are grown and then exposed to an increased shear. This acts to break flocs and therefore reduce the average floc size in the suspension. The breakage procedure breaks flocs at their weakest points, which results in smaller pieces that are smaller and more compact. This has been shown by fractal dimension analysis of polystyrene-alum flocs. There is a large amount of evidence suggesting that flocs are examples of fractal structures (Gregory, 1998, Gorczyca and Ganczarczyk, 1999, Thomas *et al.*, 1999; Bushell *et al.*, 2002). The floc fractal dimension (D_f) can indicate the openness of the internal floc structure with a higher fractal dimension indicating a more compact structure. Flocs formed at a high shear rate (G_{av} = 300 s^{-1}) were small and had a fractal dimension of 2.65 whilst floc size increased and the fractal dimension was reduced to 2.4 when the flocs were formed at a much lower shear rate (G_{av} = 50 s^{-1}) (Spicer *et al.*, 1998). More compact structures indicate that primary particles may have more attachments with one another or repulsive forces between these particles is at a minimum. Floc restructuring during breakage is one mechanism for allowing primary particles to become closer to one another so that floc internal bonds break and re-form at more favourable points within the floc where the attractive force is greater or the repulsive force lower.

Similarly, compaction is thought to explain why an optimum coagulant dose exists in terms of floc strength. For charge neutralisation coagulation mechanisms, the optimum floc characteristics should be seen when the repulsion forces between primary particles are low. However, a slight amount of charge repulsion allows attached particles to re-arrange into more compact structures rather than attaching at the first contact (Waite *et al.*, 2001). In water and wastewater operations, charge neutralisation is generally achieved by the addition of charged metal ions of an opposite charge. Increasing or decreasing the ratio of the charged coagulant will therefore change the balance of the charge within the floc above or below an optimum (Bache et al., 1991).

Table 5.5
Floc strength estimates using a variety of different methods.

Method	Type of Floc	Floc Size, μm	Floc Strength Estimate, N m^{-2}	Reference
Oscillating multigrid mixer	Alumino-humic flocs	238 182 143 120	0.08 0.16 0.29 0.42	Bache et al. (1999)
Oscillating multigrid mixer	Rice starch aggregated at isoelectric point (IEP)	1100	1.0	Bache et al. (1997)
	Latex aggregated at IEP	600	0.9	
Micromechanics	Polymer A – calcium carbonate flocs	25	100	Yeung and Pelton (1996)
	Polymer B – calcium carbonate flocs	10	1000	
Micromanipulation	Latex aggregated at IEP	2.5	5.3 μN*	Zhang et al. (1999)
	Latex aggregated by Brownian motion	1.7	3.1μN*	

* data unavailable to normalise to N m^{-2}

A generally held conception within the water and wastewater industry is that the addition of polymer acts to increase floc structural characteristics by aiming to increase floc size, strength, settleability and filterability (Bratby, 1980). From the limited data presented in this review, this statement appears to be at least partially true in terms of floc strength and size. Sewage flocs and alumino-humic flocs were seen to increase in size with the addition of a polymeric floc aid. However, only the alumino-humic flocs showed improved resistance to shear, implying an increase in floc strength. The difference in floc strength with and without polymer is likely to be some reflection of the binding mechanisms of the polymer to the primary particles of the floc. For a range of polymers (anionic, cationic and non-ionic), floc strength was seen to decline for biological flocs in their response to increases in shear (Lee and Liu, 2001; Leentvaar and Rebhun, 1983). No mechanistic explanations were given although one possible hypothesis was a toxicity effect from the polymer on the biological component of the floc. The polymer may act to kill bacteria or prevent bacterial adhesion mechanisms within the floc, thus reducing the overall bonding capacity in the floc through changes in concentration and character or extracellular polymeric substances. However, some of this previous research goes against intuitive and practical experience of polymers used in water and wastewater treatment. Much more work is required in this field to more adequately quantify the effects of polymeric flocculants.

During the removal of solid particles, enmeshment and bridging of particles within the precipitated coagulant matrix are thought to be the principal binding forces holding the floc together. These types of bond are considered much stronger than van der Waals attractive forces formed during charge neutralisation (Bache et al., 1997). This is supported by the experimental data shown in Table 5.5. From the available data, there are considerable differences in floc strength for different types of aggregate. Flocs composed of coagulant and particles were one to two orders of magnitude stronger than charge neutralisation flocs.

Flocs formed from waters of high colour and high NOM content are widely recognised as being fragile structures when compared to other flocs (Bache et al., 1997). The floc strength data available confirm this, with these flocs being weaker than flocs formed by charge neutralisation at their IEP. The weakness of humic flocs is explained by Bache et al. (1999) to be due to the fact that charge neutralisation is the main removal route NOM removal preventing stronger bridging bonds from forming. This in part explains the weakness of organic flocs but does not help explain why IEP flocs are considerably stronger than humic flocs. If both flocs are formed by charge neutralisation, then van der Waals forces between primary particles will dominate. The reduced floc strength of humic flocs therefore suggests an increased degree of repulsion within the floc matrix. This makes sense because organic matter is a complex mixture of different organic compounds (Goslan *et al.*, 2002). These organic molecules have differing degrees of charge and hydrophobicity (O'Melia *et al.*, 1999). Therefore, whilst some molecules and parts of molecules may be effectively charge neutralised by coagulants, regions of similarly charge will exist that increase repulsion within the floc.

FLOC STRENGTH SUMMARY

There have been a number of different approaches taken in measuring floc strength. Some workers consider that floc strength is indicated by the size a floc reaches at the end of its growth phase, whilst others consider the force required to break already formed flocs. Most research has concentrated on exposing flocs to increased shear in a containing vessel by the application of stirring, ultrasonification and oscillation. More recently microscopic techniques have been developed that relate the energy required to pull apart or compress individual flocs until breakage.

Whilst there is strong evidence showing an empirical relationship between the applied shear and the final floc size distribution, there are a number of problems associated with comparing the data from one study to another. The interpretations of floc breakage models need to be further analysed. These models ascribe floc breakage as either erosion or fragmentation based upon the rate of floc breakage with increasing shear relative to the eddy size in the system. These interpretations are open to debate given the incomplete understanding of how eddies interact with flocs during floc breakage. Different impeller geometries and containing vessels give rise to different shear patterns and maximum shear levels experienced by flocs that result in different floc breakage behaviour. For this reason a simple consideration of how floc size changes relative to its initial size may give the best global indication of how strong flocs are.

Some general trends have emerged from floc strength tests that show that floc strength increases with decreasing floc size. The addition of polymer decreased floc strength for biological flocs and increased floc strength for chemical flocs.

Floc strength is a difficult parameter to measure accurately due to the inherent fragility and complexity of floc structures. As yet, there is no detailed understanding of the internal composition and bonding that occurs within a floc aggregate such as the number of bonds, locations of points of attachment and internal attractive and repulsive forces within the structure. Therefore many of the floc strength and breakage models developed so far have not reached a thorough mechanistic level.

MEASURING FLOC STRUCTURAL CHARACTERISTICS

The aggregation of fine particles and colloids into larger assemblages is a process that can occur both naturally and artificially. The resulting aggregates that form are known as 'flocs' which are best described as being highly porous, irregularly structured and loosely connected aggregates composed of smaller primary particles (Dolfing, 1987, Huang, 1994 and Kim *et al.*, 2001). In natural aqueous environments examples of floc formation include the transport and deposition of particulate matter in estuaries (Manning and Dyer, 1999), the assemblage of marine particles (e.g. plankton, organic matter, faecal material and minerals) into large aggregates known as marine snow (Ransom *et al.*, 1998) and the colloidal aggregates that are present in most natural surface waters (Gregory, 1997). However, industrial processes that require the separation of solids from liquids may be enhanced by an artificially induced floc formation stage. This includes bioprocess, chemical and mineral processing industries and at water treatment works (WTW) and wastewater treatment works (WWTW) (Zhang *et al.*, 1999).

The size and structure of flocs are considered fundamental to the operation of industrial unit processes (Waite, 1999). In water and wastewater treatment processes, the aim is to remove impurities from water in the form of solid particles. Once the solid particles are produced, they may be separated from water using sedimentation, flotation, filtration and thickening techniques (Rebhun and Lurie, 1993). The physical characteristics of the floc are therefore fundamental in determining their removal efficiency. For example, large compact flocs have a high settling rate that results in a treated water of low turbidity during settlement (Wilen *et al.*, 2003), whilst large and porous flocs aid filtration due to high permeability (Bushell *et al.*, 2002)

Quantifying floc characteristics is made difficult due to the highly irregular three-dimensional structure of flocs and their inherent delicate nature. In addition, the characteristics of flocculated aggregates have been shown to change depending upon the physical and chemical conditions prevailing in the flocculator (Farrow and Warren, 1989). However, a review of the literature shows there is a comprehensive amount of work on the evaluation of floc structures. The principal aim of this review is to assess how floc structural characteristics can be measured.

COAGULATION AND FLOCCULATION

An in depth review of coagulation and flocculation theory is beyond the scope of this review, for a more rigorous and thorough treatment readers are referred to Gregory (1989) and Amirtharajah and O'Melia (1990). However, in order to understand the importance of floc structure it is firstly necessary to briefly review current knowledge of floc structure and formation. In aqueous systems, floc aggregates are composed of smaller sub-units. In most cases the primary particles are often of sizes between 1 nm and 1 μm consequently fall into the colloidal size range.

There is some confusion in the literature as to the precise definitions of coagulation and flocculation and there appears to be a certain amount of interchanging between the two (Jefferson and Parsons, IN PRESS). However, there is some consensus that the two should be treated as separate and for the purposes of this review the following definitions are used from Cornwell and Bishop (1983) and Gregor *et al.* (1997). Coagulation is the process of chemically changing colloids so that they are able to form bigger particles by coming close to one another. This may be achieved by particle destabilisation through double layer compression, enmeshment, chemical reaction or chemical sorption. Flocculation is the process of transferring coagulated colloids into contact with each other to form larger aggregates (flocs).

The exact process of particle destabilisation and the subsequent colloid aggregation is complex. It is generally considered to be a two stage process of particle transport and particle attachment (Thomas *et al.*, 1999). The schematic in Figure 5.7 shows a simplified view of the steps involved. Floc formation is considered a balance between aggregation and breakage (Biggs and Lant, 2000).The rapid initial formation of microflocs is dominated by aggregation, however the importance of floc breakage increases until a steady state floc size is reached.

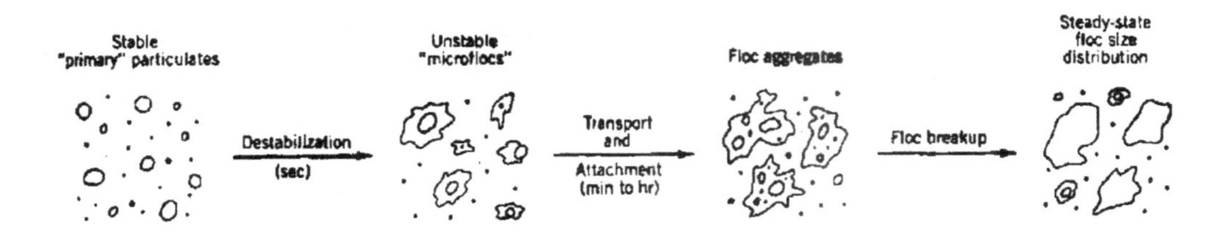

Figure 5.7 The steps of particle transport and attachment for aggregating particles (adapted from Montgomery, 1985).

Floc Structural Properties and their Measurement

The evaluation and quantification of floc structural characteristics is made difficult due to the highly irregular three-dimensional structure of flocs and their inherent delicate nature and porosity. However, there is a great deal of information in the literature on methods for the quantification of flocs. The following section deals in turn with measuring floc size, shape and fractal dimension.

FLOC SIZE

Many different measurements have been chosen as the representative characterisation of floc size. A simple measure of floc size is the floc longest dimension. On its own this measurement is of limited use as it only gives an indication of floc size in one dimension. A more common approach is to find the longest dimension of the floc in both the horizontal (d_{hor}) and vertical (d_{vert}) planes as shown in Figure 5.8 (Farrow and Warren, 1989 and Manning and Dyer, 1999). This also allows an indication of the floc height: width ratio and gives an indication of floc shape.

Typically, when referring to floc size a floc equivalent diameter measurement is made (Cousin and Ganczarcyk, 1998). The use of equivalent diameters allows the particle to be defined as a sphere or circle that is in some way equivalent to the particle. Such a standardised measurement allows a comparison to be made between very irregular forms. However, unless the particle being measured is a sphere, then each of these different diameters will take a different value for the same particle. Rather than an absolute value, equivalent diameters should be used for comparative purposes. For this reason it is important that the choice of equivalent diameter remains the same when comparing floc size. Dharmarajah and Cleasby (1986) list fifteen different characteristic diameters that are used to quantify non-spherical particles. Some of these are not applicable to the measurement of flocs because they would damage the fragile aggregates. This precludes the use of sieve diameters, which involves passing the aggregates through a sieve and determining the smallest mesh size that will allow the particle through. The most common size measurements that are used in the literature for flocs are summarised in Table 5.6.

As microscopy has been widely applied in particle sizing (Allen, 1997 and Aguilar *et al.*, 2003) most of the floc diameters in Table 5.6 are from two-dimensional images. As with all two dimensional measurements of complex and irregular three dimensional structures, there are difficulties in getting representative size data from a single measurement. Additionally, the results strongly depend upon the orientation of the floc presented to the researcher because a single non-uniform shape has an infinite number of linear dimensions, so it is only when these results are averaged for a large number of particles that a meaningful number can be found. The British Standard for microscope counting suggests that a minimum of 625 particles should be sized in order to get a representative size distribution (BS3406, 1963).

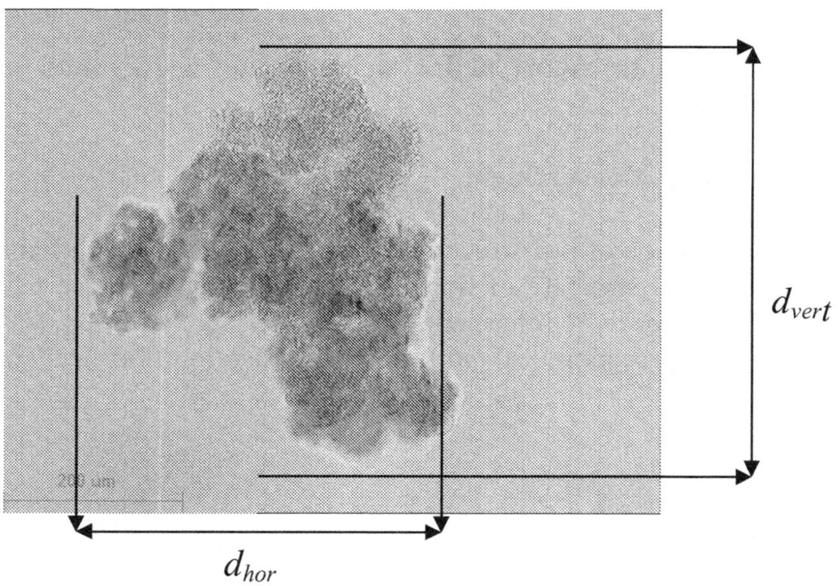

Figure 5.8 The maximum dimensions in the horizontal and vertical planes for a typical floc.

The diameters based upon 2 dimensional images (such as the projected area diameter (d_a), the Martin's diameter (d_M) and the Feret's diameter (d_f)) are known as statistical diameters because they are only an acceptable indication of particle size distribution if enough measurements are made. The situation is complicated because particles have a tendency to orientate themselves on slides such that they present their maximum area (Allen, 1997). This means that the dimension perpendicular to the viewing plane is generally the smallest and is often neglected. Therefore there is a tendency for statistical diameters based upon 2 dimensional images to be larger than those based upon 3 dimensions. Martin's and Feret's diameters also rely upon the random orientation of the floc in the plane parallel to the viewing direction if only a single measurement per floc is taken from a fixed direction. This confers an advantage on the use of the projected area diameter (d_a). However the advent of image analysis tools has enabled the quick estimation of a range of single measurements to be taken from around the same floc from any number of different directions such that an average value can be used, removing the need for random orientation in this plane. It should also be noted that if the projected area diameter is obtained from a randomly orientated particle in all 3 dimensions (d_p) then the diameter should be representative of the particle in all 3 dimensions as opposed to only 2 and thus provide a more accurate representation of the overall floc size. Indeed, if any of the other statistical diameters are obtained from the projected area of randomly orientated particles then they too should be representative of the three dimensional floc.

Particle size estimates based upon volume are particularly useful for settlement purposes. The settlement of flocs is a particularly important operational parameter because increased rates of floc settlement results in better solids removal in settlement tanks. In laminar flow, particles fall randomly so orientation should average out over a range of measurements. In non-laminar conditions, particles tend to orientate themselves to resist motion, so the free falling diameter found is smaller than the Stoke's diameter. It is therefore recommended that laminar conditions are applied in order to find a more representative indication of particle size.

Table 5.6
The most common equivalent diameters used for characterising floc aggregates (taken from Dharmarajah and Cleasby, 1986 and Allen, 1997).

Floc Diameter	Description	Diagram	Equation for Calculation
Perimeter diameter, d_c	The diameter of a circle with the same perimeter (P) as the measured particle.	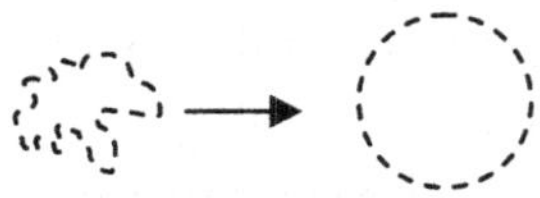	$d_c = \frac{P}{\pi}$
Projected area diameter1, d_a	The diameter of a circle with the same projected cross-sectional area (A) as the floc measured in a stable orientation.		$d = 2\sqrt{\frac{A}{\pi}}$
Projected area diameter2, d_p	The diameter of a circle with the same projected area as the floc measured in a random orientation.		
Surface diameter, d_s	The diameter of a sphere having the same surface area (S) as the floc.		$d_s = \sqrt{\frac{S}{\pi}}$
Volumetric diameter, d_v (OR equivalent spherical diameter)	The diameter of a circle with the same volume (V) as the floc measured.	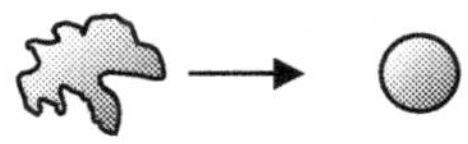	$d_v = \sqrt[3]{\frac{6V}{\pi}}$
Surface-volume diameter, d_{sv}	The diameter of a sphere with the same surface area to volume ratio as the floc.		$d_{sv} = \frac{d_v^3}{d_s^2}$
Free-falling diameter, d_f	The diameter of a sphere having the same density and free-falling speed as the floc in the same fluid at the same density and viscosity.	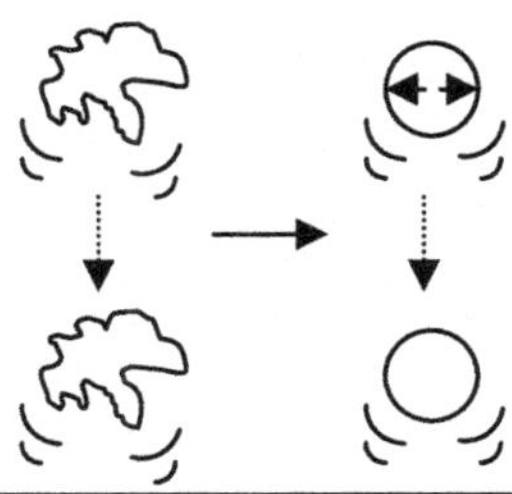	
Stokes diameter, d_{st}	The diameter of a free falling particle in the laminar flow range (where Re < 0.2).		$d_{st} = \frac{18\mu v}{\rho_f - \rho}$
Feret's diameter, d_F	The (mean) value between pairs of parallel tangents to the projected outline of the particle.	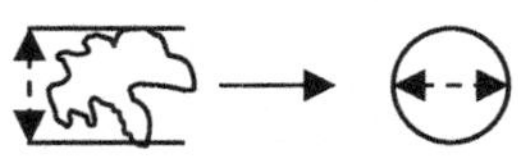	-
Martin's diameter, d_M	The length of the chord parallel to a fixed direction which splits the floc projected area into two equal parts.	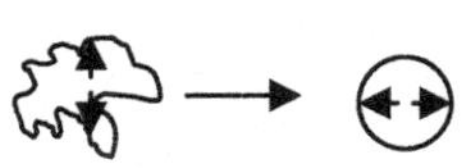	-
Circumscribing diameter, d_{sc}	The diameter of the smallest circle that circumscribes the outline of the projected floc.	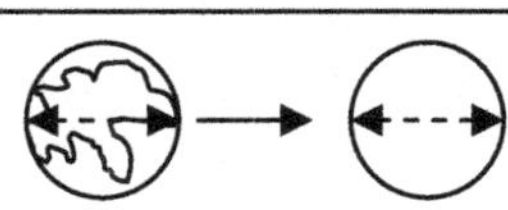	-
Inscribing diameter, d_I	The diameter of the biggest circle that fits inside the outline of the projected floc.	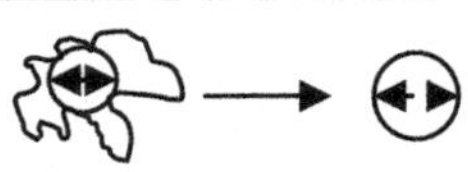	-

Floc Shape

An indication of floc shape may be provided by sphericity and circularity shape factors. The indices measure how much a particle varies from a sphere or a circle (Equations 5.16 and 5.17). The sphericity factor is a function of the volumetric diameter (d_v) and the surface area diameter (d_s).

$$Sphericity = \left(\frac{d_v}{d_s}\right)^2 \quad (5.16)$$

$$Circularity = \frac{P^2}{4\pi A} \quad (5.17)$$

Circularity is related to the perimeter (P) and the projected area of the particle (A). A value of close to zero indicates a shape approaching a straight line, whilst a value of 1 indicates the shape is a perfect sphere or circle. The shape factors may be of use to show the change of particle shape under differing conditions. For example, Cousin and Ganczarczyk (1998) compared the affect of salinity on activated sludge flocs. A reduction in the value of the circularity shape factor with increasing salinity indicated that the flocs were becoming more elongated at higher salt concentrations.

METHODS FOR DETERMINING FLOC SIZE AND SHAPE

Most of the methods for determining floc strength rely upon some measurement of floc size before and after an energy input. Techniques for quantifying and measuring floc size and shape parameters are made difficult due to the inherent irregularity of floc structures in both two and three dimensions. Most efforts have been to size flocs from magnified images captured from cameras (Wang *et al.*, 2002). Two fundamental difficulties arise when using such an approach. The first is which comparable floc characteristic(s) should be measured and the second is how the flocs are prepared prior to being measured.

Farrow and Warren (1993) have divided some of the methods used for characterising aggregate particle size into a number of separate categories (Table 5.7). Physical sizing techniques such as sieving are inappropriate for aggregates due to their delicate nature. Different workers have used a variety of different methods for the determination of aggregate size. It is important to ensure that the extraction, preparation and measuring technique:

1. measures a representative sample or sub-sample of the original floc suspension
2. does not damage, break or change the flocs
3. does not encourage further aggregation

Table 5.7
Some of the methods used for obtaining aggregate size.

General Floc Sizing method	Floc Size found from:
Microscopy	(1) observation of static floc size
	(2) observation of dynamic floc size
Photography and image analysis	(1) observation of floc static size taken from suspension
	(2) observation of floc dynamic size under turbulent conditions
	(3) observation of floc dynamic size under laminar flow
Light scattering	(1) back/front scattering of light by floc particles
Transmitted light	(1) light transmitted through floc suspension
Individual particle sensors	(1) optical sensing of flocs
	(2) electrical sensing of flocs

Microscopy

Microscopy is one of the most widely used technique for measuring particle size (Allen, 1997; Aguilar *et al.*, 2003). Microscopy has been used for decades as a method for sizing and counting flocs (Li and Ganczaryck, 1986; Droppo *et al.*, 1996). Before the advent of image analysis gaining useful floc size data was laborious and highly dependent upon the skill of the microscopist. A non-biased selection of flocs is required that is representative of the flocs contained within the sub-sample presented to the researcher. Aggregate size is estimated by reference to a graduated eye piece graticule or by placing flocs in cells with background grids or scales of a known size (for example a plankton counting chamber). This also requires good practice on behalf of the microscopist.

In most instances, carefully dropping a small sample of the suspension onto a microscope slide or into a measuring cell on a slide is suitable for particle size analysis under a microscope (Spicer and Pratsinis, 1996; Wang *et al.*, 2002). In some studies, a cover slip is placed over the sample floc sample as it removes depth of field problems (Cornelissen *et al.*, 1997; da Motta *et al.*, 2001). However, it is difficult to accept that this approach does not change the flocs due to their delicate nature. The compression from the cover slip is likely to considerably change floc structure.

The problem associated with all techniques where flocs must be removed *ex situ* from the suspension arises from the method of aggregate extraction and preparation prior to being sized. As reported in Farrow and Warren (1993), Camp (1968) used a dipped tube technique, whereby a hollow glass tube was submerged in a flocculated suspension and sealed at the top and then removed. The tube was then placed horizontally beneath the microscope for analysis. This method provides a good representative sample of the flocs without much rupture. However, in the tube flocs can settle onto one another and cannot be distinguished from one another. Therefore this cannot be a true representation of the actual size distribution. Wang *et al.* (2002) use a similar method to withdraw flocs for analysis using a sampling tube. The contents of the tube were carefully emptied into a microscope cell previously filled with water. By ensuring the tube was large enough and the cell filled with water no floc breakage was seen. In addition, the dilution upon entering the cell should prevent the flocs from falling onto one another. In addition it is important that the technique must not allow further agglomeration to take place during the sample preparation, therefore as dilute suspensions as possible are favoured.

Cousin and Ganczarczyk (1998) and Gorczyca and Ganczarczyk (1999) have used an agar solution (1.5-4 %) to solidify activated sludge flocs and alum flocs in suspension in a Petri dish. A thin layer of the solidified suspension containing an equal distribution of flocs can then be viewed and measured under a microscope. This technique aims to ensure the flocs randomly orientate themselves in the suspension and overcome the non-random orientation of flocs settling onto the slide surface (Farrow and Warren, 1993). Effectively a static measurement is giving a three dimensional representation of the floc. However, it is not clear what affect the agar has floc structure. A further method from Gorczyca and Ganczarczyk (1999) embedded cubes of agar-solidified alum floc suspensions in a hardening resin. In this way, very thin sections (2 μm thick) could be cut through the cubes to allow floc internal structure to be viewed under a microscope. However, when compared to the agar technique smaller flocs were seen suggesting the hardening resin has a considerable affect on floc structure. Indeed, for both techniques there must be some question as to whether you get random orientation of flocs in the thin agar suspension of the Petri dish.

The use of wide-mouthed pipettes is a common method for floc extraction (Li and Ganczarczyk, 1986 and Spicer and Pratsinis, 1996). Individual flocs may be selected in this way or a larger sub-sample of the suspension may be removed. The latter option is preferred and is more widespread because it requires no bias on behalf of the worker for a selection of a whole range of floc sizes. The use of pipettes is a widely used technique, however Manning and Dyer (1999) suggest that they are a destructive method. They explain that their use may account for differences between floc sizes generated from *in situ* and *ex situ* methods. It is therefore useful to compare *in situ* and *ex situ* techniques. Spicer *et al.* (1998) have compared the size of polystyrene-alum flocs generated from a Malvern Mastersizer (a light scattering method discussed later in the review) using three types of sample delivery mechanisms: a) a 5 mL hand pipette, b) a syringe pump and c) a peristaltic pump. The mass-mean floc diameters after 15-20 minutes of flocculation was ~150 μm using the pipette and ~250 μm using the syringe and peristaltic pumps. This would seem to confirm that hand pipettes adversely disrupt flocs. However, the authors discuss that it was actually the result of flocs settling between sampling and measurement by the Mastersizer because the rate of delivery using to the measuring cell is very slow using the hand pipettes and that the pipettes did not adversely affect the floc size.

Microscopy has the advantage of allowing individual particles to be viewed, scrutinised and analysed at high magnification. This allows the researcher to get a feel for the structure of each aggregate under investigation and give a better indication of floc shape and irregularity. In many applications, microscopy is the only method available to find floc porosity and other shape factors. In addition, microscopy is a relatively inexpensive method. Limitations of microscopy are the small depths of field possible with light microscopes such that a particle may be entirely focused due to its 3D structure projecting into the plane of view (Allen, 1997). A can be seen from Table 5.8, the recommended 625 individual particle counts per sample is generally not seen for most studies using microscopy to investigate floc size distributions. This suggests that the microscopy work carried out to date should only be used as an indicator of the actual floc size or as a compliment to other sizing techniques because it does not meet the rigorous statistical criteria of the British standard.

Table 5.8
The number of flocs sized per sample for a number of different studies.

Type of floc under investigation	**Number of flocs measured**	**Authors**
Activated sludge	245-377	Li and Ganczarczyk (1986)
Polystyrene beads	500+	Spicer and Pratsinis (1996)
Activated sludge	70+	da Motta *et al.* (2001)
Alum flocs	~ 100	Gorczyca and Ganczarczyk (1999)

Microscopy has been and still is a widely used technique for floc sizing. Sample extraction and preparation is key to gaining accurate knowledge of floc size distributions. Simply transferring a diluted floc sample into a shallow microscope well is the quickest and easiest and most widely used method. Considerable time and effort is required needs to be invested in order to achieve the necessary accuracy and a statistically significant distribution.

Photography and Image Analysis

Image analysis is the manipulation of information within an image to turn it into a more useful form whilst digital image analysis is the manipulation of digital images using a computer (Image Pro guide, 2001). For the purposes of this review, where the term image analysis is used this refers to digital image analysis. The basic stages and requirements of performing image processing and analysis are shown in Figure 5.9. Image analysis usually requires image processing which is the conversion of one image into another. Usually this is done to improve the quality of the image for analysis. For example, if an image does not have a well defined contrast between the object and background a particle may be incorrectly sized due to a blurred boundary between the two (Chakraborti *et al.*, 2000).

The main components of a modern image analysis system are an image capture device (usually a close-coupled device (CCD) camera or digital camera) connected to a computer with an image grabber. Computer software is required for the image processing and analysis and a variety of commercial products are available. Image analysis is often combined with microscopy by mounting a CCD camera onto the microscope (Li and Ganczarczyk, 1986; Cousin and Ganczarczyk, 1998; da Motta *et al.*, 2001; Kobayashi, 2004). The advent of photography and image analysis has allowed much quicker measurements of an almost inexhaustible number of different floc size measurements to be made from floc samples when compared to traditional microscope methods (Wang *et al.*, 2002). However, it still relies upon an unbiased approach from the microscopist. Image analysis has also been used to monitor floc suspensions *in situ* (Ducoste and Clark, 1998; Chakraborti et al., 2000; Bache and Papavasilopoulos, 2003). Flocs are monitored by capturing images of a stirred suspension by focusing on a plane a short distance

(0.3 – 1 cm) behind the wall of tank containing the suspension. Calibration is achieved by focusing on a graticule suspended into the tank prior to flocculation experiments.

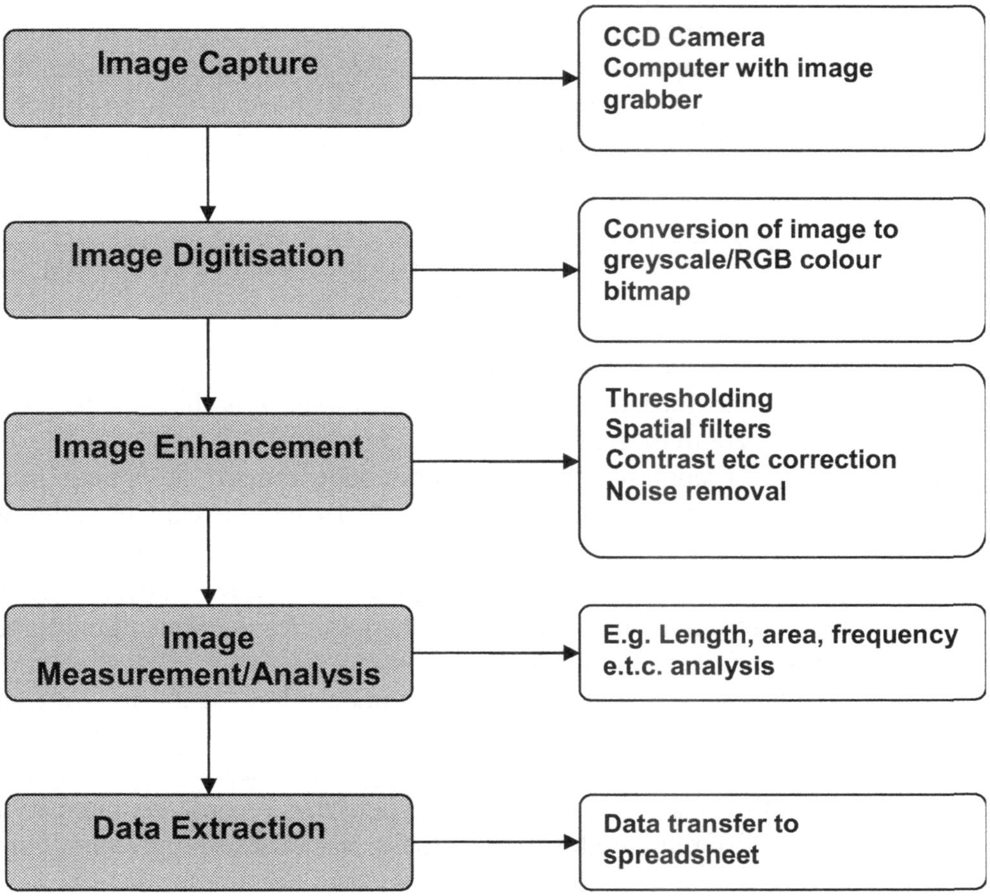

Figure 5.9 The steps involved in digital image analysis

Light Scattering

As light is passed through a suspension of particles some part of the light is absorbed by the particles whilst some light is scattered. The remainder of the light passes straight through the suspension. The way in which the suspension does this is dependent upon particle size, the nature of the particles and the suspending medium (Farrow and Warren, 1993). In light scattering particle sizing techniques, the measured scattering pattern of an applied laser is compared to the predicted scattering pattern based upon an optical model in order to generate a particle size (Selomulya *et al.*, 2001). Lorenz-Mie theory is the classical model for determining particle diameter from light scattering and is the basis for all particle sizing instruments that measure particle size in this way (Black *et al.*, 1996). The principle equation for Lorenz-Mie theory is shown in Equation 5.18

$$\chi = \frac{\pi m d}{\lambda} \tag{5.18}$$

χ is the fundamental parameter for light scattering, d is the particle diameter, m is the refractive index of the particles and the λ is the wavelength of the incoming laser.

The model assumes particles are (1) spherical, (2) the laser illuminates particles uniformly and (3) the laser beams are plane light waves. This theory works well for particles that are smaller than the cross section of the laser beam. However, as particles get larger assumptions 2 and 3 become unreliable. For example, at a typical beam size of 100 μm, the assumptions of plane light and uniform illumination do not hold at particle sizes above 10-20 μm (Black *et al.*, 1996). Fraunhofer diffraction theory is a modified version of Lorenz-Mie theory that takes this into account (Equation 5.19).

$$X = \frac{\pi d R}{\lambda f} \tag{5.19}$$

X is the fundamental parameter for light scattering, d is the particle diameter, R is the radial distance in the focal plane as measured from the optical axis, λ is the wavelength of the incoming laser and f is the focal length of the receiving lens.

It is important to note the Fraunhofer theory does not depend upon the optical properties of the particles in suspension. The theory holds true for all particles except particles with a refractive index approaching 0 or for very small particles (less than 10 μm). Fraunhofer theory considers only the light diffracted by the particles in suspension, however where a significant amount of light is transmitted through the particles or past the particles the transmitted light (also known as anomalous scattering) impacts on the results.

The most common commercial particle size instruments use light scattering to determine particle size. These instruments (such as the Malvern instruments) measure particle size by passing a laser beam through a suspension of particles. Small particles scatter light at high angles whilst large particles scatter at low angles. An array of ring detectors records the intensity of the scattered light at a range of different angles. From these responses, proprietary computer programmes iterate particle size distributions from Lorenz-Mie and Fraunhofer theory.

A certain amount of light passes through the suspension and out of the optical system. The amount of light lost is dependent upon the size and concentration of the particles in suspension. The instrument measures this in terms of the laser obscuration, in other words the amount of the laser that is scattered or absorbed by the particles. The manufacturers recommend that between 10-30 % obscuration is achieved for reliable measurements (Guan *et al.*, 1998). The concentration required to reach this range will vary for different suspensions, however Farrow and Warren (1993) suggest that solids should be < 0.03 % solids by weight. Excessive obscuration leads to significant underestimation of the laser scattering.

These techniques rely upon a constant flow of the suspension through the instrument during the measurement cycle. This feature has been harnessed to allow the development of a non-intrusive methodology for measuring dynamic floc size (Spicer *et al.*, 1998; Biggs and Lant, 2000; Chaignon *et al.*, 2002). These methods have a stirred vessel containing the aggregate suspension and are connected to the particle sizing device by plastic tubing. Intrinsic to this type of system is a requirement to pump the suspension through the optical unit of the size analyser. As has been previously been discussed, Spicer *et al.* (1998) compared 3 types of pumping techniques for delivery to the optical cell. They concluded that a continuous recycle using a peristaltic pump on the return side of the particle was the least severe technique on the flocs and allowed easy continuous monitoring of the suspension. Flocs were extracted from the recirculation zone in the flocculation vessel (a distance midway between the top of the tank and the stirrer) to ensure an accurate representation of the bulk suspension was sampled. Rattanakawin and Hogg (2001) used a similar set-up without the recirculation system. The floc samples were discarded once they had passed through the size analyser. This was done to ensure that any affect of the pump on floc size was ignored. However such a method could not be used for continuous monitoring of the flocculation process as the suspending fluid would run dry.

Biggs and Lant (2000) compared the effect of pump speed on the size distribution of activated sludge flocs for a continuous system using a Malvern mastersizer as the size analyser. They compared four different flow rates ranging from 1.7-5.5 ml s^{-1} using a peristaltic pump. They found that the optimum pump speed was at a flow of 3 ml s^{-1}, above this rate shear within the pump and tubing significantly reduced the floc size. Whilst below this flow there was a significant time lag before a representative sample was measured in the particle size analyser. Spicer *et al.* (1998) also used a flow of 3 ml s^{-1} in 6 mm internal diameter tubing (corresponding to a Reynolds number of 618) for research into flocculation of polystyrene beads. Lartiges *et al.* (1995) used a pump speed corresponding to 1.75 ml s^{-1} as this did not encourage aggregation or break-up in the pump system for research on optimising coagulation of raw river water.

Transmitted light

Another technique that has been used extensively to monitor the size and growth of floc suspensions is the photometric dispersion analyser (PDA). First described by Gregory (1985), the PDA gives a combined measurement of the particle size and frequency for a flocculating suspension. The device consists of a light source, detector and processing equipment that monitors turbidity fluctuations in the sample. Specifically the PDA measures the average light transmitted through a suspension and the rms value of the fluctuating component (Gregory and Dupont, 2001). The ratio of the two gives the flocculation index (FI), which gives a good measurement of aggregation. The PDA has been widely used in closed loop systems to measure dynamic floc size, similar to those mentioned for the light scattering techniques (Burgess and Phipps, 2000; Fitzpatrick *et al.*, 2003; Yukselen and Gregory, 2004; McCurdy *et al.*, 2004). Whilst the other sizing methods give an absolute value for floc size, the PDA gives a combined value that increases with increasing particle size and number. However, for a system containing a constant solids fraction an increase in the FI can be attributed to an increase in floc size as the larger particles have a greater signal fluctuation than smaller ones (McCurdy *et al.*, 2004). The PDA is therefore good at showing relative changes in floc size, such as during floc growth and breakage phases and giving qualitative comparisons between different treatment variables. Unlike, some of the other sizing techniques, the PDA is unable to give information on floc size distributions and the solids passing through the measuring cell must be at a high enough concentration to provide a reliable signal.

Individual Particle Sensors

Individual particle sensors measure single particles as they pass through an aperture onto an electric field (electrical sensing) or through a light beam (optical sensing). In both of these techniques, the major source of problems for particle sizing comes from the break-up of floc aggregates as they pass through the aperture of the measuring cell.

In the electric field method, particles are suspended in an electrolyte solution and then passed through an electric field. The change in resistance caused by the particle is proportional to the particle size with a small correction factor (Farrow and Warren, 1993). The Coulter counter is the most common electrical sensing technique. Leentvaar and Rebhun (1983) summarise that the Coulter counter significantly underestimates floc size when compared to optical analysis as it only measures the volume of the solid in the floc and not the effective volume of the floc including pores and water. The effect of the electrolyte solution on floc macrostructure has yet to be fully investigated at the ionic concentrations involved using electric field methods, but Cousin and Ganczarczyk (1998) have shown an increase in activated sludge floc porosity, diameter and elongation at very high salt concentration.

The optical sensing methods measure particles of a size >10 μm. The amount of light attenuated by particles as they cross a light beam is proportional to particle size. However this method is limited by a need for a low particle concentration for accurate measurement and a narrow size distribution band due to the narrow size of each aperture in these instruments and there are few examples of these instruments being used for floc sizing.

No particle sizing method is perfect for measuring floc size and each of the techniques have their own advantages and disadvantages. The tedious requirements for sample preparation and floc transfer indicate that microscopy is not the most suitable technique for finding floc size. Photography *ex-situ* gives a considerable improvement in that flocs do not need removing from the flocculating vessel but high quality images are required for accurate image analysis. Nevertheless, both photography and microscopy can give a crucial feel to the researcher of the type of particle they are dealing with and are often the only reliable method for getting floc shape factors and porosity measurements. The particle sensing instruments are not ideal due to the problems associated with only being able to measure narrow size bands at a time using electrolyte solutions in the measuring cell. On-line techniques such as those involving light scattering and the PDA allow quick measurements to made and their non-intrusive nature and ability to monitor a wide range of particle size distributions make them ideal for showing quantitative size distributions (light scattering) and qualitative changes in floc size (PDA).

Fractal Dimension

Since Mandlebrot introduced the concept of fractal theory in the 1970's, the application of fractal geometry is now a well established means of describing the complicated structure of particle aggregates (Gorczyca and Ganczarczyk, 1999, Thomas *et al*., 1999; Selomulya *et al*., 2003; Chakraborti *et al*., 2003). Fractal objects may be defined as those objects that:

1) show self similarity
2) express a power-law relationship between two variables
3) can be characterised by a non-integer fractal dimension

Self similarity is the existence of the same pattern regardless of the magnification from the which the fractal object is viewed from. In many systems exact self-similarity is not observed, but a less restrictive definition of a fractal object is that it shows statistical self-similarity. This means that on average different sections of the object look similar to one another (Kaye, 1989). The second characteristic of fractal objects is a power relationship between two variables of the object. This may be a link between area (A) and length (L) as in Equation 5.20, or the relationship between volume (V) and area as in Equation 5.21.

$$A \propto L^{D_f} \qquad (5.20)$$

$$V \propto A^{D_f} \tag{5.21}$$

Flocculated aggregates are examples of mass fractal objects. This means that both the internal structure and the surface of the aggregate exhibit fractal properties. Mass fractals are summarised by Equation 5.22.

$$M \propto L^{D_f} \tag{5.22}$$

M is the mass of particles, *L* is a characteristic measure of size and D_f is the mass fractal dimension. Gregory (1998) summarises that the choice of measurement for the size *L* does not matter as the same trends are seen so long as the choice is constant. For Euclidean objects, the one dimensional value of D_f will be 1 for a linear line, 2 for a two dimensional planar shape and 3 for a compact three dimensional shape. Fractal objects take non-integer values of D_f and are therefore said to show non-Euclidean dimensionality. Values approaching 3 for a three dimensional floc therefore indicates a high degree of compaction whilst values approaching 1 indicates a very loose and open structure. The fractal dimension can therefore give important structural information of floc compaction and the space filling nature of the aggregate. The fractal dimension of a floc may be found in a number of ways. These may be broadly categorised into techniques that use scattering (of light, neutrons or x-rays), settling and two dimensional fractal analysis using image analysis (Waite, 1999).

Scattering

The pattern in which an aggregate scatters incoming radiation gives information on the aggregate structure as a function of a length scale (Bushell *et al.*, 2002). The way in which an object scatters light can give fractal values if enough is known about the scattering properties of the material contained within the aggregate. This technique assumes:

- The primary particles that make up the aggregate are uniform in shape and size.
- The refractive index of the aggregate material is low so that the wavelength of the incident light does not become shortened.
- Light is only scattered once as it passes through the suspension of aggregates before hitting the detector. Multiple scattering should be minimised by ensuring the concentration of particles is low (Tang *et al.*, 2002).

Generally, finding the fractal value from scattering theories rely upon the power law based upon Rayleigh-Gans-Debye (RGD) scattering theory shown in Equation 5.25.

$$I(Q) \propto Q^{-D_f} \tag{5.23}$$

I(Q) is the intensity of the scattered radiation and *Q* is the wave number estimated from Equation 5.24.

$$Q = \frac{4\pi n \sin(\theta/2)}{\lambda} \tag{5.24}$$

n is the refractive index of the suspending medium, θ is the scattered angle, λ is the wavelength of the radiation in a vacuum.

The fractal dimension Df is found from the slope of the line of a log-log plot of Q(I) against Q (Wu *et al.*, 2002). For fractal objects a power law relationship exists between *Q(I)* and *Q*. This dependency is only valid when:

$$\frac{1}{R_{agg}} \rangle\rangle\ Q\ \rangle\rangle \frac{1}{R_{part}} \quad (5.25)$$

R_{agg} is the radius of the aggregate and R_{part} is the radius of the primary particle.

This is because when *Q* approaches the size of R_{agg} the relationship is affected by the edges of the aggregate whilst when *Q* approaches the size of R_{part}, light is mainly scattered by the primary particles and not the aggregate (Guan *et al.*, 1998; Waite *et al.*, 2001). The primary particles of the flocs must also satisfy independent scattering for RGD scattering theory approximations to be obeyed. The RGD approximation is deemed applicable when:

$$|m-1| << 1 \quad (5.26)$$

$$(2\pi n/\lambda)L|m-1| << 1 \quad (5.27)$$

where *m* is the material refractive index and *L* is the length of the scattering body.

Most application of scattering has been to mono-disperse systems where information is known about primary particle size and the scattering behaviour of the particles under investigation. Examples include flocs formed from particles of latex (Tang, 1999; Selomulya *et al.*, 2001), aluminium oxide (Waite *et al.*, 2001), and iron oxyhydroxide (Waite, 1999). In these cases the assumptions mentioned above are generally valid for application to floc aggregates. Application to more complex flocs typically found during water and wastewater treatment processes has been more difficult. In these instances, information on primary particle composition can be limited with little to no knowledge of particle refractive index. Furthermore the primary particles may be non-uniform and consist of particles with different refractive indices. Therefore a number of the assumptions mentioned above may not be met when analysing complex floc structures. As a practical example of this, a mixture of iron oxyhydroxide and kaolin has a considerably distinct shaped scattering curve when compared to a pure iron oxyhydroxide system (Waite, 1999). In the case of activated sludge good scattering law relationships have been seen (Guan *et al.*, 1998; Waite, 1999). This is because the very low refractive index of the bacteria in the floc allows RGD theory to be met.

A common limitation to this technique is the limited scale of investigation of the technique. Commonly, the scattering power law relationship breaks down at small floc sizes compared to the average floc size. This was seen for kaolin suspensions where the average floc sizes ranged between 200-350 μm in diameter whilst the linear portion of the linear scattering relationship applied to flocs less than 50-100 μm (Wu *et al.*, 2002). For activated sludge flocs ranging up to 400 μm in diameter, the power law cut-off was below 70 μm (Waite, 1999). The application of light scattering to larger floc systems is an area that needs further investigation to find whether the larger flocs have variable fractal dimension or interfere with the scattering.

Settlement

Using settlement as a means of determining the fractal value of aggregates is a well established technique to the more widely used small angle light scattering method. The use of sedimentation to determine floc structural characteristics takes on an extra relevance because the settling behaviour of aggregates is an important parameter for optimising the sedimentation procedure. Floc settling behaviour is dependent upon size, effective density and porosity (Tang *et al.*, 2002). The fractal structure of flocs structure can have two possible consequences on its settlement behaviour because flocs take on increasingly non-spherical forms as they grow. This may act to increase the drag on the particle when compared to a solid sphere of the same size. Conversely, the porosity of flocs can act to reduce drag by allowing advection of the suspending medium through the floc structure (Bushell *et al.*, 2002).

The following determination of floc fractal dimension from settling velocity has been taken from Miyahara *et al.* (2002). A spherical particle at its terminal settling velocity may be summarised by Stoke's law as shown in Equation 5.28.

$$v = \frac{(\rho_s - \rho_l)gd}{18\mu} \tag{5.28}$$

v is the terminal settling velocity, ρ_s is the density of the particle, ρ_l is the density of the liquid, d is the floc diamater, μ is the viscosity of the suspending medium and g is acceleration due to gravity. Whilst Stokes law may be an over-simplification to completely describe floc setting, it is generally believed that flocs settle slow enough in order for Stokes derived equations to apply (Gregory, 1998). Shape factor and drag coefficient corrections are usually added to account for the irregular shape of flocs. A fractal floc consisting of similar primary particles may be summarised by:

$$i = \left(\frac{d}{d_p}\right)^{Df} \tag{5.29}$$

where i is the number of primary particles, d_p is the primary particle diameter and D_f is the fractal dimension. The mass and volume balances of the floc are given by:

$$V_f = V_s + V_l \tag{5.30}$$

$$\rho_f V_f = \rho_s V_s + \rho_l V_l \tag{5.31}$$

Where V_f is the floc volume, V_s is the volume of solids in the floc and V_l is the volume of liquid in the floc. Combining equations 5.29-5.31 into 5.28 gives:

$$v = \frac{d_p^{3-D_f} d^{D_f-1}(\rho_s - \rho_l)g}{18\mu} \tag{5.32}$$

The slope from a log-log plot of floc settling velocity against size will therefore yield the fractal dimension (Johnson *et al.*, 1996). The fractal dimension is found from the slope of the plot with D_f being equal to the value of the slope + 1. The equation only applies when the floc Reynolds number is less than one and the flocs are fall isolated at their terminal settling velocity in laminar floc. Wu *et al.* (2002) summarise that most experimental systems measuring settling rate meet the criteria for finding the floc fractal dimension. Problems may be accounted when the floc porosity is high, in these instances advection through the floc significantly increases the settling rate over that predicted by Stokes equations. In most practical instances, porosity effects on settling are neglected and currently far from understood. Therefore, care must be made of interpreting data when the floc fractal dimension is significantly less than 2 (Gregory, 1998). In addition, measuring floc settling requires meticulous preparation and a large sample number in order to get accurate results (Bushell *et al.*, 2002).

Image analysis

The combination of microscopy and image analysis software have has been widely used to calculate floc fractal dimension (Bellouti *et al.*, 1997; Cousin and Canczarczyk, 1998; Chakraborti *et al.*, 2003). Generally, high quality images of flocs are taken and the two dimensional (2D) fractal dimension found. This may be achieved in two common ways. The first is from the relationship between floc area and length (Equation 5.20). A log-log plot of floc area against size as found from image analysis yields a line with a slope giving the fractal dimension (Chakraborti *et al.*, 2000). A second way of determining floc fractal dimension is the box counting method. The process begins by covering the floc image with boxes of a minimum size to just cover the floc, this is then repeated with smaller box sizes (Bushell *et al.*, 2002). Plotting the number of boxes needed to cover the object against the size of the box on a log-log scale gives a line with a slope equivalent to the fractal dimension (Bellouti *et al.*, 1997). Commercial software packages are able to do this analysis very quickly and easily.

The main requirement for 2D fractal dimension analysis using image analysis is for the image to be of suitable quality for commercial software packages to be able to distinguish the floc from the background (Chakraborti *et al.*, 2000). In practice this often requires considerable image correction prior to fractal analysis and works best with flocs that show good contrast with their background and are not transluscent.

To summarise, Table 5.9 highlights the major advantages and disadvantages of each of the techniques mentioned. Light scattering methods work well with small, open flocs that have low refractive indices. Bushell *et al.* (2002) state that these are precisely the type of aggregates that settlement and image analysis techniques do not apply well to because they do not settle well, the modelling of the settling of open flocs is very difficult due to permeability effects and these particles are difficult to see because of their low refractive index. However, when analysing systems of high particle concentrations, shadowing effects and multiple light scattering invalidates the scattering models. Therefore, the calculation of fractal dimension and particle size only holds when within the obscuration threshold of the scattering instrument being used (Guan *et al.*, 1998). Similarly, flocs composed of a number of different primary particles have scattering behaviour that is difficult to predict and should not be used in these instances. The technique is also limited in that fractal values are generally only found for the small flocs in the system. However, when using the settling technique the fractal relationship is always seen across the whole range of flocs under investigation. Settling is a reliable technique provided a large enough sample is measured which can make this technique very time consuming. Settling can be widely applied to most floc systems provided that flocs are relatively compact thus avoiding interference in settling from porosity and advection though the floc. Careful temperature control and quiescent conditions in the settling column must be provided in order to prevent disruption to the floc.

2D image analysis relies upon flocs that have a high degree of definition between the solid of the floc and the background. Therefore pale translucent activated sludge flocs are often not ideal for this technique. In addition sample preparation prior to image capture is an important consideration and as has been mentioned in previous sections, the preparation stage must not act to damage or interfere with the floc structure. The main advantage of the box counting method is that the fractal dimension of individual flocs are measured. This can highlight differences in floc fractal dimension within a system whilst all the other techniques report average fractal values for the whole system.

Finally, it is important to report which technique has been used to measure floc fractal dimension. This is because each technique gives a different answer and may in fact be measuring different structural properties of the floc. For example, in a comparative study by Wu *et al.* (2002) the fractal dimension of activated sludge using settling was 1.31 whilst it was 2.06 using light scattering. Furthermore, image analysis of 2D images can only give a maximum fractal value of 2 whilst the maximum is 3 for the analysis of 3D flocs in settling and scattering.

Table 5.9
The advantages and disadvantages of the techniques used for determining the fractal dimension of floc aggregates.

Technique	**Advantages**	**Disadvantages**
Light scattering	- Rapid, non-intrusive method - Lends itself well to dynamic, online analysis - Very good for analysis of small aggregates with an open structure and low refractive index - Takes a large number of readings from many aggregates in a few seconds	- Not good for polydisperse aggregates made from many primary particles - Choosing an appropriate model for scattering behaviour can be difficult - Results affected by contamination from dust etc. - Power-law relationship breaks down at large floc size
Settling	- Best for measuring fractals of compact flocs - Cheap and simple - Not prone to contamination issues - Good for aggregates of made from a number of different primary particles	- Time consuming - Finding an appropriate drag coefficient is difficult - Can get non-random orientation of falling aggregates - Careful regulation of settling column required
Image analysis	1. Best for large, open aggregates 2. Not prone to contamination issues 3. Examination of single flocs allows detailed information on variation in floc structure within a sample	- Time consuming - Requires well defined, high contrast images for accurate analysis – which flocs generally aren't

FLOC MEASUREMENT SUMMARY

A range of techniques for measuring floc structural characteristics of size, shape and fractal dimension have been presented. Of the sizing techniques, microscopy is the most time consuming requiring considerable sample preparation and analysis time in order to achieve satisfactory results. However, microscopy and photography can give an important feel for the type of floc under investigation and is also the only reliable method to determine floc shape characteristics. On-line light scattering and PDA devices show good capability for measuring a whole range of different floc types and size in a non-intrusive way. Fractal dimension analysis can be measured using three main techniques. Light scattering works best for small, open flocs of low refractive index whilst larger, flocs of low porosity and of high colour contrast are more suited to settling and 2D image analysis under a microscope.

CHAPTER 6
DIAGNOSTICS

INTRODUCTION

This chapter presents the results of the investigations into the physical properties of flocs formed by coagulation with/without additional treatment with ion exchange. The chapter focuses on the size, strength, settling rate and regrowth potential of flocs and how operational decisions impact on the properties measured. The chapter also includes a more mechanistic section which attempts to link raw water characteristics, through fractionation, to the physical properties of flocs.

MATERIAL AND METHODS

Floc size and breakage

A laser diffraction instrument (Malvern Mastersizer 2000, Malvern UK) was used to measure dynamic floc size as the coagulation and flocculation process proceeded. A PB-900 variable speed jar tester (Phipps and Bird, Virginia USA) was used with 76 x 25 mm flat paddle impellers for coagulation experiments. A 1 litre cylindrical jar was used with holding ports for the inflow/outflow tubing secured onto the sides of the jar. The suspension was monitored by drawing water through the optical unit of the mastersizer and back into the jar by a peristaltic pump on the return tube using 5 mm internal diameter peristaltic pump tubing at a flow rate of 1.5 $L.h^{-1}$. The inflow and outflow tubes were positioned opposite to one another at a depth just above the paddle in the holding ports.

Water samples were placed on the jar tester under the following conditions: 200 rpm for 1.5 minutes, followed by a slow stir phase at 30 rpm for 15 minutes. Coagulant and pH adjustment chemicals were added at the start of the rapid mix. Size measurements were taken every minute for the duration of the jar test and logged onto a PC. Each size experiment was repeated 3 times for each set of conditions. For floc breakage experiments, the jar tester was programmed as before, however after the slow stir phase the suspension was exposed to increased shear for a further 15 minutes. Separate experiments were carried out and replicated 3 times at increased shears of 30, 40, 50, 75, 100, 150 and 200 rpm.

Method development

In order to provide the necessary shear to break the floc suspensions a jar tester with controllable rpm was used. The calibration of the jar tester was checked by comparing the rpm displayed on the control panel with the real rpm as determined by frame by frame analysis of video film. As can be seen in Figure 6.1 there was close agreement between the displayed rpm and the actual rpm, implying that the jar tester could be reliably used to provide controllable shear.

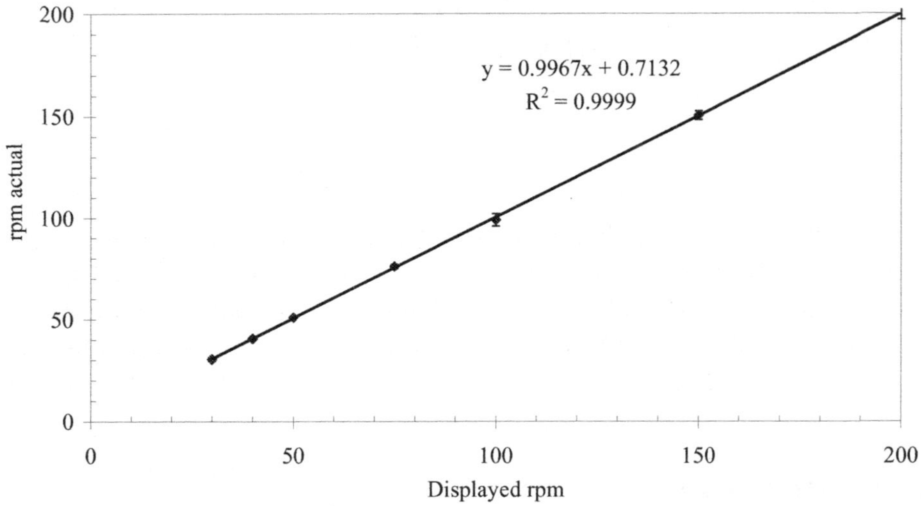

Figure 6.1 The calibration of the jar tester.

The floc sizes given by the laser diffraction instrument were verified using conventional light microscopy and image analysis software. Flocs were transferred from the jar tester using a wide mouthed pipette after the slow stir period (30 rpm) and after increased shears of 50, 100 and 200 rpm. Approximately 100 flocs from each set of rpm conditions were measured using this method.

As can be seen from Figure 6.2 there was reasonable agreement between the two sizing methods. For the small particle sizes (d_{10}) the laser diffraction instrument tended to give an increased aggregate size which was more pronounced at lower rpm. This was likely to be a reflection of the bias of laser diffraction instruments towards larger particles due to the calculation of size from a particle volume. At higher rpm there were fewer larger particles due to floc breakage and hence the bias will be reduced. The median floc sizes (d_{50}) were of a similar value for each of the techniques. The microscope technique gave floc d_{90} sizes that were larger than for the laser diffraction technique. This was a somewhat surprising result given the natural bias of laser diffraction to larger particle size. However, it was thought that flocs in the microscope cell underwent further aggregation during the transfer procedure or fell on top of one another giving rise to poor discrimination between individual flocs and hence an over estimation of size. These results show that while there were some differences in the resulting particle size using the two techniques, there was good agreement of the median particle size, suggesting that a high degree of confidence can be placed in these median values. Following from this, it was decided that the Mastersizer would be used to evaluate floc size during growth, breakage and re-growth.

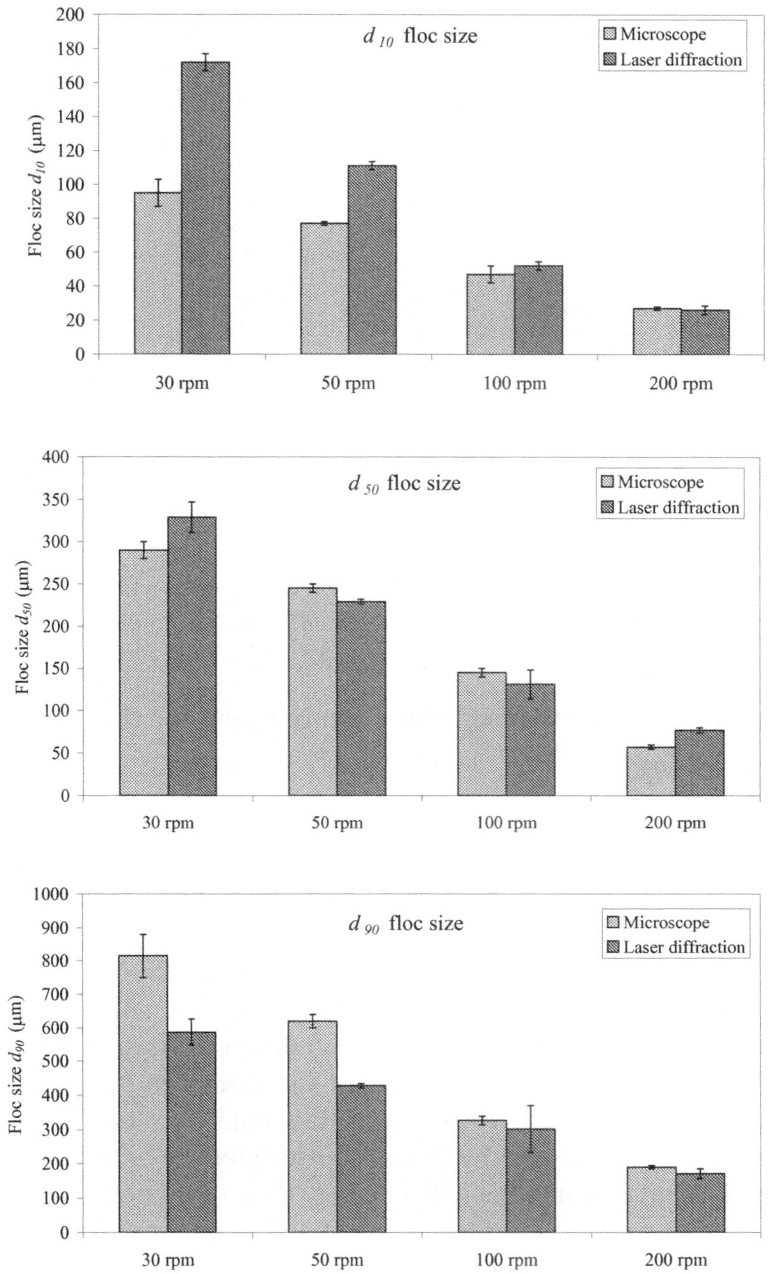

Figure 6.2 A comparison of the d_{10}, d_{50} and d_{90} floc size as measured using microscopy and laser diffraction techniques.

The Mastersizer technique was then applied to NOM flocs. The rpm of the peristaltic pump was optimized for NOM flocs in order to insure that flocs were not affected by the pump speed. Flocs were formed using a humic solution at an approximate concentration of 10 mg L^{-1} and using ferric sulfate coagulant at 8 mg L^{-1} and pump speeds of between 4 and 20 rpm were investigated and the resulting floc sizes were compared as given by the Mastersizer (Figure A.8). At rpm of below 5, the flocs settled in the tubing and measuring cell of the system, therefore giving rise to significant under-sizing of the suspension. At 6 rpm the flocs did not settle in the tubes but became stuck in the measuring cell of the Mastersizer instrument, therefore floc matter accumulated giving rise to an over-sizing of the flocs. At 7 rpm, flocs were observed to flow through the system whilst above this rpm, significant floc breakage was seen in the tubing as reflected by the rapid decrease in floc size shown in Figure 6.3. It was therefore decided that 7 rpm would be used for NOM flocs.

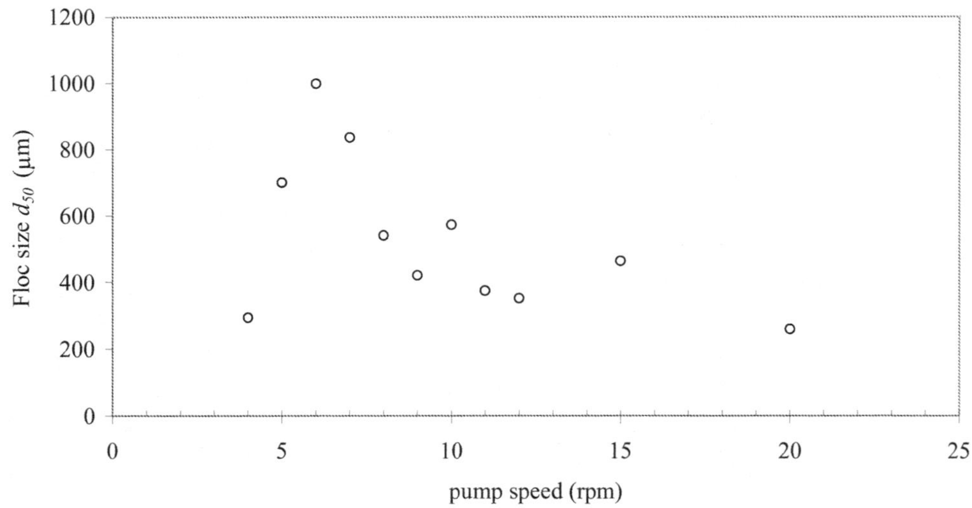

Figure 6.3 The optimisation of the peristaltic pump speed for the measurement of NOM floc size.

Structural analysis

During small angle light scattering, a light beam is passed through a sample. The particles in the sample scatter light proportional to their size and at a constant angle independent of which part of the particle is hit by the beam. Small particles scatter light at high angles, whilst large particles scatter at lower angles. The Malvern Mastersizer has an array of photo-sensitive detectors at different angles between 0.01 – 40.6 ° which detect the light scattered by the sample.

The determination of the fractal dimension of the flocs can be achieved through small angle light scattering using the Malvern Mastersizer. Raw data from the instrument is processed through a dedicated spreadsheet provided by Malvern instruments which resolves the intensity of light at each detector. The total scattered light intensity I is a function of the wave number Q, where Q is the difference in magnitude of the incident and scattered laser beam, which is given by (6.1):

$$Q = \frac{4\pi n \sin(\theta / 2)}{\lambda} \tag{6.1}$$

n is the refractive index of the suspending medium, θ is the scattered angle, λ is the wavelength of the radiation in a vacuum.

For independently scattering aggregates, I is related to Q and the fractal dimension by D_f (6.2):

$$I \propto Q^{-D_f} \tag{6.2}$$

A confirmation of the power relationship in (5) is to plot I against Q on a log-log scale. A power law relationship exists if this yields a straight line, the slope of which is used to give D_f.

The relationship only holds when the length of investigation is much larger than the primary particles and much smaller than the floc aggregates (6.3):

$$\frac{1}{R_{agg}} << Q << \frac{1}{R_{part}} \tag{6.3}$$

Settling rate

The settling column apparatus are shown in Figure 6.4. Floc aggregates were introduced into the settling column via a tapered entry port to ensure the flocs settled into the centre of the column. In order to avoid thermal currents disrupting floc settlement, the central settling column was enclosed by a water bath. The water bath was connected to a ThermoHaake K10 heat-refrigerated circulator (Hakke, Germany) to ensure a constant temperature was maintained in the column at 21.5 – 22.0 °C. The settling column was filled with de-ionised water and left to reach the required temperature. Floc images were captured using a CV M90 colour close-coupled device (CCD) camera (JAI UK Ltd, England). Image analysis software (Image Pro Plus from Media Cybernetics, Maryland USA) was used to determine floc settling velocity. The image analysis system was calibrated by suspending a 1 cm microscope graticule with 100 graduations in the centre of the settling column and focusing the CCD camera on the graticule.

Flocs were aggregated on a jar tester under the required conditions at room temperature (21-22° C). Flocs were then carefully removed from the jar using a wide mouthed pipette and re-suspended in de-ionised water held in a water bath at 21.5° C for dilution and temperature control of the flocs. Individual flocs could then be removed and placed in the settling column. Previous research has shown that particle/particle interactions can hinder settlement rates and it is therefore important that aggregates fall in isolation of one another (Thomas, 1999). Visual observation showed that little change in floc macrostructure occurred using the pipettes, however if flocs were observed to break during the transfer procedure then they were discarded.

Falling flocs were captured as image sequences. As a focused floc passed in front of the camera, the image grabber was manually triggered to take a series of 10 images. The time between each frame was set at 1 second. This meant that the distance travelled by the floc could be calculated per frame and therefore per time period, thus giving a settling velocity. The projected area of the floc presented in front of the camera was determined using the image analysis software and converted to an equivalent diameter. This standardised diameter was recorded along with its settling velocity for 100 aggregates for each set of coagulation conditions.

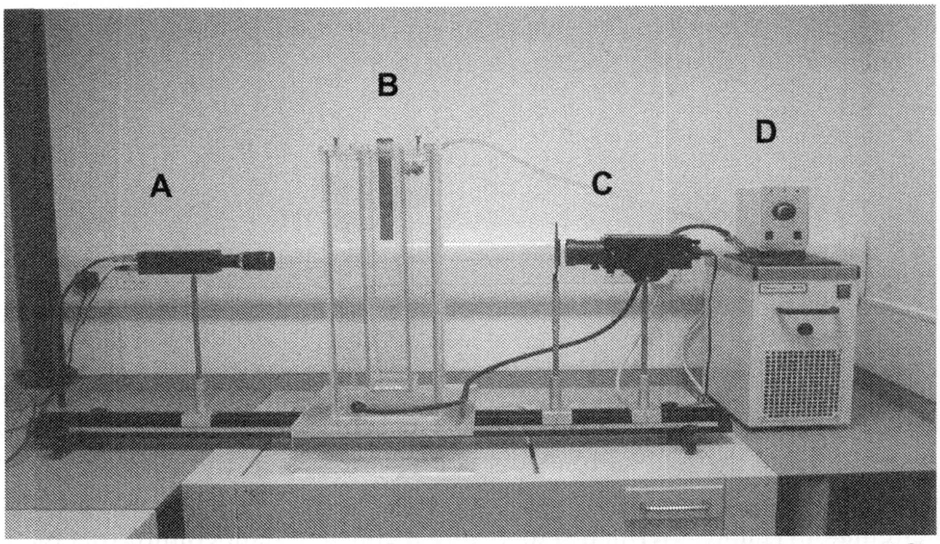

Figure 6.4 Settling column apparatus. A = CCD camera linked to PC and image analysis software; B = the central settling column contained within a sealed water bath; C = lamp for background lighting; D = temperature controlled-circulator for the water bath.

Microscopy

Qualitative floc structural characteristics were investigated using an Olympus BHB light microscope (Olympus European, Hamburg Germany) and image-analysis software (calibrated using a microscope graticule). Flocs were placed in an adapted microscope slide consisting of glass slide base with a raised rubber border with a depth of 3 mm to provide a well to contain whole flocs. The suspension was then secured by placing another microscope slide on top. Source images from the CCD camera were digitised to allow image processing to occur. The software used for floc analysis was Image Pro Plus from Datacell. All captured images were converted into grey scale images with a bit depth of 16, providing 65,536 levels on the grey scale. Size measurements were made by comparing an aggregate image to an image containing something of a known size. In this case, a 1 cm microscope graticule with 100 graduations was used for system calibration. A focused image of the graticule was captured and digitised. Calibration was then achieved by specifying the number of units the reference graticule represents and, using the software drawing tools, by placing a defining line over the length specified. Because image analysis software bases measurements on pixel size a pixel/size unit scale is reported.

The arranged system ensures that only flocs falling in the same plane as the centrally positioned graticule will be in focus. The tapered entry port directs flocs towards the centre of the column to increase the chances of images being in focus. The accurate measurement of focused flocs is dependent upon the depth of field of the image. This is the combined distance in front and behind the central focusing point (CFP) of the camera. Flocs may appear in focus even though they are not passing through the CFP, thus reducing the accuracy of any measurement made from flocs in front or behind the CFP. However, more accurate measurements may be made by increasing the magnification of the image because this reduces the depth of field. Ideally, the depth of field should be one floc diameter thus ensuring that flocs in focus will be passing through the CFP. In practical terms this is very difficult to achieve because the range of floc sizes may vary by up to 2 orders of magnitude. For the purposes of these experiments it was decided to use a relatively high order of magnification and accept some degree of error on measurements of small flocs. Very small flocs below 50 μm were ignored for this reason.

System calibration using standardised glass beads

The validity of the settling column system was checked by using standardised soda-glass spheres (Duke Scientific Ltd) of known diameter 99.1 μm ± 5.0 and density 2.45 g cm^{-3}. The beads are close to being perfect spheres, therefore a comparison of Stokes law predictions of terminal settling velocity with the value calculated from the settling column and image analysis tools will ensure that the system provides rigorous and valid results. The settling velocity of 40 glass beads were measured. The mean settling velocity of the glass beads was found to be 7.62 ± 0.54 mm s^{-1}. This compared favourably to a value of 7.7 mm s^{-1} as predicted by Stokes law using a water density of 1000 kg m^{-3} and viscosity of 0.001 Pa s^{-1}. It was therefore accepted that the system was able to accurately measure settling velocity.

Environmental scanning microscopy

Scanning electron micrographs were measured using an environmental scanning electron microscope with a peltier temperature stage. Microanalysis was performed with an EDS system from EDAX operated with a 1 micron spot size. Samples were transferred into the ESEM chamber using a wide mouthed pipette and the temperature was equalised to 2 C. The pressure in the stage was then reduced to 6 torr while maintaining the humidity close to 100%. Images were obtained using an electron beam of voltage of 15 kV.

Rheology

Description of Technique

Rheological measurements were performed with the vane technique (as discussed in Chapter 5) as it is known to successfully reduces or eliminates wall slip thus allowing accurate yield stress measurements of concentrated particulate systems (Nyugen, 1985).

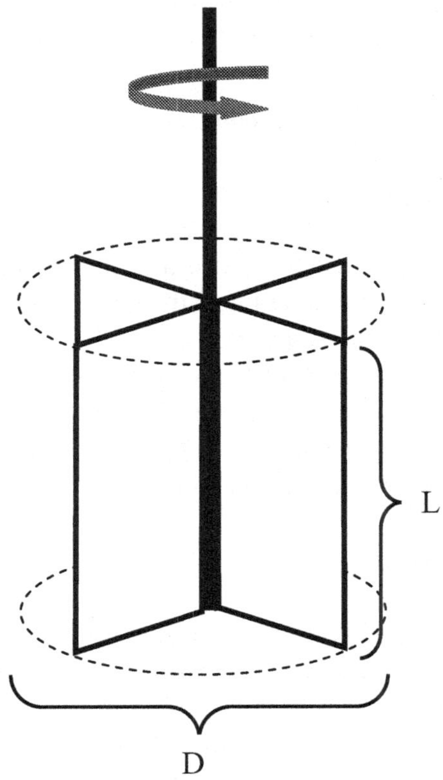

Figure 6.5 Schematic diagram of the vane geometry.

The vane consists of four perpendicular blades immersed in the suspension which is to be measured (Figure 6.5). The torque is determined at a fixed shear rate, where the maximum torque measured is used to calculate the yield stress, τ_y. The yield stress of the material is calculated using equation (6.1), Nyugen (1983, 1985)

$$\tau_y = \frac{T_m}{\pi D^2 \left(\frac{L}{2} + \frac{D}{6} \right)} \tag{6.4}$$

where, T_m is the maximum measured torque. [Nm]
L is the length of the vane. [m]
D is the diameter of the vane. [m]

The parallel plate was used for all rheological tests because of the need to identify and eliminate rheological problems such as wall slip. The parallel plate geometry assumes cylindrical coordinates and the absence of inertia for the purposes of calculation. The equations shown below are used for processing the raw data obtained from rheological experiments. The full derivation can be found in Macosko (1994). Torque is the measured parameter, while the angular velocity is imposed on the rheometer for a required strain rate. The material is subjected to a straining angle, θ dependent on the desired strain.

Strain can be calculated using equation (6.5)

$$\gamma = \frac{r\theta}{h} \tag{6.5}$$

where, γ is the strain.
r is the radius of the plate. [mm] and
h is the height from the top of the plate. [mm]

The velocity of the sample is given by equation (6.6) below and is a function of only the radius and the gap height.

$$v_\theta(r, z) = \frac{r\Omega z}{h} \tag{6.6}$$

where V_θ is the velocity of the sample. [mm/s]
Ω is the angular velocity [rad/s] and
r and h were introduced earlier in equation (6.5).

The strain rate is the differential of the sample velocity in the θ direction (shear rate = dv_θ/dz) given by equation (6.7)

$$\dot{\gamma}(r) = \frac{r\Omega}{h} \tag{6.7}$$

where $\dot{\gamma}$ is the strain rate [s^{-1}] and all other symbols have been introduced previously.

From equations (6.4) and (6.7), the strain rate and strain are shown to be functions of the plate radius. The maximum strain rate is at the rim of the plate or when r =R and is zero at the centre of the plate or at r = 0. This is referred to as the rim shear rate,γ_R and is the parameter most often used for parallel plate geometry.

A differential equation is required to determine the shear stress profile using the parallel plate geometry. Equation (6.8) below is used for calculating the shear stress as derived from a torque balance given by equation (6.9), whereby the radius has been substituted with the shear rate using equation (6.7).

$$M = \frac{2\pi R^3}{\dot{\gamma}_R} \int_0^{\dot{\gamma}_R} \dot{\gamma}^2 \tau\left(\dot{\gamma}_r\right) d\dot{\gamma}_r \tag{6.8}$$

$$\tau(R) = \frac{M}{2\pi R^3}\left[3 + \frac{d \ln M}{d \ln \dot{\gamma}_R}\right] \tag{6.9}$$

where M is the torque [g-cm] and τ is the shear rate [s^{-1}].

From the above equation, it can be seen that the torque, which is experimentally obtained data, has to be differentiated. This may result in large errors in the processing of shear stress data. Thus, equation (6.6) is rarely used by rheometer software to process raw data. For Newtonian fluids, the value of $d(\ln M) / d(\dot{\gamma}_R) = 1$. This reduces the above equation to give the apparent shear stress, which is often used to calculate the apparent viscosity, Carvalho (1994). The equation for the apparent shear stress is given below as equation (6.10). This equation is often used to approximate the shear stress of non-Newtonian materials to avoid the problems which arise from solving for stress using equation (6.10), which requires experimental data to be differentiated.

$$\tau_A(R) \frac{2M}{\pi R^3} \tag{6.10}$$

where τ_A is the apparent shear stress. [Pa]

One of the limitations of the parallel plate geometry is that the ratio of the gap height to the plate diameter should not exceed 0.1. Furthermore, the particles present in the test sample should be less than 10% of the gap height. Equation (6.10) will be used in order to simplify the calculations. The maximum expected error using equation (6.10) is about 20%.

Dynamic Oscillatory Rheometry

Oscillatory measurements allow the sample to be probed in order to determine the structure-fluid relationships of the material. The material is probed within its linear viscoelastic region, where the stress and strain have a linear relationship. Generally the material is subjected to a small sinusoidal strain at varying frequencies. The stress will oscillate at the same frequency as the strain but will be shifted by a phase angle, δ (Figure 6.6). Equations (6.11) and (6.12) give the mathematical representation of the sinusoidal strain and stress respectively.

$$\gamma = \gamma_0 \sin \omega t \tag{6.11}$$

$$\tau = \tau_0 \sin(\omega t + \delta) \tag{6.12}$$

where γ_o is the strain amplitude, τ_o is the stress amplitude and ω is the frequency.

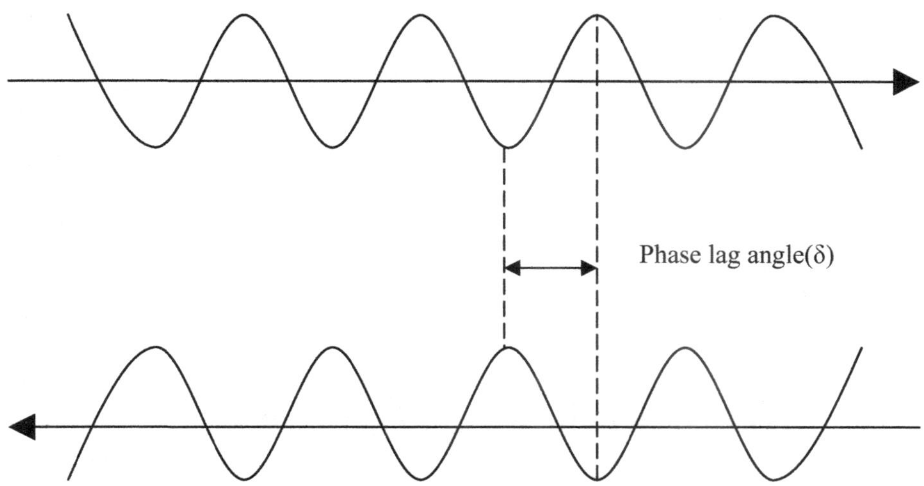

Figure 6.6 Diagram showing the input sinusoidal strain and output stress of a strain controlled rheometer.

The phase shift is a function of the viscous component of the material. A perfectly elastic material will have no shift in the phase angle since the response to the deformation will be instantaneous. Thus a purely elastic fluid will have a phase angle of 0^0. On the other hand, a purely viscous fluid (ie. Newtonian fluid) will have a phase angle of 90^0. Therefore, a viscoelastic fluid will have a phase angle, $0^0 < \delta < 90^0$.

In oscillatory tests, two important functions used to define the elastic and viscous properties of the material are the storage and loss moduli. The viscous component is defined as the loss modulus G", while the elastic component is defined as the storage modulus, G'. The relationship between the output stress and the storage and loss modulus is given by equation (6.13)

$$\tau(t) = \left[\tau_0' \sin(\omega t) + \tau_0'' \cos(\omega t)\right] \tag{6.13}$$

where τ is the shear stress. [Pa]

τ_0' is the stress amplitude, which is in phase with the strain wave.

τ_0'' is the stress amplitude, which is out of phase with the strain wave.

Equation (6.10) is obtained by decomposing the stress wave, resulting in two waves of the same frequency, whereby sin(ωt) is in phase with the input strain and cos(ωt) is 90^0 out of phase. The storage modulus, G' can be defined simply as the extent to which energy is stored and used to deform the molecule, while the loss modulus, G" is the extent to which energy is lost due to viscous dissipation. G’ and G” can be defined mathematically by equations (6.14) and (6.15).

$$G' = \frac{\tau_0'}{\gamma_0} \qquad (6.14)$$

$$G'' = \frac{\tau_0'''}{\gamma_0} \qquad (6.15)$$

Substituting equation (6.14) and (6.15) into equation (6.13) will relate the stress to the dynamic moduli, G’ and G” given by equation (6.16)

$$\tau(t) = \gamma_0 \left[G' \sin(\omega t) + G'' \cos(\omega t) \right] \qquad (6.16)$$

where G’ is the in-phase or storage modulus and
G” is the out-of-phase modulus or the loss modulus.

During dynamic frequency sweeps, the material is subjected to a small sinusoidal strain for a specified range of frequencies. This is a suitable test for studying the elastic and viscous behaviour of a fluid. From equations (6.12) and (6.13), it can be shown that the loss tangent is given by equation (6.17), Macosko, (1994)

$$\tan \delta = \frac{G''}{G'} \qquad (6.17)$$

The loss modulus is defined as the measure of energy dissipated per cycle of deformation per unit volume, Nyugen (1985). The greater the amount of energy loss by viscous dissipation, the larger the magnitude of tan δ. The loss modulus and storage modulus are related to the viscous component of the complex viscosity and the dynamic viscosity by equations 6.18 and 6.19.

$$\eta'' = \frac{G'}{\omega} \qquad (6.18)$$

$$\eta' = \frac{G''}{\omega} \qquad (6.19)$$

where η’ is the dynamic viscosity. [Pa.s]

η'' is the imaginary component of the complex viscosity which is also known as the out of phase viscosity [Pa.s].

The magnitude of the complex viscosity is given by equation (6.20)

$$|\eta^*| = \left(\eta'^2 + \eta''^2\right)^{1/2} = \left[\left(\frac{G''}{\omega}\right)^2 + \left(\frac{G'}{\omega}\right)^2\right]^{1/2} = \frac{|G^*|}{\omega} \tag{6.20}$$

Small angle oscillatory measurements are a useful method for characterising material structure breakdown. The idea to use small amplitudes is to ensure the material is kept within the elastic region, while relatively low frequencies are used to reduce any contributory effect which oscillatory shear may have on structural breakdown . Dynamic strain sweeps will be the main tool used for this analysis. The torque is measured over time for a sinusoidal strain input at a fixed frequency. Figure 6.7 below is a typical result of a dynamic strain. The material is probed within the linear viscoelastic region where the stress and strain have a linear relationship (ie. the material is deformed elastically). Figure 6.8 shows the linear relationship between stress and strain. The yield stress of the material can be obtained by determining the point at which the material begins to plastically deform that corresponds to the end of the linear region of the material.

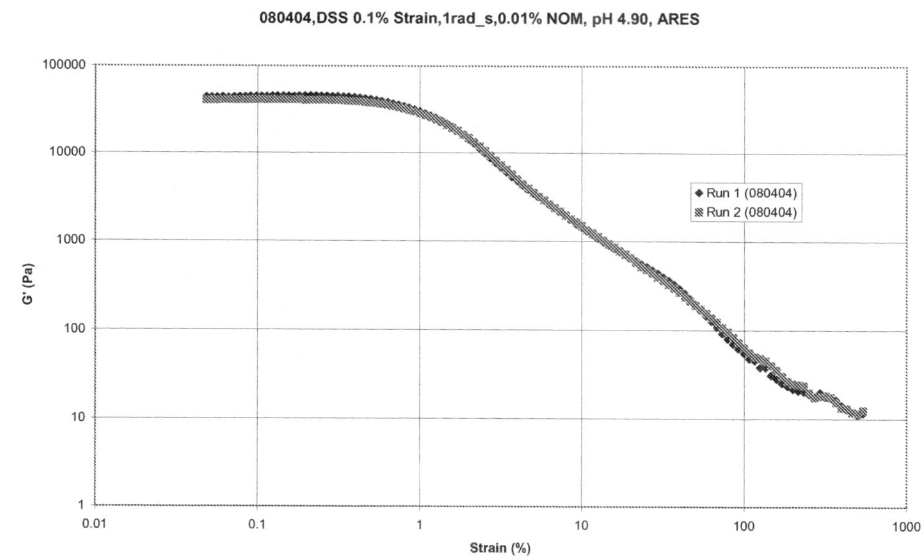

Figure 6.7 Plot of G' as a function of strain for dynamic stress sweep experiments using different gap heights.

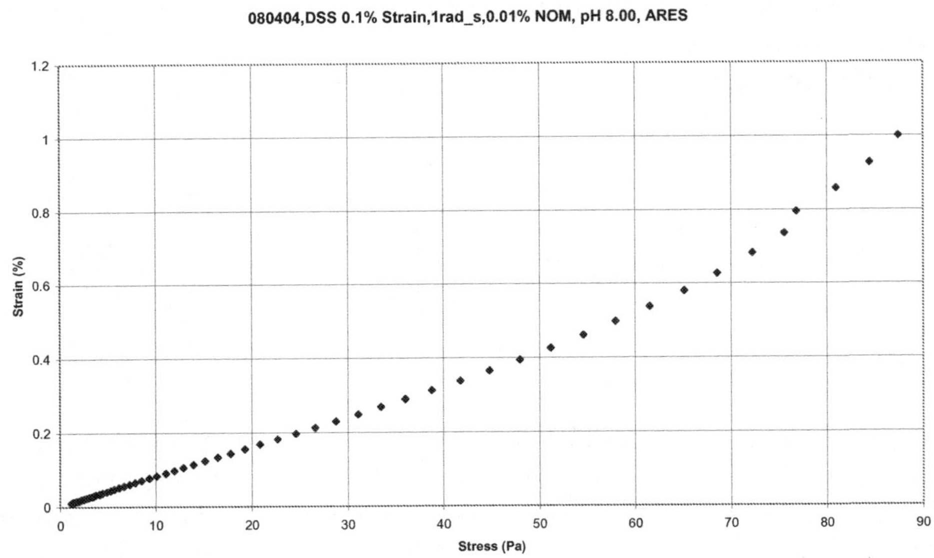

Figure 6.8 Plot of shear stress versus strain.

Method development

Originally experiments were planned with alumina and haematite to simulate the two coagulant systems of principle concern to Chapter 4. Unfortunately the zeta potential of the haematite surface remained less than 20 mV across a relatively wide range of pH which resulted in the haematite being difficult to disperse and hence gave a relatively high shear yield stress in that pH region. Also, for these experiments the samples were prepared at high pH and the pH was adjusted from high to low. It appears that the strength of the network was maintained at a high level at pH levels below the iep, ie the material was not redispersing, indicating that the particle bonds formed due to the reduction in electrostatic repulsion are not easily broken when the electrostatic interaction is then increased. This was confirmed by experimental data which showed that for samples prepared at low pH, the shear yield stress values obtained at the same pH below the iep were much lower.

The alternative approach was to use metal oxides as a nucleus and to deposit a coating of ferric hydroxide as a surface layer. This had the advantage of being a better approximation to the practical system but the disadvantage that it required greater preparation time and introduced some variability with respect to the extent of coverage and the strength of attachment of the ferric hydroxide flocs to the underlying oxide. Alumina substrates were rejected as the ferric coatings did not appear to substantially alter the IEP or the yield stress of the system and consequently did not appear to simulate ferric over alumina very well. Ultimately, silica was chosen as the underlying oxide while the optimal level of coating arrived at was 2.5 % w/w.

Adhesion measurements

Adhesion measurements were performed using a Nanoscope IIIA atomic force microscope (Digital instruments, Santa Barbara, CA) with silicon nitride nanoprobe cantilevers (Nanoscope version 4.42r4). Detailed description of the force measurement made using the atomic force microscope can be found elsewhere (Durham *et al.,* 1991) but brief explanation is given below. The probe is brought towards the surface of the sample where the interfacial interaction is measured based on the deflection of the cantilever. Upon contact, the probe is loaded to a preset force value, and then withdrawn. Once the elastic energy in the cantilever exceeds the adhesive force, separation occurs and the probe returns to a position of zero deflection. The adhesion force is the calculated from the maximum deflection in the retract portion of the force curve.

Samples were prepared by pre soaking the test fraction on either molecularly smooth mica or an ion exchange resin bead. Where appropriate the cantilever was then modified by attaching a resin bead using a micromanipulator following the method of Ducker *et al* (1991). The spring constant of the cantilevers was determined by measuring the resonance frequency of loaded and unloaded cantilevers with an accuracy of 0.3±0.03 $N.m^{-1}$ and was measured with each new tip prior to force measurement. Once prepared the samples were placed in the liquid cell of the microscope and filled with low alkalinity water of similar quality to the raw waters under investigation. The measurement was performed at least 40 times at four separate locations on each sample to provide an average value of the adhesion force.

RESULTS AND DISCUSSION

Impact of season

Raw water characteristics

The water under investigation was characteristic of a moorland catchment being of high colour and NOM concentration, low turbidity (<3 NTU) and low alkalinity (<10 mg/L as $CaCO_3$). It can be seen that the general water characteristics do not significantly change across the periods of measurement (Table 6.1). SUVA was seen to increase by 24-30 % in the summer and autumn months. DOC removal was seen to increase with increasing coagulant dose other than in summer when high removal was seen for a low dose. There was a drop in organic removal rates in autumn and winter at optimum doses suggesting these waters were more recalcitrant to treatment. As can be seen in Figure 6.9, fractionation of these waters shows there to be only small changes in the character of the water throughout most of the year. The hydrophobic fractions (the humic and fulvic acids) were the dominating components throughout the year, accounting for between 60-70 % of the total DOC. The largest change in the fractional composition was seen in winter when the fulvic acid fraction (FAF) was seen to increase by 20 % when compared to the other seasons. This was commensurate with a similar decrease in the humic acid fraction.

Table 6.1
The seasonal change in raw water characteristics over a one year period and the coagulant dose used and DOC removals achieved.

Seasonal water	DOC ($mg.L^{-1}$ as C)	UV_{254} (m^{-1})	SUVA ($L.mg.m^{-1}$)	Coagulant dose ($mg.L^{-1}$ Fe)	DOC removal (%)
Spring	11.4	52.6	4.6	14.4	81
Summer	8.9	50.8	5.7	8.0	88
Autumn	10.2	60.2	5.9	12.2	75
Winter	9.2	40.1	4.4	10.0	63

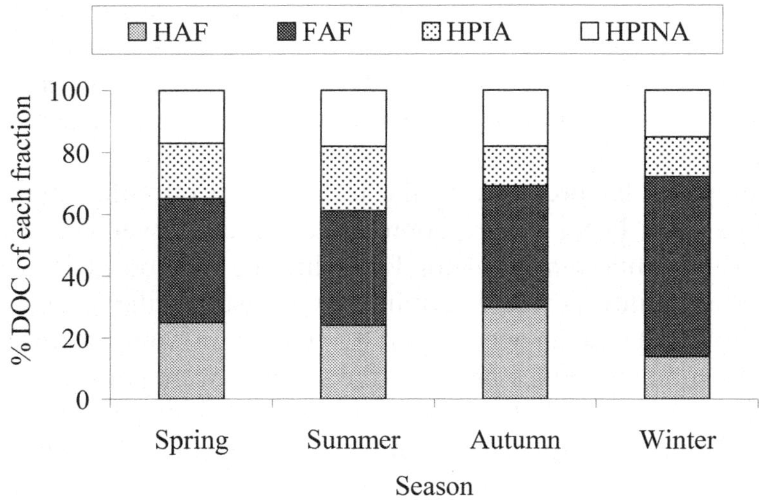

Figure 6.9 A demonstration of the change in composition of NOM with season for the Albert Reservoir, Yorkshire.

Floc size and strength

The size profiles of flocs upon exposure to increasing levels of shear on a jar tester are shown in Figure 6.10 for the seasonal waters. The d_{50} floc size after 15 minutes of slow stir was of the following general trend: autumn (median floc size 764 μm ± 34) > summer (724 μm ± 65) > spring (649 μm ± 38) > winter (594 μm ± 45). However, the overlap of adjacent error bands suggest this size order was indicative only. The rate of floc formation was quicker for the spring, autumn and winter waters than for the summer water which had the lowest coagulant dose. In comparison to other studies using similar floc sizing techniques, these flocs were 6-7 times larger than activated sludge flocs (Biggs and Lant, 2000) and 5-6 times larger than kaolin coagulated with polyaluminium chloride (Lin *et al.*, 2002). There has been no previous work using a similar laser based diffraction technique to observe the size of Fe-NOM flocs, however the range of floc sizes found were of a similar size to metal-NOM flocs observed using CCTV with image analysis (Bache and Rasool, 2001).

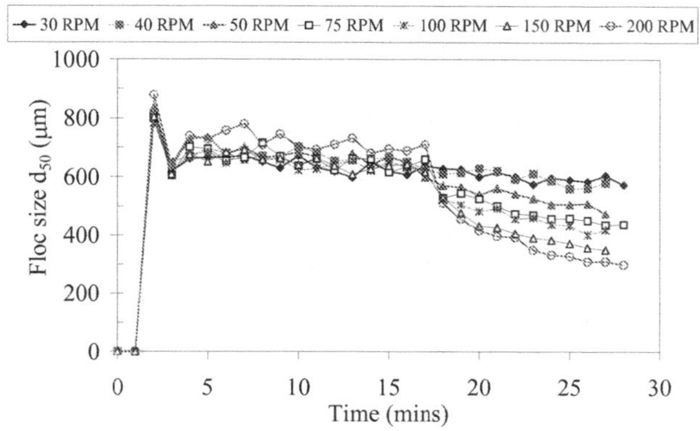

Spring (14.4 mg/L Fe)

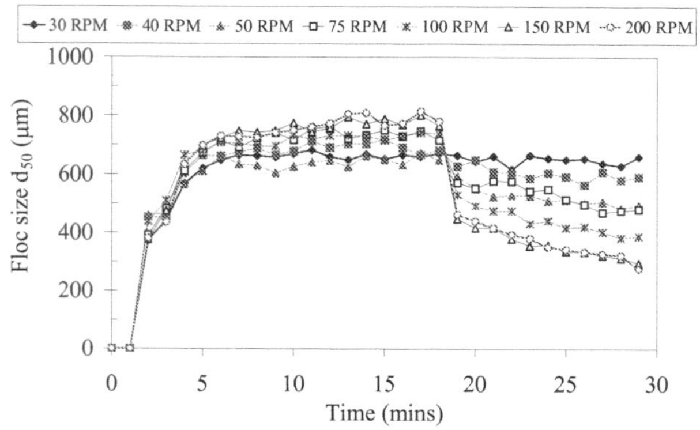

Summer (8 mg/L Fe)

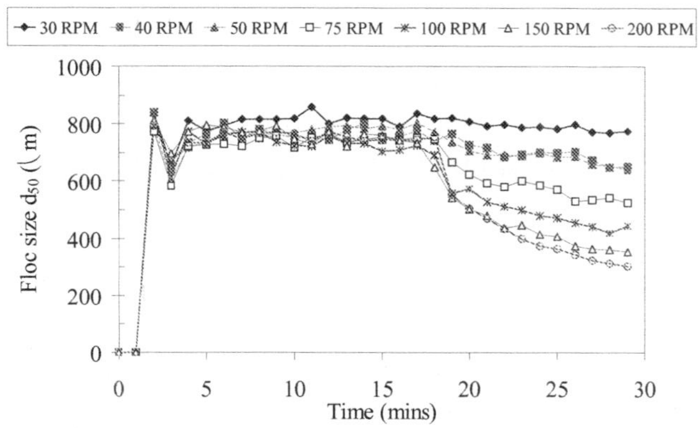

Autumn (12.2 mg/L Fe)

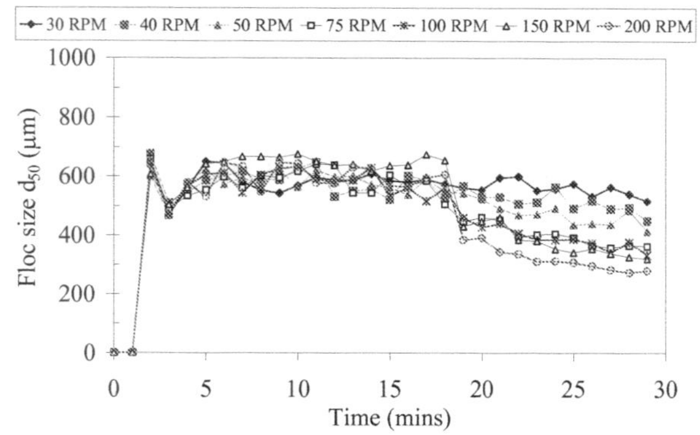

Winter (10 mg/L Fe)

Figure 6.10 The growth and breakage behavior of flocs with increasing shear from three seasonal periods in 2003.

With increasing rpm on a jar tester the steady-state floc size was shown to decrease. At high rpm there was an initial large-scale decrease in floc size followed by a more gradual decline. At low rpm there was a more gradual decline in floc size until a steady state size was approached after 15 minutes exposure to shear. The size after 15 minutes at increased rpm are shown in Figure 6.11. The rate at which a floc suspension degrades with increasing rpm gives an indication of floc strength. A steeper slope suggests a floc suspension will undergo a larger proportional reduction in floc size when compared to a degradation curve with a shallower slope. The summer and autumn floc suspensions had similar degradation slopes (0.47 and 0.49 respectively) whilst the initially smaller spring and winter flocs degraded at a significantly lower slope (0.34 and 0.31 respectively). However it should be noted that the autumn flocs remained larger than the other seasons until very high shears (>100 rpm) when flocs converged on a similar microfloc size.

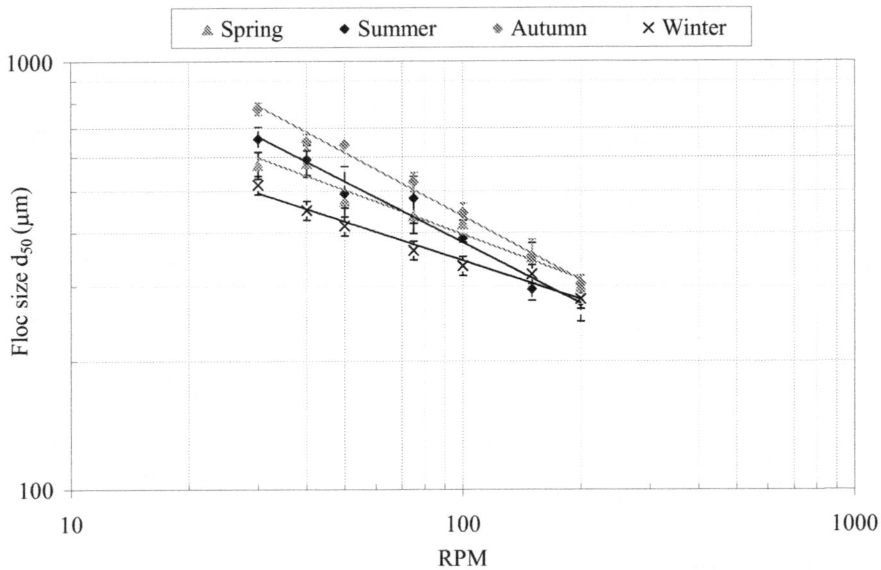

Figure 6.11 The rate of degradation of NOM flocs with increasing shear for the seasonal waters.

Settling rate

An example of settling behaviour of Fe-NOM flocs is shown in Figure 6.12 for flocs formed from the summer and autumn water. Due to the relative scatter and crossover of the data, it was not possible to present all of the data for all of the seasonal waters on the same graph. For this reason, a review of the settling data for organic flocs is summarised in Table 6.2 as typical settling velocities obtained from the regression line through the data points. The floc settling data showed that the autumnal water flocs had generally faster settling rates.

Table 6.2
The settling velocity and fractal dimension of different types of flocs.

Type of floc	Settling velocity range (μm/s)	
	Small floc (400 μm diameter)	Large floc (1000 μm diameter)
Spring	290	728
Summer	267	680
Autumn	423	891
Winter	315	763

These experiments have established a seasonal change in floc characteristics with the autumnal flocs showing the most marked difference in structure. Explaining the reasons for the resulting floc structure was made difficult due to the different initial NOM concentrations and coagulant doses for each of the seasonal waters. The different DOC removals achieved with optimum coagulant doses based on NOM removal and as used at the WTW reflect a seasonal change in NOM. During winter and autumn the organics became more recalcitrant to removal as reflected by lower percentage DOC removals. However, there was no obvious correlation between the floc growth, size and breakage and the coagulant dose or the initial DOC's of the water. It was therefore likely that the seasonal NOM composition had a considerable impact on the floc structure. The increase in the FAF seen during winter may be a factor that influenced the reduced floc size seen during this season when smaller flocs formed. However, significant changes in floc structure were seen during the other seasons which could not be explained by the small scale changes in NOM fractions observed during these times. It was therefore difficult to interpret how the specific fractions impacted on floc structure.

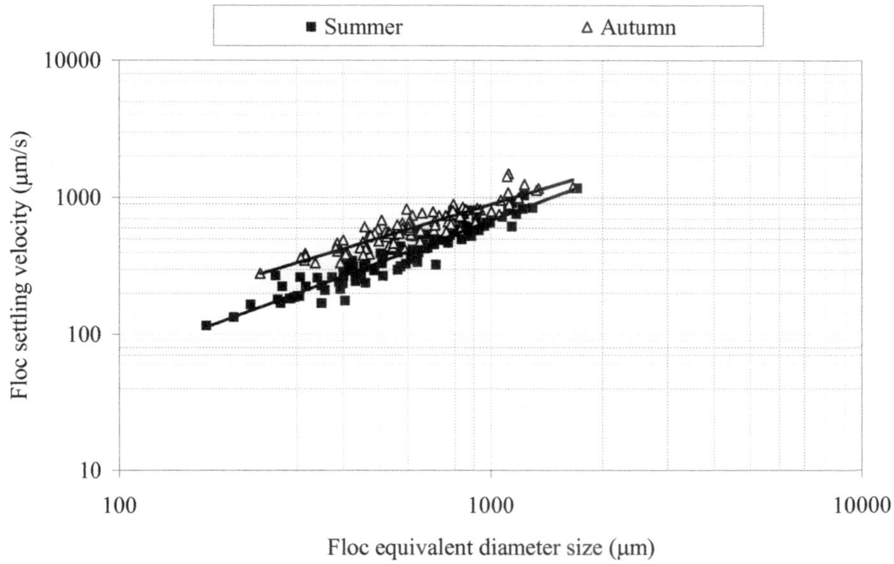

Figure 6.12 An example of the settling behavior of NOM flocs in relation to their size.

While the specific reactions of NOM fractions with coagulant chemicals has not been fully understood, a possible explanation for the observed floc characteristics was SUVA. The autumn and summer waters had the highest SUVA and this was concurrent with flocs that were initially large whilst the low SUVA, high FAF winter water gave rise to small weak flocs. Large flocs are indicative of strong flocs (Bache and Papavasilopoulos, 2003 and Yukselen and Gregory, 2004). This is because in order to maintain a large floc size for a given shear these flocs must have higher resistance to breakage and should therefore be considered stronger. Higher SUVA values are indicative of an increase in hydrophobic, highly charged and high molecular weight NOM molecules (Edzwald and Tobiason, 1999). It is likely that an increase in these charged components will react better with coagulants and become better incorporated in the floc when compared to lower charged molecules. This may help to explain why waters give periodic operational problems at WTW. However, given the previously mentioned differences in initial DOC and coagulant dose this impact needs further investigation.

Comparison of chemical options

Coagulation tests

Optimization of the coagulant conditions through preliminary jar testing revealed that the highest removals were obtained at different water pH depending on the choice of coagulant (see Section 4 – process comparison).

Size

A qualitative visual comparison of typical flocs formed from each of the systems is shown in Figure 6.13. It can be seen that the flocs formed using Fe based coagulants were darker in appearance than Al flocs. For the NOM flocs, the MIEX® pre-treated flocs had a more rounded and compact appearance than the other treatment options which appeared to show a more dendritic structure.

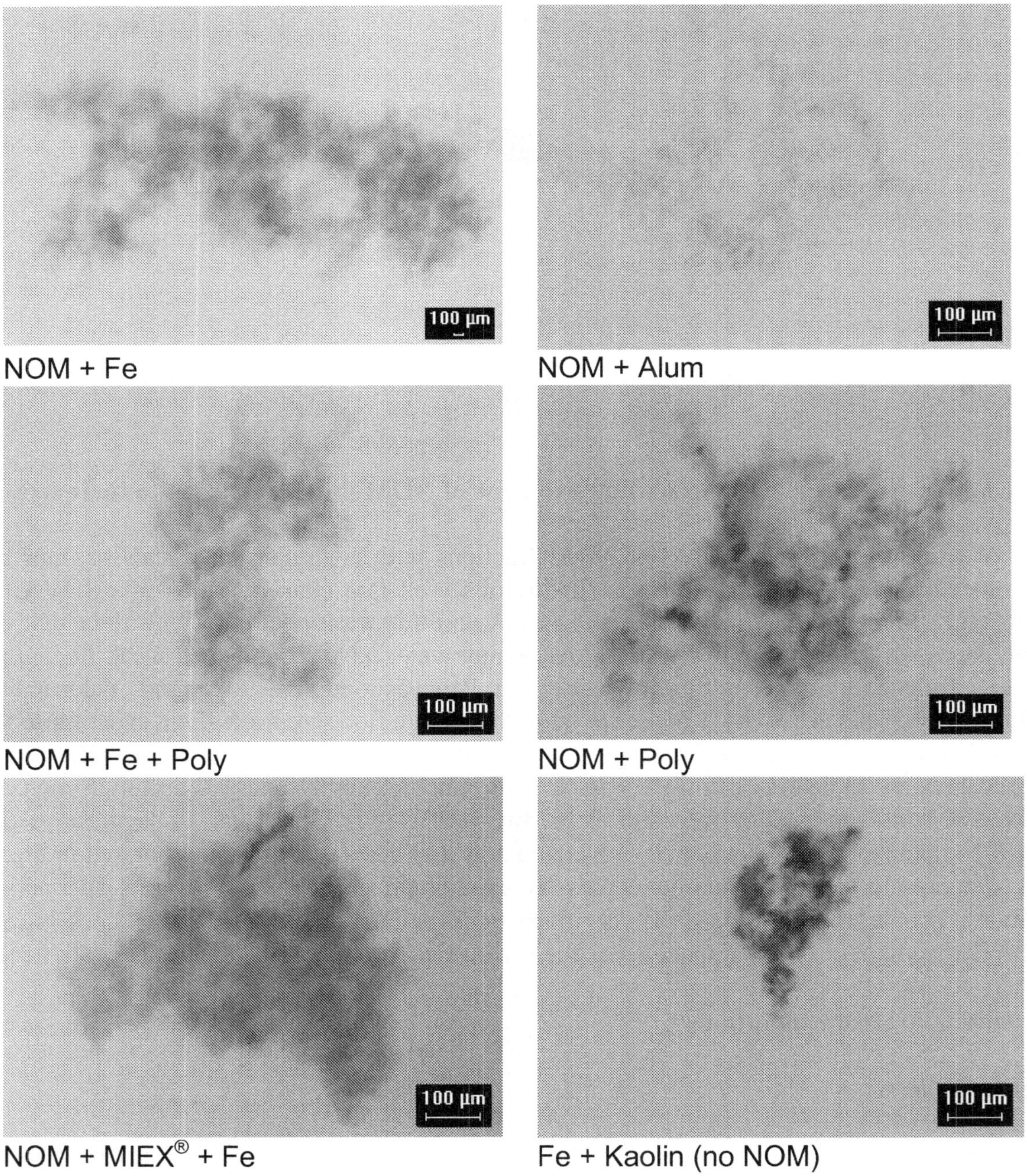

Figure 6.13 Microscope images of typical examples of each type of floc.

Strength

Comparison of the floc growth profiles demonstrates that, as expected, the floc size is highly dependent on chemical choice (Figure 6.14). The d_{50} floc size after 15 minutes decreased according to the following coagulant system: MIEX® + Ferric (1000 ± 200 μm), pDADMAC (840 ± 74 μm), Ferric (720 ± 84 μm), pDADMAC + Ferric (573 ± 65 μm), Alum (463 ± 30 μm) and kaolin + Ferric (334 ± 15 μm).

The response of the floc size to increased levels of shear after 15 minutes of slow stir revealed a two stage decrease in floc size (Figure 6.15). During the first two minutes after exposure to the increased shear there was a large scale decrease which is then followed by a gradual exponential decline. The extent of each phase is dependent on coagulant choice, initial floc size and the imposed shear. For instance in most systems no rapid decline is observed until the shear rate exceeds 50 rpm. The MIEX® pre-treated flocs showed a more inconsistent breakage profile, this was thought to result from these flocs sticking and blocking the measuring cell of the laser diffraction instrument; properties that were not seen for the other flocs. Examination of the floc size distributions before and after the exposure to 15 minutes of increased shear revealed a basic pattern between all the different systems (Figure 6.9). The exposure to increased shear did not substantially alter the shape of the distribution but caused a shift in particles sizes and a slight decrease in the peak height suggesting that the effect of shear was consistent across all size classes.

For all of the flocs under investigation, a straight line relationship was seen between the applied shear (rpm) and the floc size on a log-log scale with an R^2 correlation co-efficient between 0.93-0.99 (Figure 6.15). The slopes of the lines increased according to the following order: Alum (0.29), pDADMAC + Ferric (0.32), Ferric (0.47), MIEX® + Ferric (0.54), pDADMAC (0.74) and Kaolin + Ferric (0.77). The coefficients reported here suggest that degradation of most of the flocs were dominated by fragmentation mechanisms whilst the kaolin and pDADMAC flocs were more balanced between fragmentation and erosion mechanisms. For the NOM flocs it was observed that the flocs that were initially larger remained larger until shears of above 100 rpm when floc sizes tended to converge on a common floc size of between 200-300 μm. The exception to this being the MIEX® pre-treated flocs which remained larger at all shear levels.

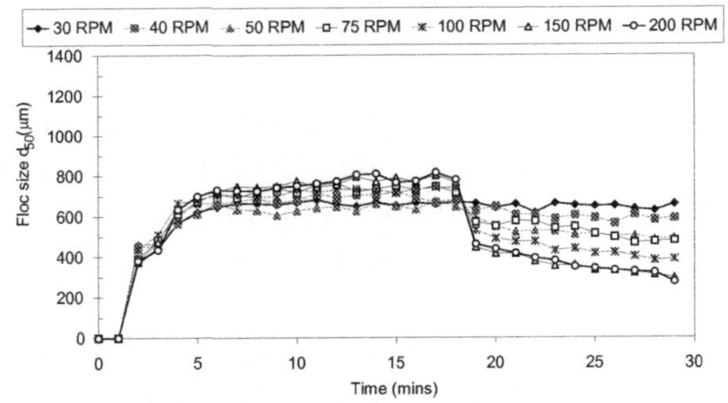

NOM + Fe

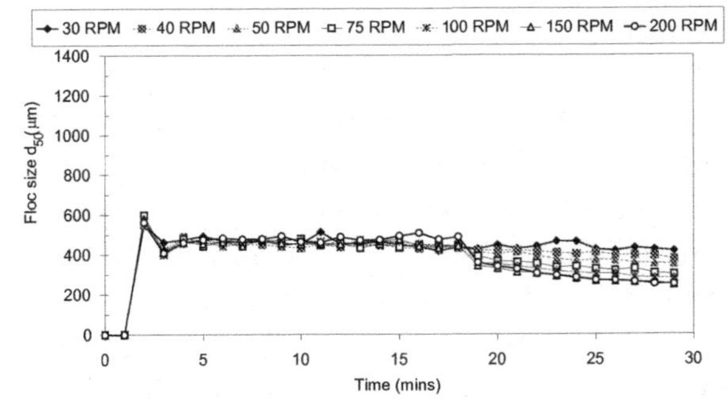

NOM + Al

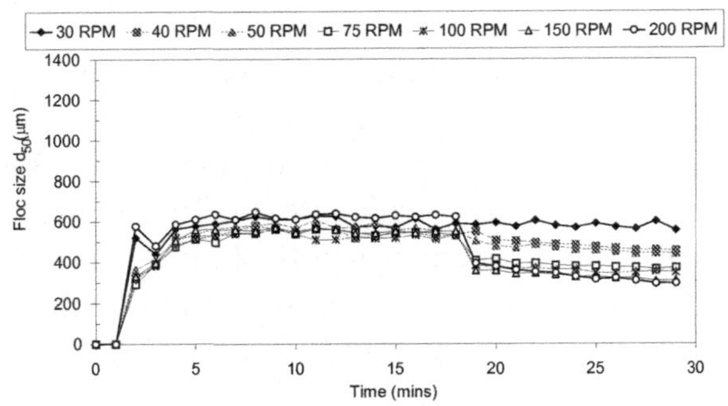

NOM + Fe + Poly

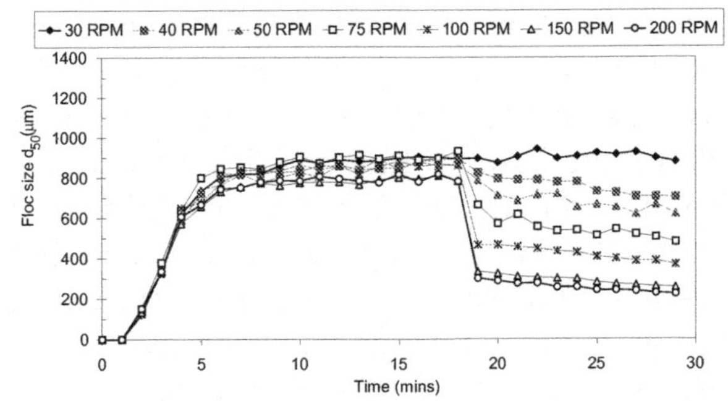

NOM + Poly

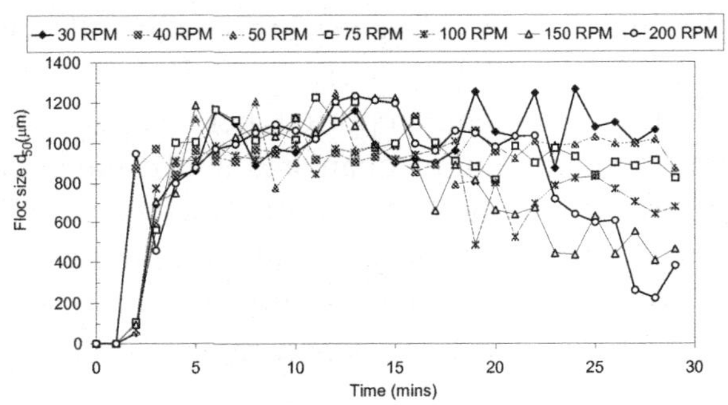

NOM + MIEX® + Fe

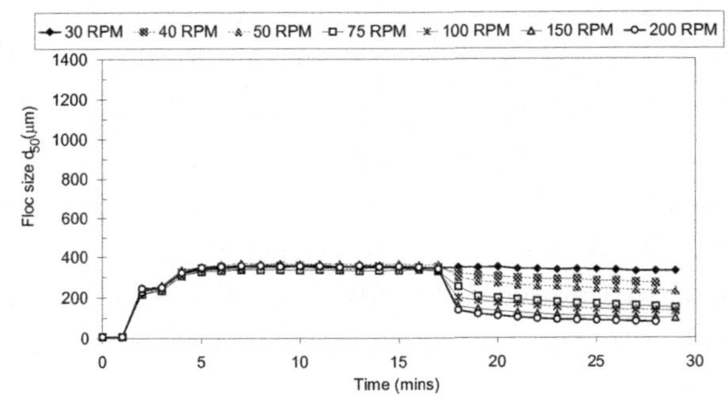

Fe + Kaolin (no NOM)

Figure 6.14 Floc growth and breakage patterns for the different coagulated systems

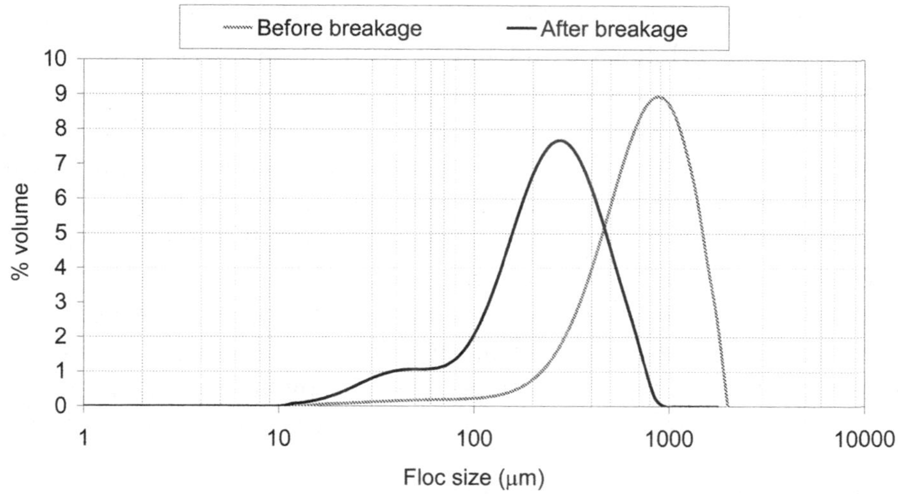

Fe + NOM system (typical of all floc systems investigated).
Figure 6.15 Typical size distributions before and after breakage at 200 rpm

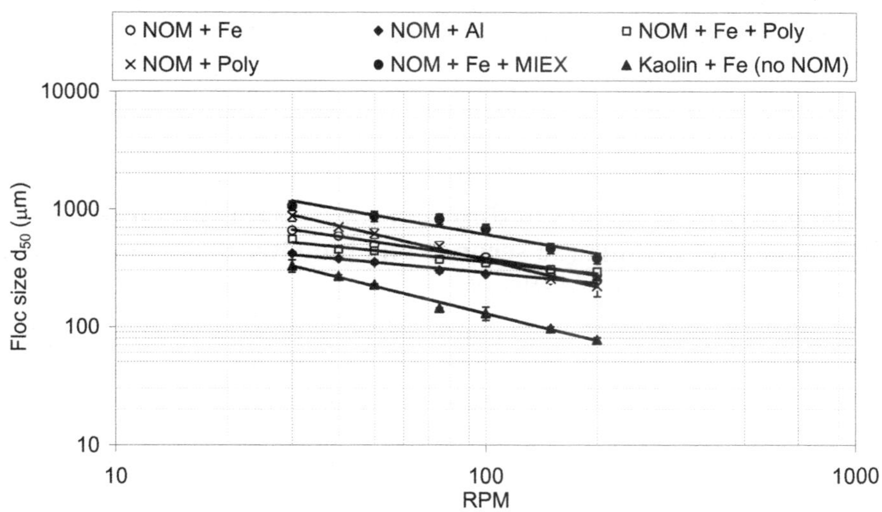

Figure 6.16 Strength profiles for the different systems.

Regrowth

After the slow stir phase, Fe-NOM flocs had reached a d(50) floc size of 700 μm and pDADMAC-NOM had a median floc size of 800 μm (Figure 6.17a). This was approximately double the size of the Al-NOM flocs. There was very close agreement between experimental repeats with floc size values varying by less than 5 %. In all cases, floc size was immediately reduced following an increase in shear. The size of the d(50) flocs after 15 minutes exposure to shear was 300 μm for the ferric sulfate coagulant whilst the alum and pDADMAC flocs were broken into smaller floc sizes of around 200 μm. As the shear was reduced again the flocs begin to re-grow, however the NOM flocs could not re-grow to anywhere near their previous size regardless of the coagulant used. The ferric sulfate and pDADMAC flocs re-grew to a d(50) size of 400 μm. Alum flocs increased in equivalent diameter to a value approaching 300 μm. Analysis of the floc PSD in Figure 6.17b show a shift in the major peak after breakage to under half the original value. There was also a considerable increase in the smaller peak at floc sizes between 10-100 μm. After the re-growth phase, this smaller peak was reduced with a commensurate shift to the right of the major peak.

The strength factor values in Table 6.3 show that the larger polymer flocs had a greater proportional size change after increased shear such that the associated strength factor was 25 compared to 40 for alum and ferric NOM flocs. The re-growth of ferric sulfate flocs was poor with a recovery factor of only 18, this was considerably lower than a value of 40 for alum and 32 for pDADMAC.

The effect of reducing the period of exposure to high shear is shown in Figure 6.18. The extent of floc breakage was much less at the reduced shear period of 30 seconds for all of the coagulants with the floc strength factor increasing by an additional 15-25 from the longer shear period (Table 6.3). Ferric and alum flocs were better able to resist shear with floc strength factor values of 56 and 64 respectively. The polymer flocs had a floc strength factor of 40 when exposed to the increased shear. At the short shear time, floc re-growth was seen to improve considerably with the recoverability factor between 66-71 for all the NOM flocs. The PSD data again showed a reduction in the flocs around 10-100 μm after floc re-growth for all of the flocs.

Table 6.3
The strength and breakage factors of different sized NOM flocs formed from three different coagulants after a short (30 seconds and long breakage period (15 minutes).

	Strength Factor		Recovery Factor	
	Long break	*Short break*	*Long break*	*Short break*
Alum	40	56	40	66
Ferric sulfate	40	64	18	67
PDADMAC	25	40	32	71

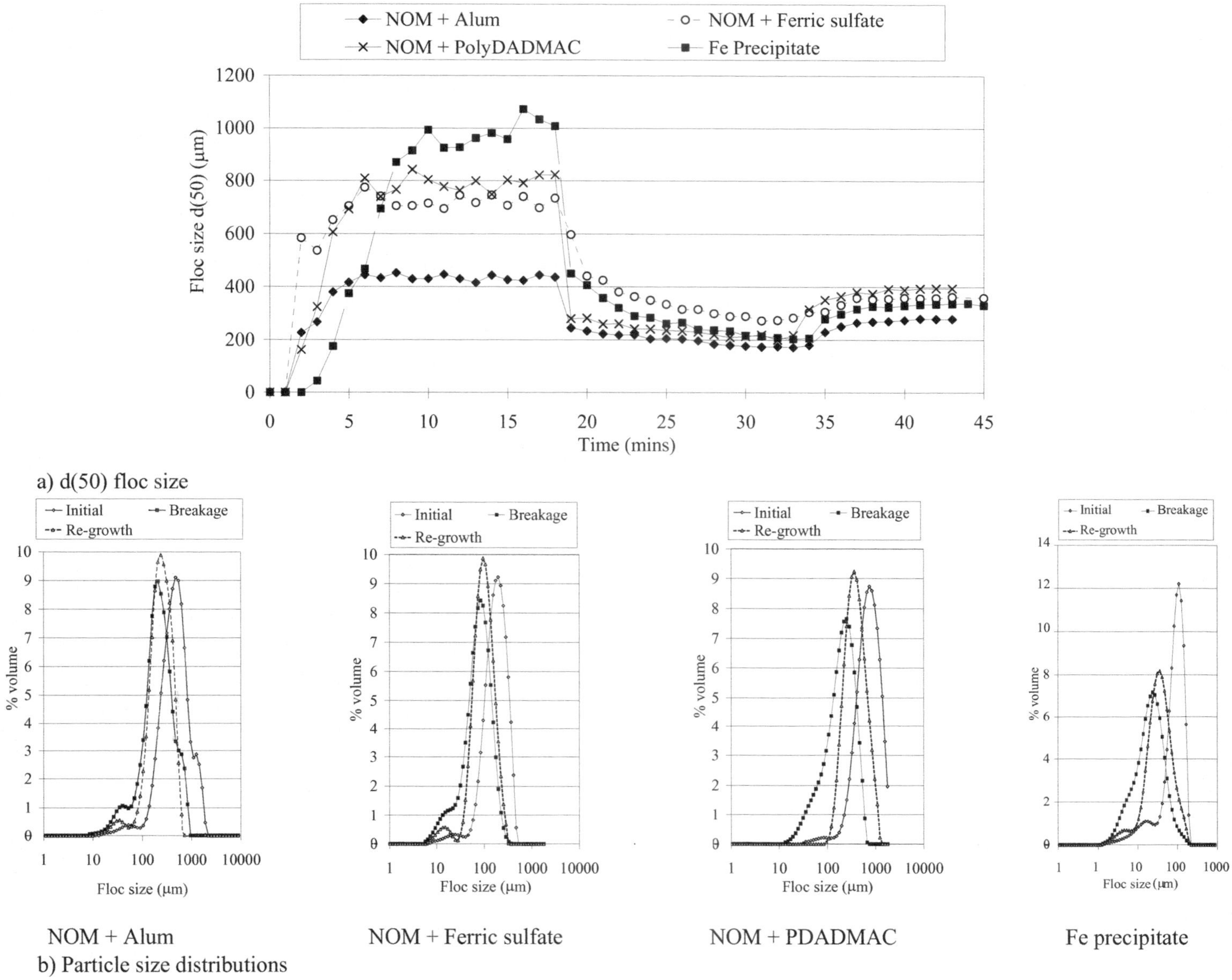

Figure 6.17 Long shear period – a) the breakage and re-growth profile of NOM flocs formed from three different coagulants (alum, ferric sulfate and pDADMAC). b) the particle size distributions of the NOM flocs after growth, breakage and re-growth after 15 minutes high shear time.

Different starting suspensions. The median size of the flocs formed from ferric sulfate precipitate were 1000 µm after the initial slow stir phase (Figure 6.18). This was considerably larger than the Fe-NOM flocs which reached 700 µm. The pattern of the breakage and re-growth behavior of the flocs was similar to that seen previously. After exposure to 15 minutes of shear, the Fe-NOM flocs were broken to a median size of 300 µm. The Fe precipitate flocs were smaller with a median equivalent diameter of 200 µm. The Fe precipitate flocs had a low floc strength factor of 21, well below the value of 40 for Fe-NOM flocs (Table 6.4). Once broken, the Fe precipitate flocs had very poor recoverability with a re-growth factor of less than 20. The latex coagulated in salt solution reached a median floc size of 150 µm after the initial slow stir phase (Figure 6.18). On exposure to increased shear the flocs were broken to a d(50) size of 25 μm with a low strength factor of 18. Once the original slow stir speed was restored, the flocs showed extensive recoverability with the flocs re-growing above their initial size to a median size of 200 μm. Two more subsequent breakage/re-growth cycles were therefore investigated which showed that there was a progressive drop-off in floc size and recoverability following these cycles (Table 6.4).

Table 6.4
The strength and breakage factors of flocs formed from different initial suspensions after a short (30 seconds) and long breakage period.

	Strength Factor				Recovery Factor			
	Long break			*Short break*	*Long break*			*Short break*
NOM	40			64	18			67
Precipitate	21			59	16			81
Latex	*cycle 1*	*cycle 2*	*cycle 3*	-	*cycle 1*	*cycle 2*	*cycle 3*	-
	18	20	24		139	84	64	

The short shear period did not break the suspensions to the same degree as for the longer shear with the floc strength factor remaining at around 60 for all of the suspensions. The recoverability of flocs improved relative to the longer shear period. The precipitate flocs had significantly higher recovery when compared to all of the flocs investigated at this short shear period with a re-growth factor of 81 compared to 56 for Fe-NOM flocs. The PSD data of both the long and short shear period showed similar changes in aggregate size as seen for the NOM flocs.

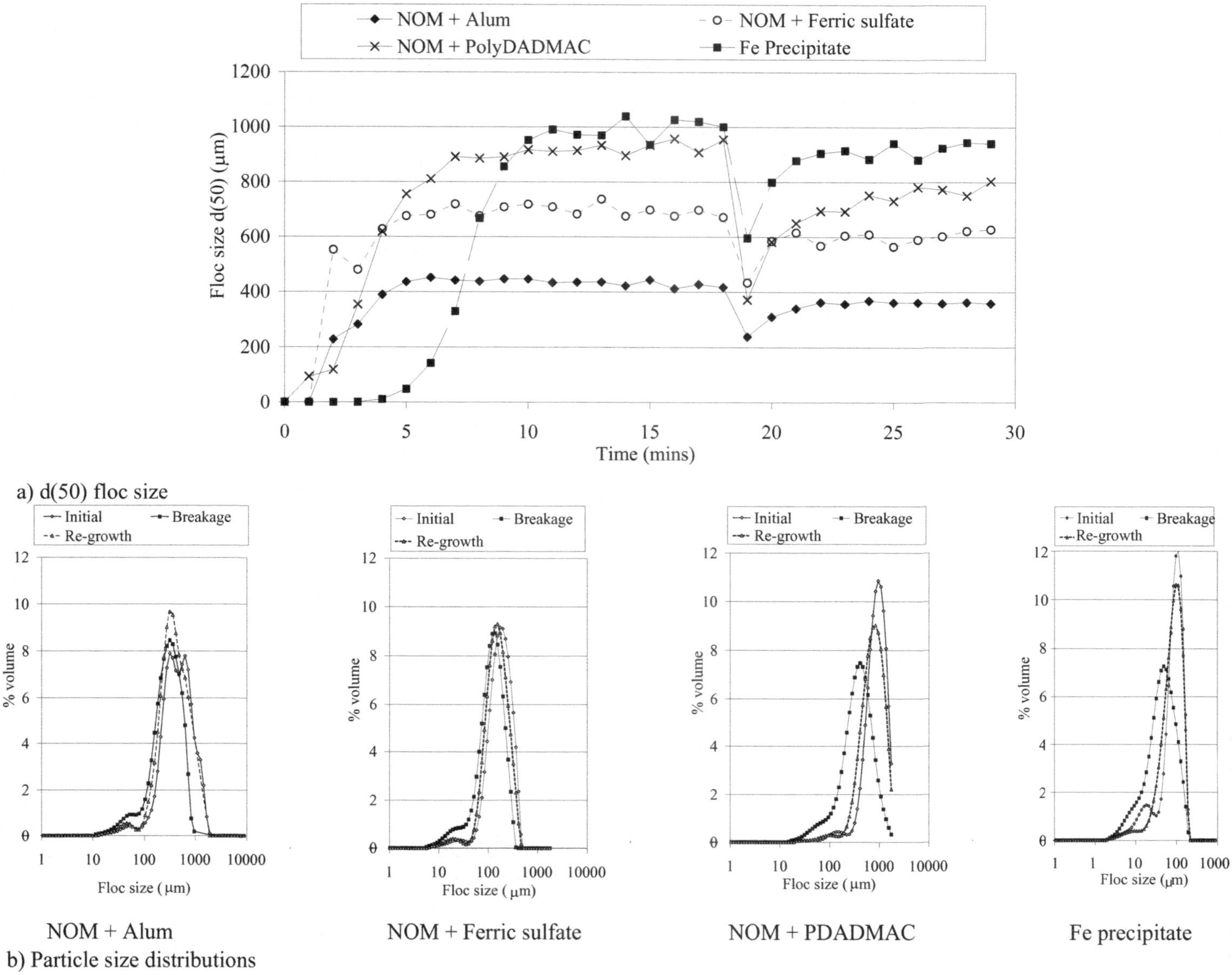

Figure 6.18 Short shear period – a) the breakage and re-growth profile of NOM flocs formed from three different coagulants (alum, ferric sulfate and pDADMAC). b) the particle size distributions of the NOM flocs after growth, breakage and re-growth after 30 seconds high shear time.

Settling rate

The settling characteristics of the floc systems demonstrated a characteristic log-log relationship between floc size and settling velocity (Figure 6.19). One data set missing was for the alum flocs. The contrast difference between the pale, translucent flocs and the background could not be precisely defined which did not allow the image analysis software to accurately size the floc which prevented the alum data being included in the study. This was despite image analysis enhancements and changes in background lighting conditions. Visual observation of the other data revealed two distinct groups of settling rates. The higher settling rate group included the kaolin + ferric (no NOM) and MIEX® + ferric flocs whilst the slower settling group included the ferric, ferric + pDADMAC and pDADMAC flocs. For instance, the settling rate of a typical 500 μm floc generated from the MIEX® + ferric system was 977 μm s^{-1} and 1079 μm s^{-1} for the kaolin compared to 382 μm s^{-1} for the ferric system, 323 μm $s^{-1}$for polymer and 348 μm s^{-1} for the ferric + polymer (Table 6.5). Closer inspection of the data showed that 95 % confidence limits on individual value estimates as predicted by the regression line gave wide intervals. However, the confidence limits of the two identified groups did not cross over for the range of floc sizes investigated, showing that there was a statistical difference between these two sets.

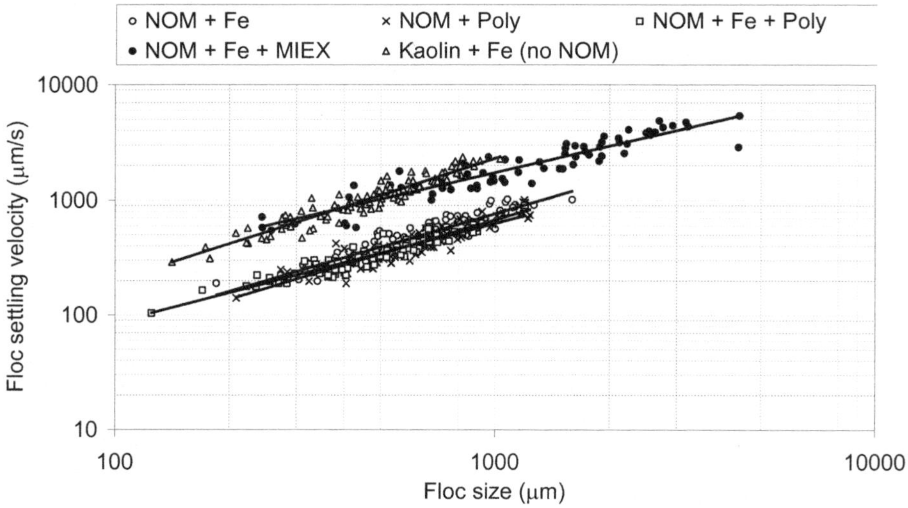

Figure 6.19 Settling rates of the different systems under investigation.

Table 6.5
Settling velocity ranges for the different floc systems.

System	Settling velocity range (μm s^{-1})	
	500 μm floc (95 % confidence limits)	1000 μm floc (95 % confidence limits)
NOM + Fe	382 (276-531)	793 (570-1101)
NOM + Al	N/A	N/A
NOM + poly	323 (241-458)	682 (509-973)
NOM + poly + Fe	348 (270-447)	659 (509-851)
NOM + MIEX + Fe	977 (641-1489)	1698 (1135-2616)
Kaolin + Fe	1079 (807-1445)	2333 (1734-3140)

The derivation of floc fractal dimension from the scattered light intensity (I) as a function of wave number (Q) are shown in Figure 6.20 for NOM flocs and different starting suspensions after 15 minutes slow stir when the flocs had reached their maximum steady state size. The linear portion of the curves suggests that there were mass fractal scaling relationships within the aggregates. For the NOM flocs it was evident that at values of log $Q > -1.75$ and log $Q < -3.5$ the data began to deviate from the linear region as the Porod and Guiner regimes were entered respectively. This gives an upper aggregate size ($1/R_{agg}$) in the region of 3 μm. This suggests that the floc fractal dimension as obtained from light scattering behavior cannot be applied to the larger flocs that dominate NOM systems where the median size was 700 μm. For latex flocs, log Q was < -3.2 giving an upper aggregate size of 1.6 μm, over five times the value of the primary particle size but still much smaller than the median particle size of 150 μm.

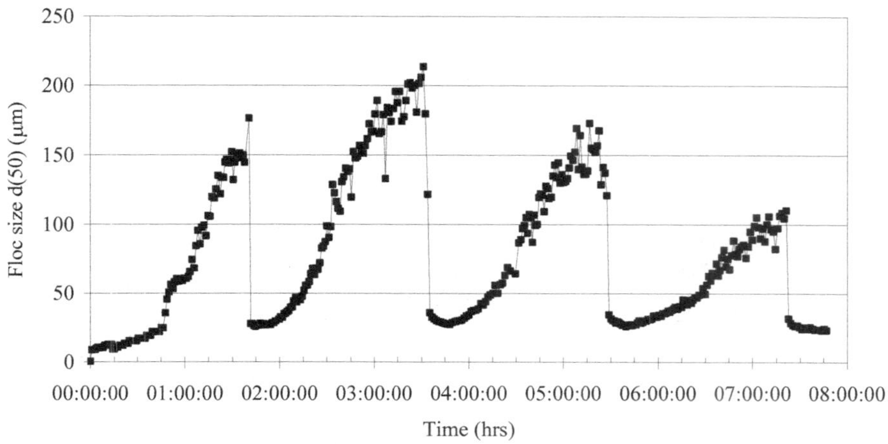

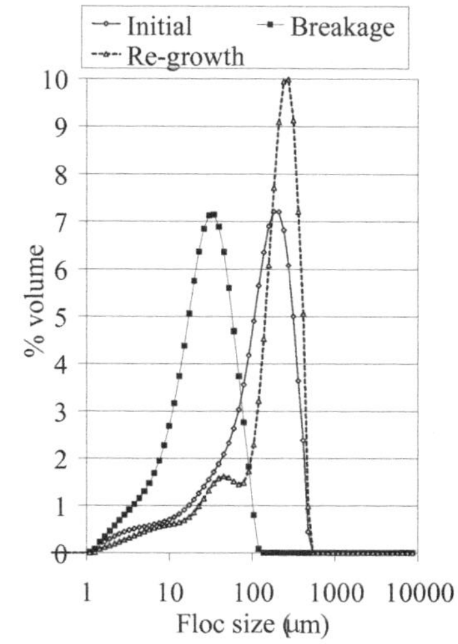

a) Latex beads (one break/re-growth cycle) b) Latex floc size distributions

Figure 6.20 a) the growth, breakage and re-growth of latex beads coagulated in a 1M NaCl solution after 3 breakage/re-growth cycles. The shear period was 15 minutes. b) The floc size distributions of latex beads after growth, breakage and the initial re-growth cycle.

Similar curves to those shown above were formed for the coagulating suspensions as they were grown, broken and subsequently re-grown. This demonstrated how the degree of floc compaction changed with different shearing patterns for the small flocs in the system. There was a considerable difference in NOM floc fractal dimension for the different coagulants (Figure 6.21). Prior to breakage, ferric-NOM flocs had the lowest D_f of 2.25 whilst alum flocs appeared to have a higher degree of compaction with a *Df* of 2.45 with the polymer flocs still more compact at a D_f of 2.55. On exposure to high shear, the D_f value of the alum and ferric NOM flocs did not change considerably. PDADMAC flocs had a considerable reduction of D_f to 2.3 when broken and once the original shear had been returned to, floc compaction was seen to increase for all of the flocs (Figure 6.22).

After 15 minutes slow stir the D_f was 2.15 for the coagulant precipitate. D_f increased to 2.6 for Fe precipitate flocs after breakage and then reduced to 2.25 during re-growth. The cyclical breakage and re-growth of latex coagulated in NaCl showed that floc aggregates reached a D_f of 1.67 after the initial growth phase (Figure 6.23). The degree of compaction increased to *Df* of 2.05 following the first breakage. Once the original slow stir was returned a D_f of 1.7 was reached. Further breakage cycles increased D_f to 2.0 and then fell to a constant value of around 1.7 when the flocs re-grew.

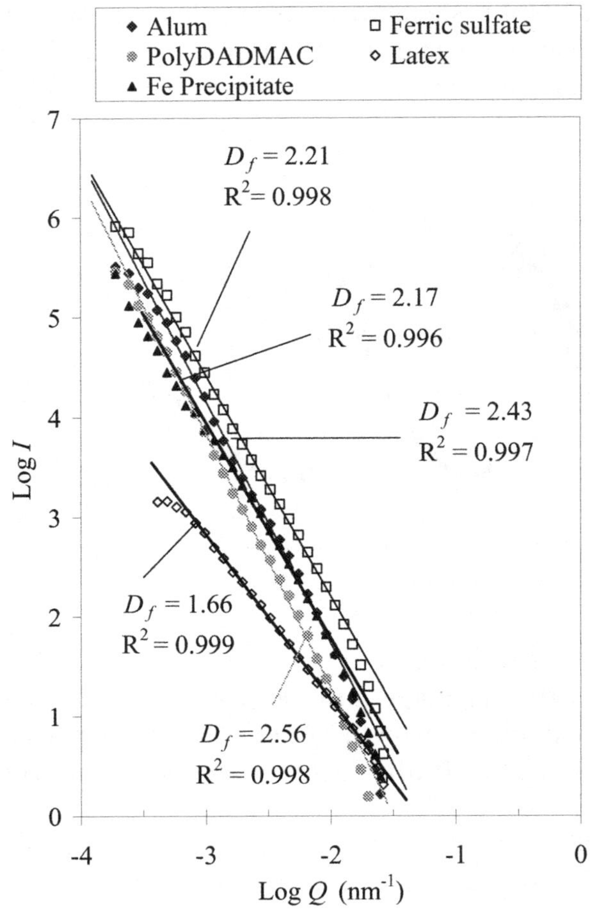

Figure 6.21 The relationship between the scattered light intensity (*I*) and the wave number (*Q*) on a log-log scale and the determination of the fractal dimension for different flocs after the initial growth phase for a) NOM flocs coagulated with different coagulants and b) different initial starting suspensions.

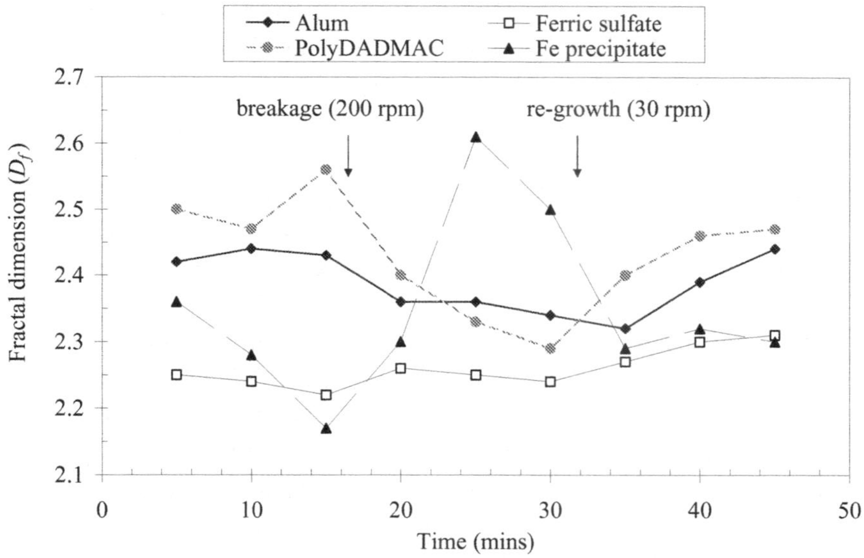

Figure 6.22 The change in fractal dimension of NOM flocs and Fe precipitate flocs with time for different coagulants after growth (30 rpm), breakage (200 rpm) followed by a return to the initial 30 rpm for re-growth.

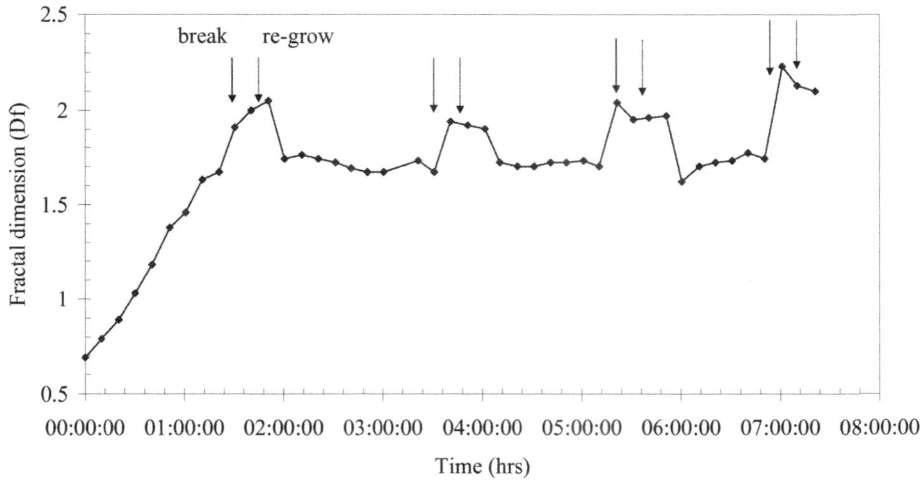

Figure 6.23 The change in fractal dimension with time of latex flocs coagulated in 1 M NaCl for cyclical breakage and re-growth regimes.

Discussion

A ranking of the floc properties from the different coagulated systems at optimized coagulant doses revealed that the different treatment options gave rise to some very different floc properties (Table 6.6). Both the increased size of the initial and broken flocs of the MIEX® + Fe treatment suggested that these flocs were the strongest and the NOM floc strength was of the following order: MIEX® + Fe > Polymer > Fe > Fe + Poly > Al. The MIEX® + Fe flocs also showed improved floc settling characteristics over the systems. The marked difference in the properties of MIEX® pre-treated flocs over the other treatment systems was surprising. For example, after ion exchange pre-treatment the flocs had a settling rate close to that of kaolin flocs despite the density of the kaolin used in this study being 2600 kg m^{-3} whilst humic material is estimated to have a density of 1750 kg m^{-3} (Hossain and Bache, 1991). The use of ion exchange for the removal of NOM has been previously been shown as a good alternative to coagulation for NOM removal (Bolto *et al.*, 2002), particularly of recalcitrant organics (Fearing et al., 2004b). This is of current importance given the evermore strict control of disinfection by-products in North America, Australasia and Europe. This study has presented evidence showing that ion-exchange pre-treatment combined with conventional coagulation can give rise to improved NOM floc structure, the knowledge of which may be of potential operational consequence.

Previous trials have indicated that as the organic fraction in a NOM-Fe floc decreases the floc size and settling rate are seen to increase to a maximum at 0 NOM concentration. Given that the organic concentration was reduced by 70 % after ion exchange this effect was thought to explain this phenomenon at least in part. However the increase in size and settling rate of the MIEX® + Fe flocs in this study was considerably above that seen previously suggesting that other factors were also important.

Analysis of the HPSEC chromatograms showed that the ferric coagulant preferentially removed the high and medium MW compounds whilst the ion exchange preferentially removed the medium to low MW compounds. This meant that after MIEX® pre-treatment only the high MW organics were incorporated into the floc matrix. Without ion exchange pre-treatment the full range of MW compounds were contained within the floc matrix and it is this difference that provides an explanation for the change in properties when compared to the other treatment options. The Fe, Fe + poly and Al treatments showed good all round removals for all organics other than the small MW compounds, whilst the polymer used alone also showed poor removal of the medium MW organics. The polymer flocs were the next biggest flocs to the MIEX® + Fe flocs. Polymer bridging may in part explain the increased floc size and strength seen for the polymer flocs (Gregory and Dupont, 2001). However, the absence of the medium to low MW compounds in the floc matrix was also likely to be a contributory factor given the differences in floc structure seen for the MIEX® + Fe floc. The specific nature of these medium and low MW compounds was not identified in this study, but the nature of organics has been shown to have a significant impact on floc structure (Vilge-Ritter *et al.*, 1999). However, further investigation of medium and low MW NOM compounds is necessary in order to see whether this was a site specific effect or is applicable to all NOM rich waters.

The use of ferric and alum based coagulants have traditionally been favored as the principal chemicals of choice at WTW throughout the world (Johnson and Amirtharajah, 1983). It is therefore useful to compare the properties of the NOM flocs formed from these coagulants. The results of this work have shown that the ferric based coagulant gave better floc structural characteristics than alum in terms of floc size and resistance to shear. The main coagulation mechanisms for the removal of NOM using traditional metal salts is the precipitation of metal-NOM complexes by charge neutralization and the adsorption of NOM onto precipitates of either metal hydroxide or the complexation products (Dempsey *et al.*, 1984; Dennett *et al.*, 1996; Gregor, 1997; Vilgé-Ritter *et al.*, 1999). There is little evidence in the literature to explain the difference in floc properties of the alum and ferric systems. The situation is complicated as a myriad of different molecules exist which can form different initial micro flocs with each coagulant. Previous comparisons of ferric and alum coagulants suggest that they operate in a similar manner and are of a similar coagulant efficacy (Johnson and Amirtharajah, 1983). In addition the precipitates of the metal hydroxides have similar surface areas and particle size

(Bache, 2004). Speciation products for Al and Fe consist of similar hydrated monomers, dimers, trimers and larger polymers which operate in a similar way (Cheng and Chi, 2002). In this study the alum and ferric operated at optimum removals at pH 6.5 and pH 4.5 respectively, typical values seen previously for NOM removal using alum and ferric coagulants (Dennett et al,, 1996; Edzwald and Tobiason, 1999; Armitharajah and O'Melia, 1999). Coagulation pH may therefore play a significant role in the floc structure as it has been seen to give rise to more loosely connected NOM aggregates with increasing pH (Vilge-Ritter, 1999). Increased deprotonation of organic molecules at high pH gave rise to increased electrostatic repulsion within Fe-NOM flocs leading to a more open and a weaker floc as a result. However, this effect has not been previously seen for Al-NOM flocs when lowering the coagulation pH from the optimum conditions led to slow forming small flocs with poor settling characteristics (Gregory and Carlson, 2004). It is therefore at present unclear why the two coagulants give rise to different floc characteristics, however the results of this study lead to the recommendation of the use of ferric coagulants over alum for NOM coagulation due to the improved removals and floc structure seen.

Polymers are traditionally used in combination with conventional coagulants as an aid to improve floc structure by increasing floc size, strength and settleability (Lee *et al.*, 1998; Bolto *et al.*, 2001). In the current study the Fe-NOM flocs coagulated with and without pDADMAC showed that the polymer reduced the strength of the floc with a corresponding decrease in size and settleability. Added after the primary coagulant polymers work by bridging together primary particles (Bratby, 1980). However, quaternary ammonium polymers such as pDADMAC that are of high MW and charge density operate better as a primary coagulant operating like conventional coagulants rather than a floc aid (Bolto *et al.*, 2001). In terms of floc structure, the use of polymer as a primary coagulant did improve floc strength and size but offered no improvement in floc settling rate and crucially showed much reduced organic removal than the other treatment options. Therefore it was established that the high MW cationic polymer used in this study offered no overall benefit as a floc aid or as a primary coagulant.

The response of flocs to increases in shear is of particular operational significance. After they have been formed, flocs are exposed to a range of shear stresses such as high shear regions around impellers of flocculators or transfer over ledges and weirs (McCurdy *et al.*, 2004). Flocs must resist these stresses in order to prevent breakage. It was observed that the initially larger NOM flocs tended to break proportionally greater than the smaller NOM flocs (as reflected by a higher value of the degradation curve) but still maintained larger broken floc sizes until the flocs approached a common value at very high shear. This is in agreement with data presented by Bache and Rasool (2001) for a range of Al-NOM flocs formed from low alkalinity, low turbidity and highly colored waters at different water treatment works.

Table 6.6
Ranking of different NOM floc properties from the different treatment options

Parameter	Highest				Lowest
DOC Removal	Fe	≈Fe + Poly	≈MIEX + Fe	Al	Poly
Size	MIEX + Fe	Poly	Fe	Fe + Poly	Al
Strength	MIEX + Fe	Poly	Fe	Fe + Poly	Al
Re-growth	MIEX + Fe	Al	Fe	≈Fe + Poly	≈Poly
Settling rate	MIEX + Fe	Fe	≈Poly		≈Fe + Poly

≈ means values were similar to previous system

The results of this work have shown that floc re-growth was limited with the exception of latex. The recoverability of flocs gives some indication of the floc internal bonding structure. Previous research has shown that flocs formed by charge neutralization should give total recoverability (Chaignon *et al*, 2002). This work has shown that this was the case for latex beads coagulated in an ionic solution during one breakage/re-growth cycle. Subsequent cycles saw a drop off in recoverability, perhaps as a result of impurities entering the system during the seven hours required for these experiments.

The irreversible breakage of the NOM flocs and Fe-precipitate flocs was seen as evidence that the flocs formed in these systems were not dominated by pure charge neutralization mechanisms and were therefore held together by chemical rather than physical bonds. For polymers, breakage of polymer chains at high shear rates is thought to reduce floc bridging capability during re-growth, whilst the reasons for irreversible floc breakage for hydrolyzing coagulants has yet to have been satisfactorily explained (Yukselen and Gregory, 2004). It was therefore difficult to give a chemical explanation for the reduced re-growth of Fe-NOM flocs when compared to alum given the poorly understood speciation chemistry of these coagulants during NOM coagulation. However, from a physical standpoint, the reduced breakage of the Fe-NOM flocs resulted in a lower frequency in floc number when compared to the Al-NOM and pDADMAC-NOM flocs which may have resulted in fewer possible particle-particle contacts to allow for possible aggregation during the re-growth phase. The limited re-growth of Fe precipitate flocs suggests that sweep flocs show limited re-growth. This agrees with the findings of other research showing that sweep flocs formed from natural waters and Fe-kaolin flocs give poor re-growth after breakage (McCurdy *et al*, 2004). A reduction in the time exposed to a similar shear showed both a significant reduction in floc breakage and an increase in floc re-growth, an effect also seen by Yukselen and Gregory (2004). This shows that flocs can resist a certain amount of shear. This results from a proportion of flocs remaining outside of floc breakage zones around impellers during the short breakage periods. Under longer breakage periods, the probability of flocs entering breakage zones is much higher.

In this study, the degree of floc compaction was measured using fractal dimension analysis. It was evident that the flocs under investigation exhibited scattering behavior typical of mass fractal objects. The fractal values of NOM flocs were in the region of 2.2-2.6 for the coagulants under investigation using SALLS. As a direct comparison, this agrees with the D_f of 2.2 seen for flocs formed from a raw water of high organic content (5-6 mg L^{-1}) using a similar technique (Kim *et al*, 2001). The increased fractal dimension of pDADMAC flocs over the other floc types indicated that polymer bridging gives rise to the most compact structures. This does not agree with findings found for activated sludge flocs and mineral flocs which show a general decreasing of fractal dimension as polymer bridging is increased (Wu *et al*, 2002; Boller and Blaser, 1998). In addition, there was no significant difference in the fractal value of Fe-NOM flocs and Fe precipitate flocs. This was surprising given the differences seen in the floc growth and breakage behaviour. Organic material has previously been shown to both stabilize and de-stabilize particles dependent upon the organic concentration (Walker and Bob, 2001). It therefore seems highly likely that adsorbed NOM within a floc should provide either increased attraction or repulsion leading to differing degrees of compaction.

A possible explanation for these inconsistencies being that in the current study the scale of investigation of the fractal dimension analysis was three to four orders of magnitude smaller than the average size of the flocs under investigation. This is a finding consistent with a number of other studies (Yan *et al*, 2000; Selomulya *et al*, 2001; Tang *et al*, 2002). Using SALLS, the fractal scaling law breaks down at increased floc size as the light becomes scattered by the whole aggregates rather than the primary particles (Guan *et al*, 1998). Larger flocs may also have inconsistent fractal dimensions such that they exhibit degrees of compaction from one another. Therefore, the findings of the floc compaction work must be limited to a small proportion of the flocs in each distribution. Specifically for coagulant-NOM flocs, little is known about the exact nature and homogeneity of the primary particles in comparison to the other systems. It was therefore concluded that the adopted fractal protocol was not sensitive enough to help explain the observed differences in NOM flocs for model systems such as small spherical beads, the technique seemed to provide more consistent results. The D_f value of 1.67 for latex aggregated in NaCl was similar to previous values found from latex aggregated in McIllvaine buffer (D_f= 1.8) (Tang, 1999) and latex aggregated in potassium nitrate (D_f= 1.7-1.8) (Selomulya *et al*, 2001).

Current understanding suggests that flocs become more compact on exposure to increases in shear as they break at their weak points and re-arrange into more stable structures (Spicer *et al*, 1998; Selomulya *et al*, 2001). The NOM flocs did not follow this expected change and the reverse was seen, which was further evidence for the limited applicability of SALLS to NOM flocs. The Fe precipitate and latex flocs followed this expected change in compaction as indicated by the change in fractal dimension such that D_f was seen to increase by 0.2-0.3 as flocs were broken into smaller and more compact structures. An increase in floc compaction provides a further explanation for the limited re-growth seen for most of the flocs. As chemical bonds are broken and flocs re-arrange into more compact structures bonding sites that were previously available become unavailable as they re-form within the broken aggregate. Whilst the same conclusions cannot be drawn for NOM flocs using the current experimental procedure, further research using other fractal dimension analysis tools (such as floc settling) may provide this information. For physical bonding (as seen for latex), floc compaction still occurs after breakage but the opportunities for floc re-formation are much greater because van der Wals forces can form anywhere on the latex particle surface such that further particle-particle contact between exposed regions can lead to physical bonds forming, thus producing similar sized flocs during re-growth.

The results of this work have shown that NOM flocs have different properties dependent on which chemical coagulant is used. The resulting structure seems to be determined by mechanisms of coagulation with bridging flocs forming the largest flocs. Furthermore, it appears that the bonds holding NOM flocs together are not purely physical bonds given the limited re-growth seen. The use of SALLS was not applicable to the NOM flocs even though a straight line relationship was seen between log Q and log I. On a wider scale, SALLS is recommended for model and uniform particles, but even then can only be applied to the small flocs in a system.

The implications of this part of the work is that coagulation regimes can be tailored to potentially meet criteria for downstream processes as well as the primary coagulation and flocculation stage. For example in unit processes which are likely to expose NOM flocs to very high shear (ie an equivalent to >150 rpm) such as dissolved air flotation the control of floc size is not important as larger flocs will be broken to the same size regardless of any increase in the relative strength of smaller flocs. This is because the average velocity gradient in a jar tester at 150 rpm is less than 100 s^{-1} (Bouyer *et al.*, 2001), whilst in a DAF unit estimates are around 1000 s^{-1} (Bache and Rasool, 2003). This high shear will break even the strongest flocs to similar sizes. Whereas for low shear settling processes coagulation regimes that generate large flocs should be preferred due to the operational benefits during solid-liquid separation as increased floc size gives rise to increased sedimentation rates (Aguilar *et al.*, 2003) and improves particle transport mechanisms during removal in filtration (Cleasby *et al.,* 1999).

DOC: coagulant ratio

Raw water characterisation and coagulation tests

Preliminary optimisation tests across a range of pH for a coagulant dose of 8 mg L^{-1} showed that for the Fe based coagulant used in this study a pH of between 4 and 4.75 gave a combined optimum for NOM removal and residual turbidity after settlement (Figure 6.24). It was therefore decided to control the pH around 4.5 for all subsequent floc experiments. This pH also corresponds to the operational pH as used by the WTW from where the water was abstracted. As the Fe:DOC ratio changed a number of significant changes in the coagulated suspensions were observed (Table 6.7). The residual DOC remained around 1 mg L^{-1} for all of the treated waters whilst the UV_{254} was seen to increase from 0.01 to 0.06 cm^{-1} from a ratio of 0.33 to 3.80. Assuming that all of the NOM removed during coagulation is encorporated into the floc then the percentage of carbon in the dry floc mass increased from 0 to 35 % with increasing carbon fraction. The zeta potential of the raw water was -17 mV which did not change in the cases where it was necessary to dilute the raw water. The zeta potential of the coagulated suspensions changed from being very positive (11 mV) at high Fe:DOC ratios to being negative at low Fe:DOC ratio (-11 mV). At ratios of between 0.75 and 1.75 the charge was just above zero (1-3 mV).

Table 6.7
Characteristics of coagulated suspensions for each of the different Fe:DOC ratios.

DOC to Fe ratio (mg:mg)	Residual DOC (mg L^{-1})	Residual UV_{254} (cm^{-1})	Dry floc mass (mg)	mass of carbon in floc		Zeta potential of coagulated water (mV)	Settled turbidity (NTU)
				(mg)	(%)		
0	0	0	24.8	0	0	+ 11.0 ± 0.7	0.55
0.33	0.71	0.01	31.5	1.6	5.1	+ 11.6 ± 0.7	0.73
0.75	1.11	0.02	36.0	4.8	13.3	+ 3.8 ± 0.7	1.14
1.00	1.32	0.02	39.2	6.2	15.8	+ 1.8 ± 0.7	1.39
1.75	0.68	0.03	29.4	6.9	23.5	+ 3.7 ± 0.7	1.49
3.80	1.31	0.06	16.8	6.3	36.5	- 11.6 ± 0.7	3.98

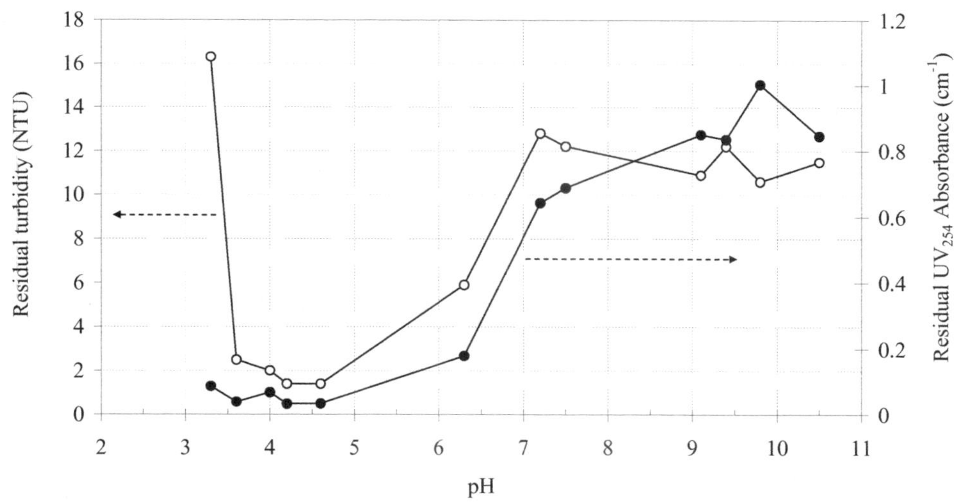

Figure 6.24 Optimisation of NOM removal with pH for a coagulant dose of 8 mg L-1 as Fe for UV254 and turbidity on the raw water (ratio 1:1).

Size

Comparison of floc size was made using the median equivalent volumetric diameter (d_{50}), however the same trends were seen for the whole range of floc sizes investigated. The general trend showed that as the ratio of organic matter increased, floc size was seen to decrease (Figure 6.25). Both the 0 and 0.33 flocs approached a d_{50} plateau maximum approaching 1000 μm. The 0.75 to 1.75 ratio flocs reached a similar maximum size of between 700-800 μm. A substantial reduction in floc size was seen for the 3.8 flocs as the maximum size reached approached 500 μm. The growth rate of the flocs was much quicker for the floc ratios of 1 and 0.75 with the maximum size being reached between 6-8 minutes. Above and below these ratios, floc growth was significantly slower, taking over 10 minutes to reach a plateau.

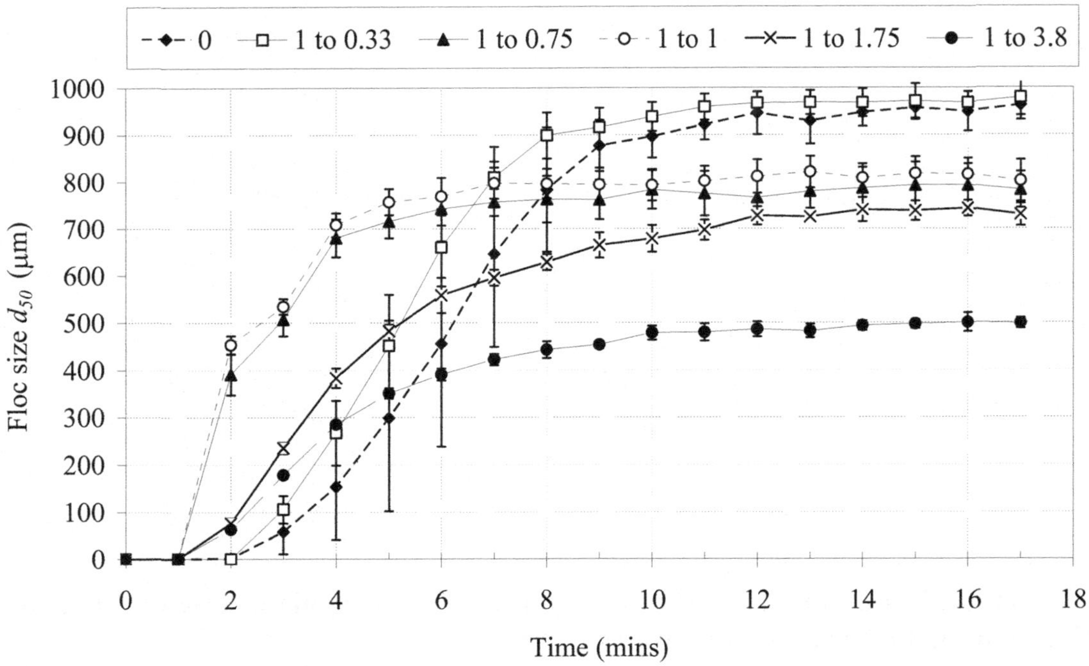

Figure 6.25 The growth rates of flocs for each of the different Fe:DOC ratios.

Microscopy

Typical floc images for each of the coagulant to organic ratios are shown in Figure 6.26. From a qualitative analysis of these images it can be seen that in the absence of NOM and at a Fe:DOC ratio of 1 to 0.33 the flocs appeared more solidly composed with fewer internal pores and softer edges. As the carbon fraction increases the flocs became more dendritic with more internal pores.

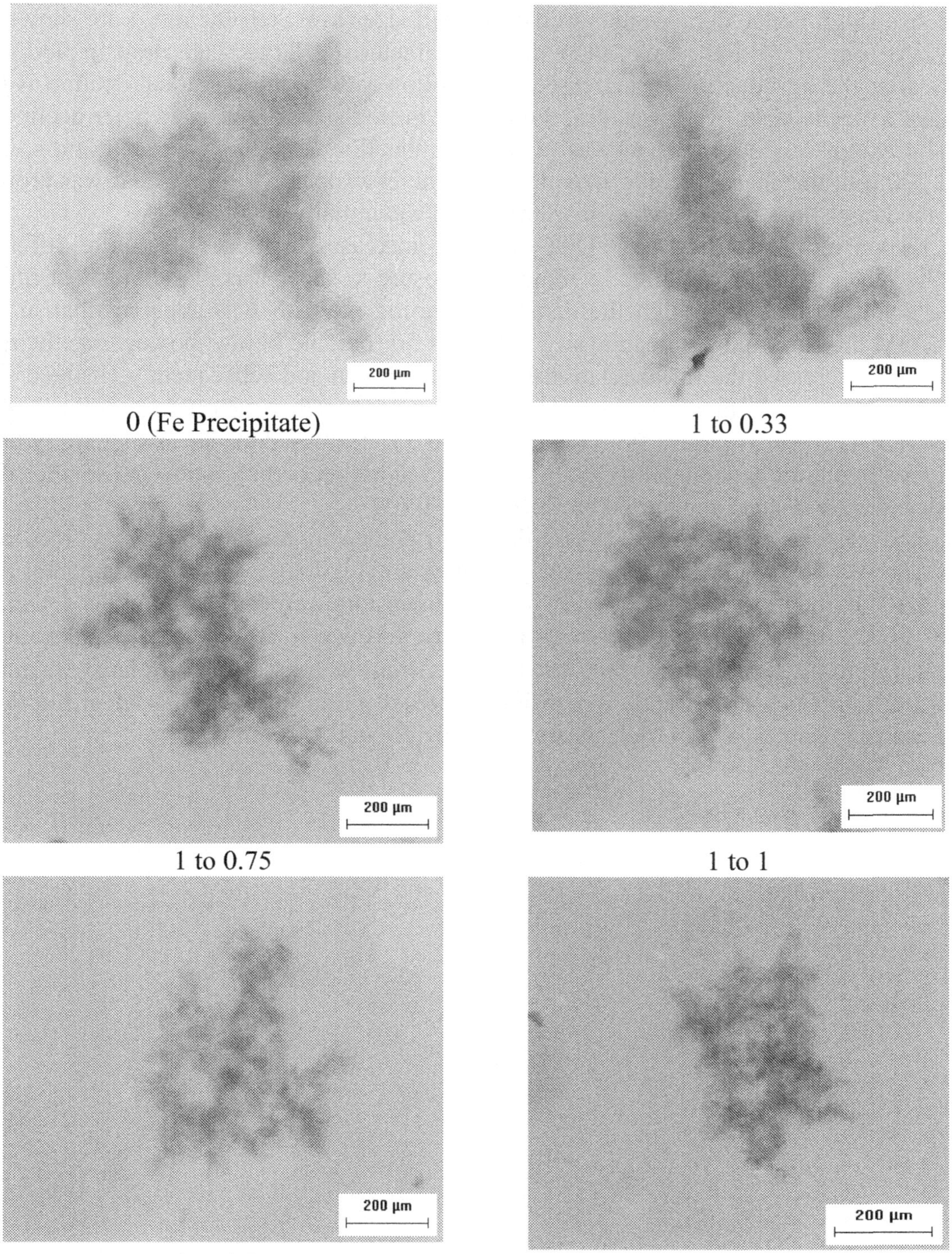

Figure 6.26 Microscope images of typical flocs from each of the different Fe:DOC ratios.

Strength

After the flocs had reached their steady state size during the slow stir period, they were exposed to increased shear. A comparison of the type of breakage profiles seen for flocs of an initial large size (high Fe:DOC ratio) and an initial small size (low Fe:DOC ratio) are shown in Figure 6.27. At rpm of 75 and above, there was a characteristic large scale drop in floc size immediately after the introduction of increased shear followed by a more gradual decline as the flocs approach a steady state. At lower rpm, the change in floc size follows a steadier decline in floc size. The extent of floc breakage was greater for the flocs of a high Fe:DOC ratio. For example, at 200 rpm the change in floc size after 15 minutes exposure to high shear was around 800 μm for the larger flocs whilst it was only 300 μm for the initially smaller flocs.

The response of the different Fe:DOC ratios to increases in shear were quite different (Figure 6.27). When the floc size after 15 minutes exposure to shear was plotted against rpm a straight line could be drawn through the data on a log-log scale. It was apparent that in the absence of NOM the flocs did not change size at rpm of 50 rpm and below, however as the rpm was subsequently increased the linear relationship was then seen and subsequently changed size rapidly with increasing rpm dropping from 1000 μm to 250 μm. This was reflected by the value of the slope (-1.17). The next largest flocs (at a ratio of 0.33) decreased in size at a similarly high rate as indicated by the slope value of -0.90. For the flocs at higher carbon ratios and smaller floc size the slopes were less steep and varied between -0.53-0.65.

The floc systems that showed a significant difference in fractal dimension from one another also showed differences in their breakage behaviour (Figure 6.27b). It was apparent that the broken size of the flocs with a higher fractal dimension were larger than flocs with a significantly lower fractal dimension at increased shears between 30-100 rpm. For example between 30-150 rpm the 0 ratio flocs (high fractal dimension) were over 200 μm larger than the 1, 1.75 and 3.8 ratio flocs (low fractal dimension). The exception to this being when the flocs approached common values at very high shears between 150 and 200 rpm.

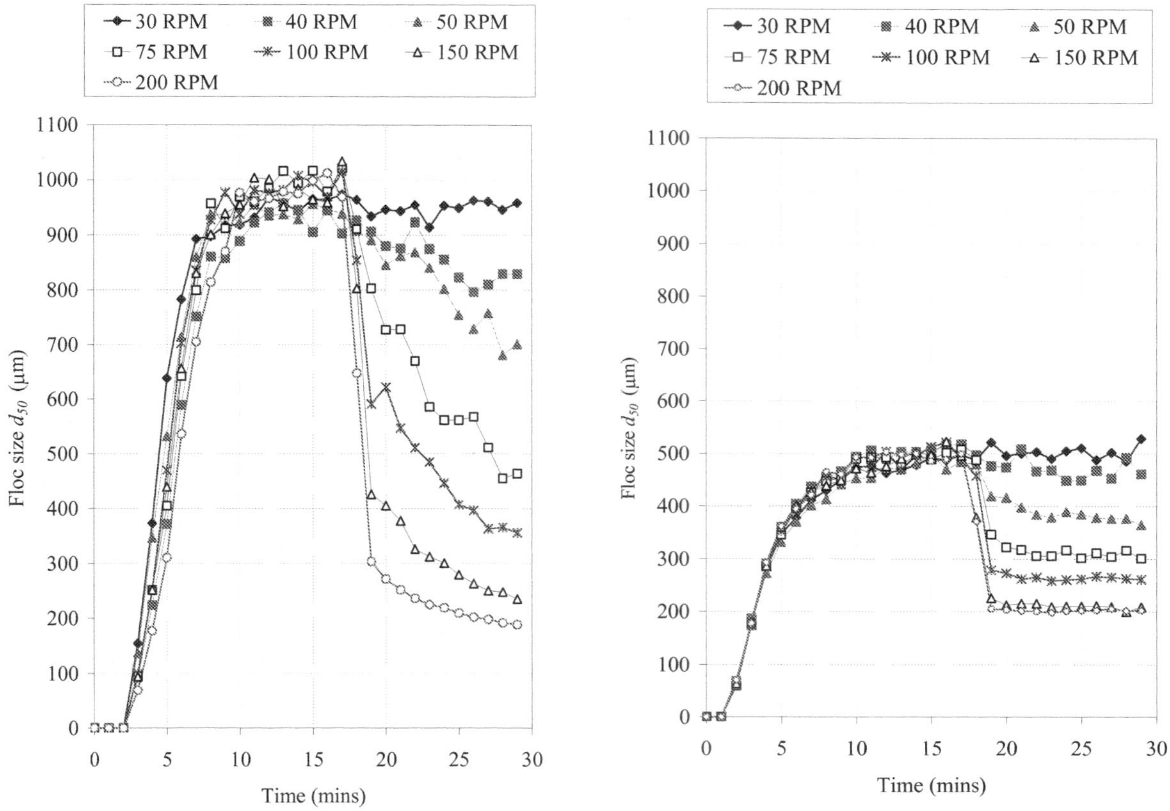

a) High Fe:DOC ratio b) Low Fe:DOC ratio

Figure 6.27 Typical breakage profiles for a low and high DOC to Fe ratio with increasing RPM.

Settling rate

The settling rates for each of the Fe:DOC ratios are shown in Figure 6.28. The first point of note is a general decrease in the settling rate of similar sized flocs as the carbon fraction increases. For example, the settling rate of a floc of 1000 μm in equivalent diameter was around 1000 μm s^{-1} for the 0 and 0.33 ratio flocs whilst at higher organic ratios of 1.75 and 3.8 the settling rate was around 600 μm s^{-1}. The improved settling rate of flocs with lower carbon content was further supported by a decrease in the settled turbidity at lower Fe:DOC ratios (Figure 6.29). Secondly, there was a general reduction in the slopes of the data with increasing carbon proportion. This was confirmed by calculation of the floc fractal dimension. Model two reduced major axis regression was used to calculate the slopes of the lines through the data. There was good linear correlation for all of the data sets with R^2 values between 0.73-0.93. From these slopes D_f was calculated which confirmed a general tendency for D_f to decrease with increasing carbon fraction with the fractal dimension decreasing from 2.20 ± 0.04 to 1.78 ± 0.04 from the lowest to highest Fe:DOC ratio (Figure 6.29). A statistical comparison of the values of the slopes showed that there were statistical differences (two-tailed t-test, 190 df, P = 0.01) between the slope of 0 (Fe precipitate) and all of the slopes at and above ratios of 1. Further significant differences were seen between ratios of 0.33 and 3.8; 0.33 and 1.75; 0.75 and 3.8; 0.75 and 1.75.

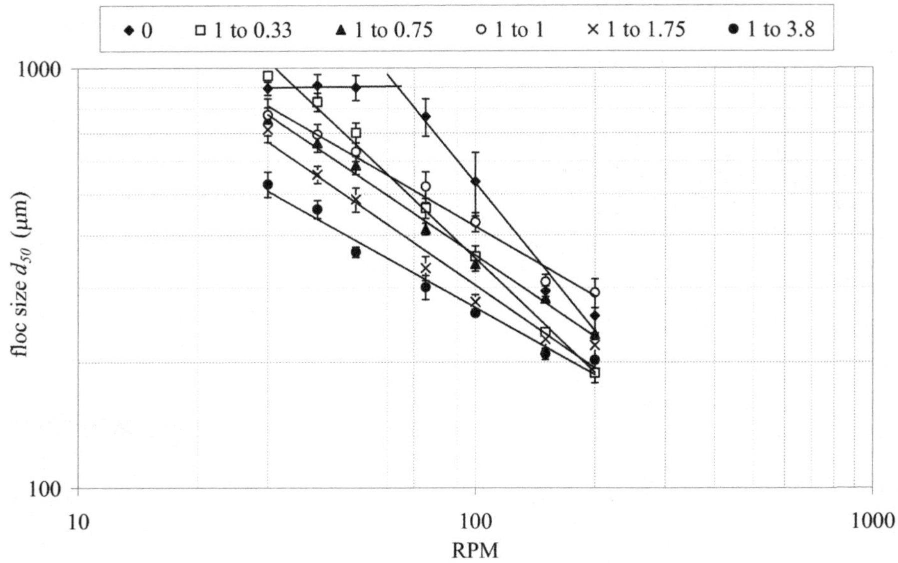

a) Floc breakage rates with increasing rpm for all the Fe:DOC ratios

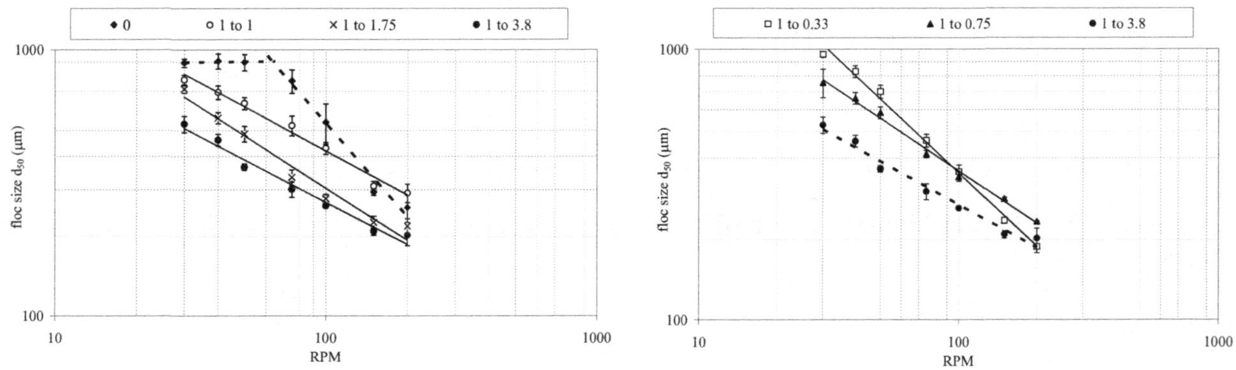

b) floc breakage rates for flocs with significantly different fractal dimensions. The heavy dashed line shows the data which has a significantly different fractal dimension from the other data.

Figure 6.28 The steady state breakage size with increasing RPM for different Fe:DOC ratios.

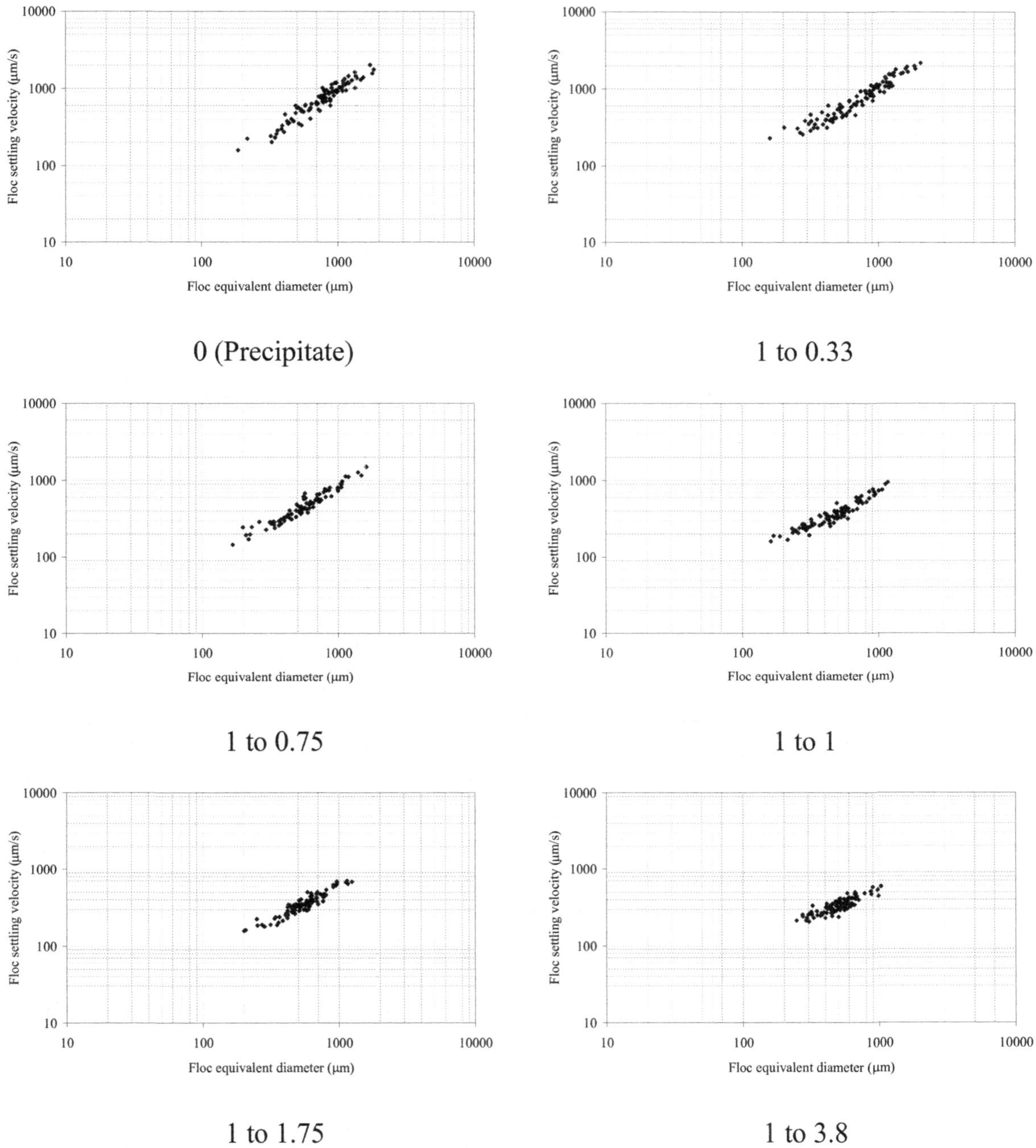

Figure 6.29 The settling rates of flocs for each of the different Fe:DOC ratios.

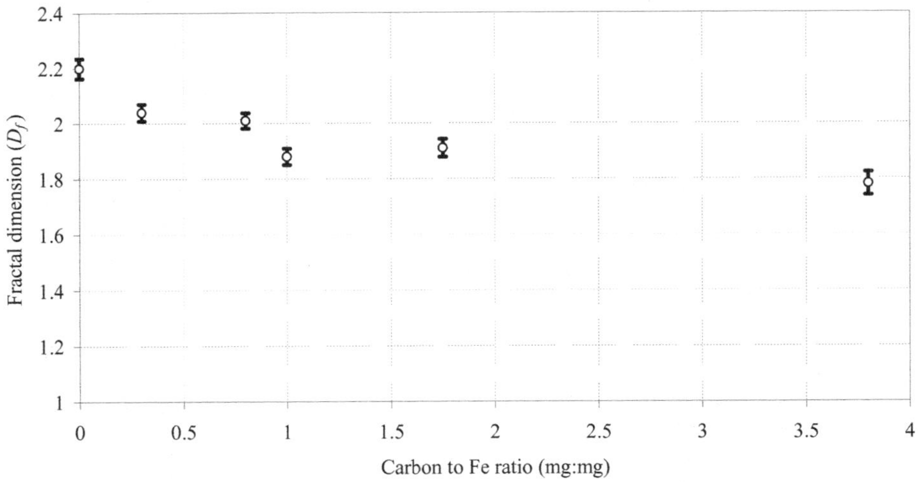

Figure 6.30 The change in fractal dimension (D_f) with increasing carbon to Fe ratio (with standard error bars).

Rheological measurements

In order to better understand the impact of changing DOC:coagulant ratio a series of parallel experiments were conducted to ascertain the impact of changing NOM concentration onto a fixed substrate concentration. The experiments required large quantities of NOM and as such were conducted on locally available sources (Wanneroo [WANt] or Hope Valley [HVt]).

The zeta potential versus pH data for Wanneroo (WANt) as a function of dose are presented in Figure 6.31. The addition of WANt shifts the iep of alumina to lower pH with increasing dose. In addition, the magnitude of the shift in iep increases with increasing NOM concentration. These data demonstrate that the NOM adsorbs to both positive and negative alumina surfaces and can cause charge reversal, indicating specific adsorption to the surface. For low doses the gradient of the curves is largely the same as that obtained for the pure alumina. This indicates that, apart from the shift in iep, the charging characteristics of the surface as a function of pH are largely unchanged. The gradient reduces slightly with increasing dose and conductivity measurements obtained during the zeta potential measurements suggest this decrease in gradient may be partially due to an ionic strength effect. The presence of significant levels of background salts in WANt means that the ionic strength of the suspensions increases with increasing NOM dose.

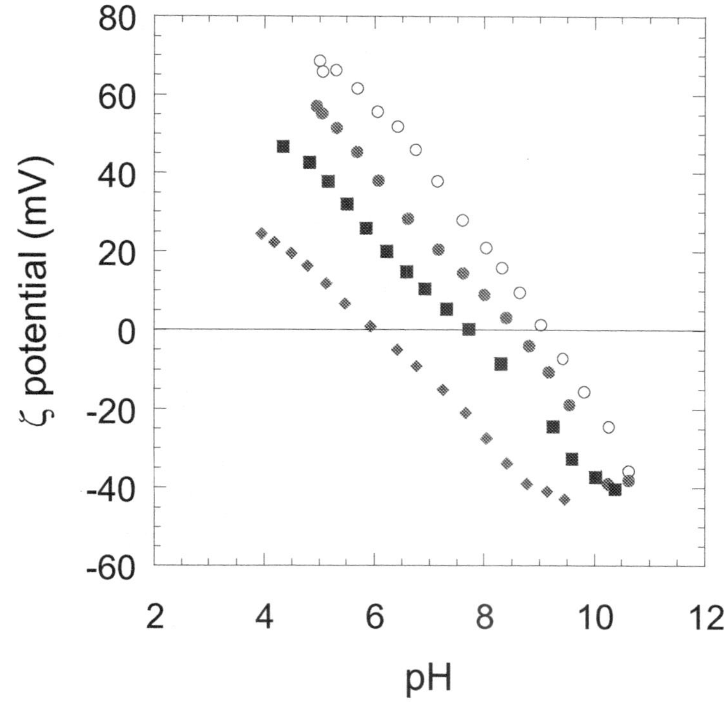

Figure 6.31 The zeta potential behaviour of AKP-30 alumina as a function of pH and [Wanneroo NOM]. The alumina volume fraction is 0.020 and the background electrolyte concentration is 0.010 M NaCl. ○, [WANt] = 0.000 % w/w TOC; ●, [WANt] = 0.025 % w/w TOC; ■, [WANt] = 0.050 % w/w TOC; ◆, [WANt] = 0.100 % w/w TOC.

Figure 6.32 shows the effect of WANt on the shear yield stress of an AKP-30 alumina suspension as a function of pH and dose. The position of τ_{ymax} is shifted to lower pH which is consistent with the electrokinetic data. There is also a reduction in τ_{ymax} and a broadening of the curves with increasing dose which correlates with the slight decrease in gradient of the zeta potential versus pH presented in Figure 6.31.

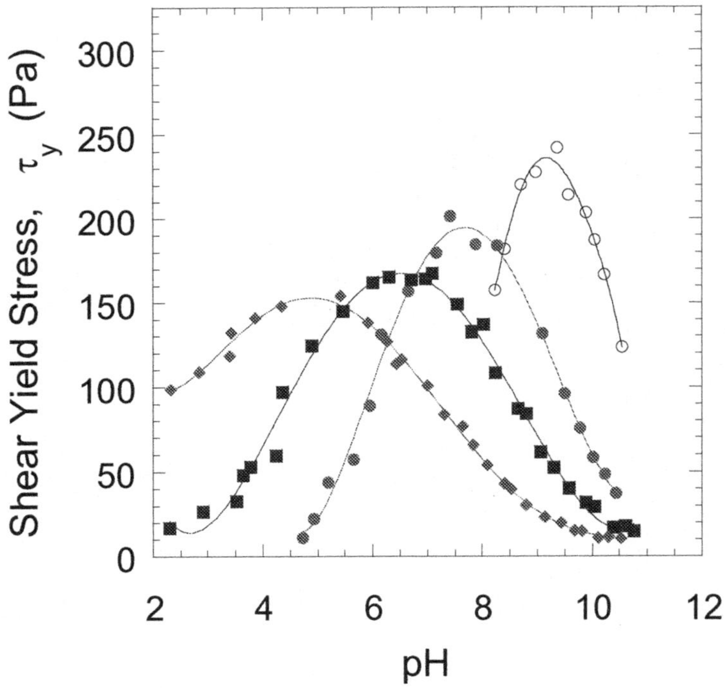

Figure 6.32 Shear yield stress of AKP-30 in the presence of Wanneroo NOM (WANt) as a function of pH and dose. The alumina volume fraction was 0.250 and the electrolyte concentration was 0.010 M KNO_3. ○, [WANt] = 0.000 % w/w TOC; ●, [WANt] = 0.025 % w/w TOC; ■, [WANt] = 0.050 % w/w TOC; ◆, [WANt] = 0.100 % w/w TOC. (lines are purely to help guide the eye)

The reduction in τ_{ymax} is caused by a reduction in the effective van der Waals attractive force at the iep. At this pH there are no electrostatic interactions and, assuming only DLVO forces apply, only van der Waals forces are present. The van der Waals force is dependent on a fundamental property of the substrate known as the Hamaker constant, A_H, and the average distance between the surfaces. As the Hamaker constant is largely unaffected by the small amount of material adsorbed (Buscall 1993), the apparent van der Waals force experienced by interacting particles can only be reduced by increasing the average distance between the particles by effectively preventing their close approach (Leong *et al.* 1993b). The NOM in this case acts as a steric barrier; preventing close approach of the particles, reducing the apparent van der Waals attractive force, and thereby reducing τ_{ymax}. The increasing reduction of τ_{ymax} with increasing dose is due to increasing surface coverage, and therefore increases in the effective steric layer thickness created by the NOM.

Figure 6.33 shows that when increasing doses of WANt is added to alumina with 2.5% v/v solids $Fe(OH)_3$ present, the NOM reduces the shear yield stress and shifts the iep of the surface to lower pH. This is analogous to the system with no $Fe(OH)_3$ present, as previously presented in Figure 6.31. Clearly, the NOM has a similar effect on both systems although significantly higher concentration is required in the presence of $Fe(OH)_3$, due to the increased surface area and the increased number of charged sites. The systems with low NOM to $Fe(OH)_3$ ratios are clearly dominated by the $Fe(OH)_3$ coating which reduces the shift in iep and maintains higher yield stresses. NOM at higher levels can reduce the effect of the $Fe(OH)_3$ by reducing the shear yield stress (Figure 6.34).

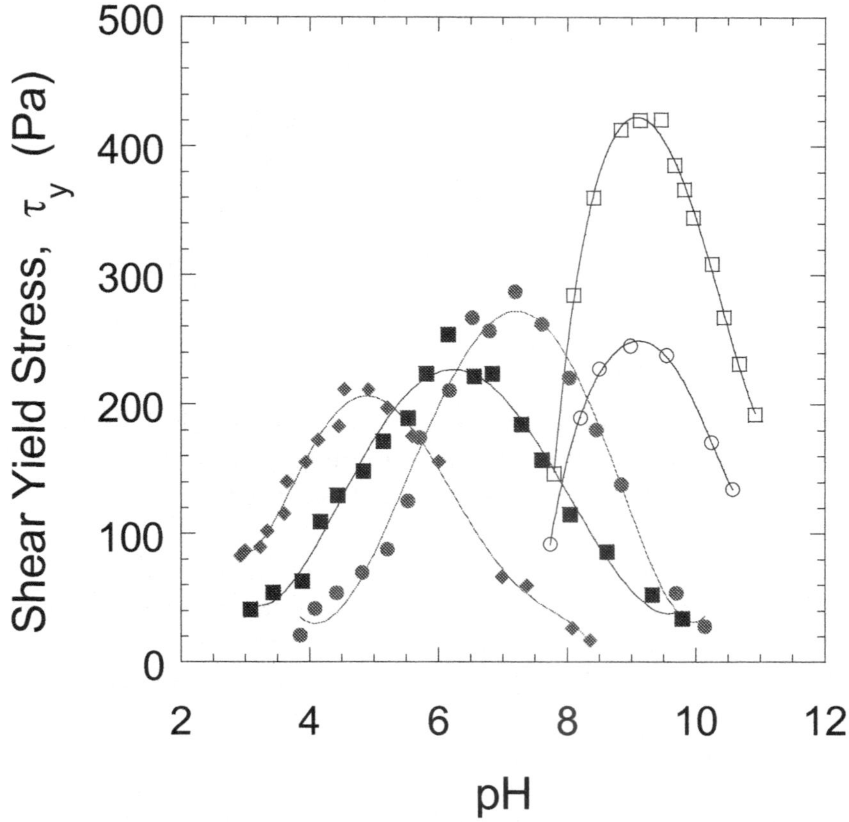

Figure 6.33 Shear yield stress behaviour of AKP-30 alumina in the presence of 2.5 % v/v ferric hydroxide and WANt as a function of pH and NOM dose. ○, no additives; □, 0.0 % w/w WANt; ●, 0.1 % w/w WANt; ■, 0.2 % w/w WANt; ◆, 0.5 % w/w WANt.

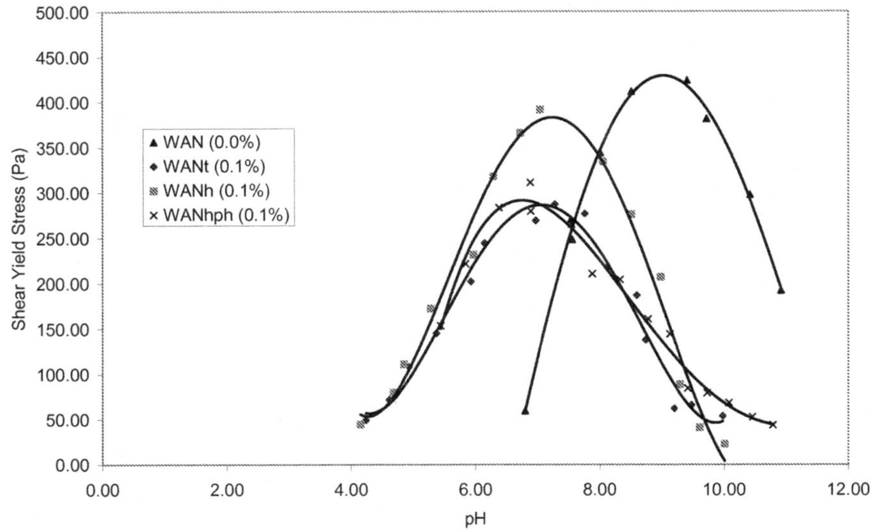

Figure 6.34 Shear yield stress of ferric hydroxide coated AKP-30 in the presence of Wanneroo NOM (WANt) and fractions as a function of pH. The solids volume fraction was 0.250 and the electrolyte concentration was 0.010 M NaCl.

The zeta potential versus pH data for Hope Valley (HVt) as a function of dose are presented in Figure 6.35. Compared to the WANt data, HVt data show much greater shifts in the iep for a given dose of NOM and a significant reduction in the gradient of the curves. In addition, at the highest dose of HVt the negative surface potential is increased considerably. As stated above, increases in electrolyte concentration cause compression of the double layer and reduction in surface potential. These effects often result in reduction in the gradient of zeta potential versus pH curves. (Johnson *et al.* 1999b; Wiese and Healy 1975). However, conductivity measurements suggest there is minimal difference in the electrolyte concentration between the alumina suspensions, with or without HVt added.

Figure 6.36 shows the effect of adding HVt, on the shear yield stress of an AKP-30 alumina suspension as a function of pH and dose. The position of τ_{ymax} is shifted to lower pH which is consistent with the electrokinetic data. There is no significant reduction for HVt at low doses, although there was significant reduction for HVt at the highest dose. In comparison to WANt, HVt is the most hydrophilic NOM sample and produces the greatest reduction in τ_{ymax} at the highest dose. However, this sample produces the lowest reduction in τ_{ymax} at the low doses used. The electrokinetic data proves that HVt adsorbs and causes changes to the iep of the surface and the mobility of the particles at all doses. The overall reduction in zeta potential across the pH range also indicates the molecules are extended from the surface even at low doses. In addition, the reduction in the zeta potential versus pH gradient would normally result in a broadening of the shear yield stress peak, however this does not occur at the two lowest doses.

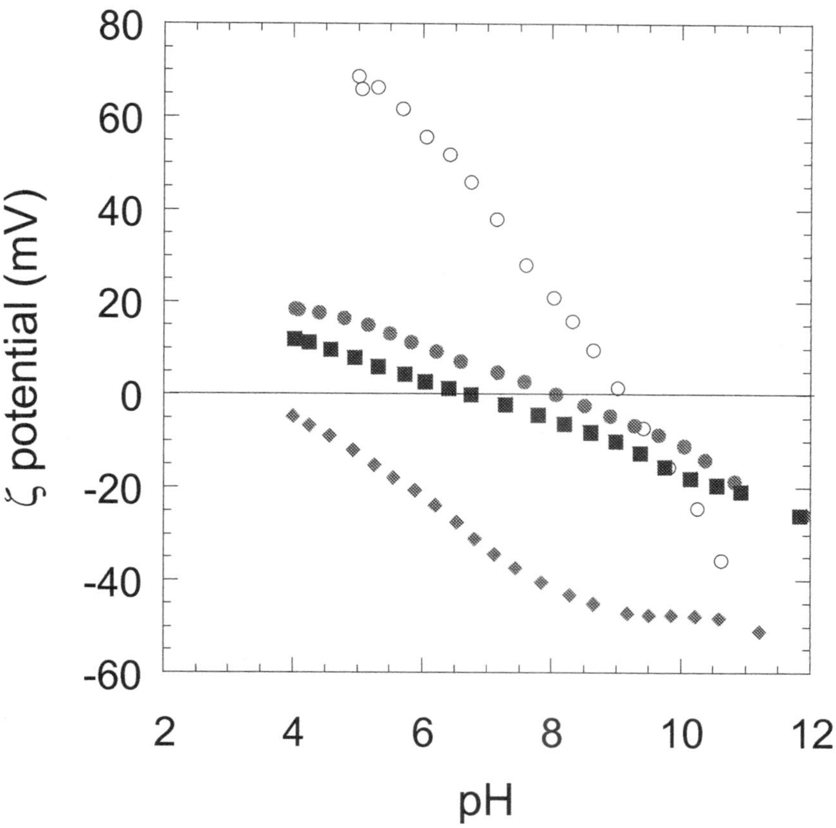

Figure 6.35 The zeta potential behaviour of AKP-30 alumina as a function of pH and Hope Valley NOM concentration. The alumina volume fraction is 0.020 and the background electrolyte concentration is 0.010 M NaCl. ○, [HVt] = 0.000 % w/w TOC; ●, [HVt] = 0.028 % w/w TOC; ■, [HVt] = 0.056 % w/w TOC; ◆, [HVt] = 0.112 % w/w TOC.

The lack of reduction in the τ_{ymax} at low HVt dose and substantial lowering at high dose is of great interest. It implies that at the low doses, whilst the HVt NOM adsorbs and thus reduces the iep, it does not form a substantial steric barrier. One possible explanation for these seemingly anomalous results is the formation of a "soft" NOM layer. This is possible if the adsorbed molecules are sparsely and weakly adsorbed and able to migrate laterally on the oxide surface. As the particles approach each other the adsorbed molecules are pushed aside, and the interaction essentially reverts to that between bare alumina surfaces, thus giving a similar value for τ_{ymax}. This would also explain the lack of broadening in the shear yield stress curve to correlate with the reduction in zeta potential gradient. At higher doses of HVt, the surface concentration reaches a critical point at which the NOM layer becomes more contiguous. This prevents the lateral movement of NOM molecules, thereby producing a much "harder" steric barrier and reducing the shear yield stress markedly. In essence, the shear plane at low dose is ill defined due to changes in the hydrodynamic properties at the adsorbed layer. At higher concentrations, the adsorbed layer increases in density and becomes firm and the shear plane becomes better defined.

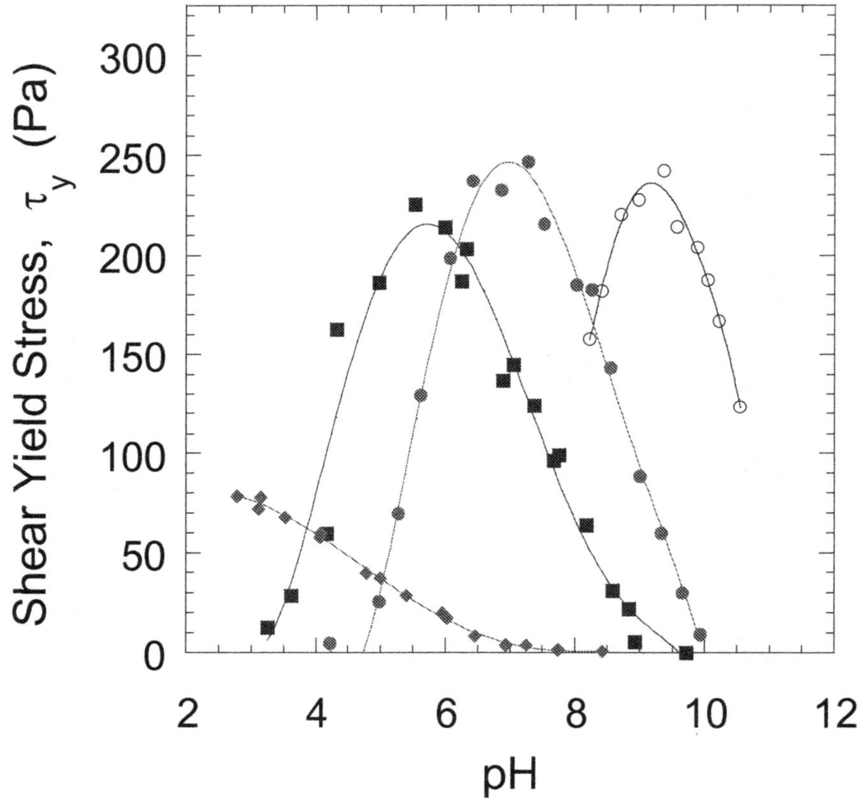

Figure 6.36 Shear yield stress of AKP-30 in the presence of Hope Valley NOM (HVt) as a function of pH and dose. The alumina volume fraction was 0.250 and the electrolyte concentration was 0.010 M KNO_3. ○, [HVt] = 0.000 % w/w TOC; ●, [HVt] = 0.028 % w/w TOC; ■, [HVt] = 0.056 % w/w TOC; ◆, [HVt] = 0.112 % w/w TOC. (lines are purely to help guide the eye)

Discussion

The results presented in this work show a definite change in floc structure with decreasing coagulant to organic ratio. The overall trend shows a decline in floc structure with increased organic carbon proportion in the floc. The proposed model for the observed changes in floc structure is attributed to the interaction of carbon within the floc matrix. When particles are coated with a layer of organic material and then subsequently flocculated with an inorganic salt, the resulting suspensions either resist aggregation or form flocs that have a lower fractal dimension than particle suspensions flocculated in the absence of organic matter (Amal *et al.*, 1992; Waite *et al.*, 1993; Wilkinson *et al.*, 1997). In these instances, organic material stabilises the particulates through steric effects and electrostatic stabilisation from the negative charge on ionised acidic groups on the organic molecules. Primary particles cannot get as close to one another as before such that they do not aggregate at all or they produce more open and loosely connected structures. In the case of NOM coagulation, there is strong evidence suggesting that the principal mechanism for NOM removal at pH below 6 is complexation of negatively charged NOM with soluble metal species of positive charge onto insoluble precipitates (Dempsey *et al.*,

1984; Dennett *et al.*, 1996; Vilge-Ritter *et al.*, 1999). The further removal of NOM in solution subsequently occurs by adsorption onto metal-NOM precipitate particles (Gregor *et al.*, 1997).

The proposed mechanistic model for the impact of NOM on floc formation for the coagulation conditions of this study is shown in Figure 6.37. In the absence of NOM, ferric hydroxide precipitates form the primary particles of the floc, as the NOM concentration increases the formation rate of Fe-NOM precipitates increases and the adsorption of NOM becomes important. At low NOM adsorption, the organic molecules only partially cover the primary particles which allows the organics to attach to individual particles at numerous sites. At higher NOM concentration increased adsorption of organics prevents this and lateral interactions between organics results in large portions of the molecules sticking out into the solution forming micelle type structures (Amal *et al.*, 1992). During further agglomeration, primary particles at low NOM concentration are able to get close enough to each other for strong attractive bonds in the primary energy zone of classical DLVO electrostatic double layer theory to become important, whilst at high NOM concentration particles are unable to get close enough to one another such that steric repulsive forces dominate the floc properties. Instead, bridging bonds form between NOM molecules and primary particles which give rise to more open and fragile flocs. There was a strong demonstration that this occurred from the observed shift in fractal dimension when the organic component in the floc was high. In addition, microscope images provide qualitative indication that a more open structure was produced. Further supporting evidence for this comes from similar observations seen using solid latex and iron oxide particles in the presence of humic acid (Amal *et al.*, 1992; Walker and Bob, 2001). This mechanistic model provides an explanation for changes in floc size, settling rate and breakage behaviour of Fe-NOM flocs, which will be subsequently discussed.

Across the range of ratios investigated a reduction of 500 μm was seen in the median floc sizes. This initial steady state size is a good indicator of floc strength. Both Bache and Papavasilopoulos (2003) and Yukselen and Gregory (2004) show that for a given shear condition, increased floc size is as a result of increased floc strength. This is because floc size is a balance between growth and breakage (Biggs and Lant, 2000). Larger flocs must therefore have higher resistance to breakage in order to reach a bigger size and should therefore be considered stronger. This gives flocs with a low Fe:DOC ratio (0 and 0.33) increased floc strength over the 0.75, 1 and 1.75 ratio flocs, which in turn are stronger than the 3.8 ratio flocs. The decrease in floc size with increasing organic content in the floc agrees with work carried out on sludge characterization by Dulin and Knocke (1989). The average diameter of flocs in the sludge increased from 10 to 15μm when comparing flocs formed from raw water with and without humic acid. It should be noted that the floc sizes were one order of magnitude smaller than the flocs investigated in this study, however it was the microfloc (small clusters of primary particles) size that was being measured from the sludge samples. It therefore seems that the bridging bonds formed when the organic fraction in the floc is high gives rise to weak flocs, whereas when the particles are able to get close to one another at low carbon fraction strong and large flocs form.

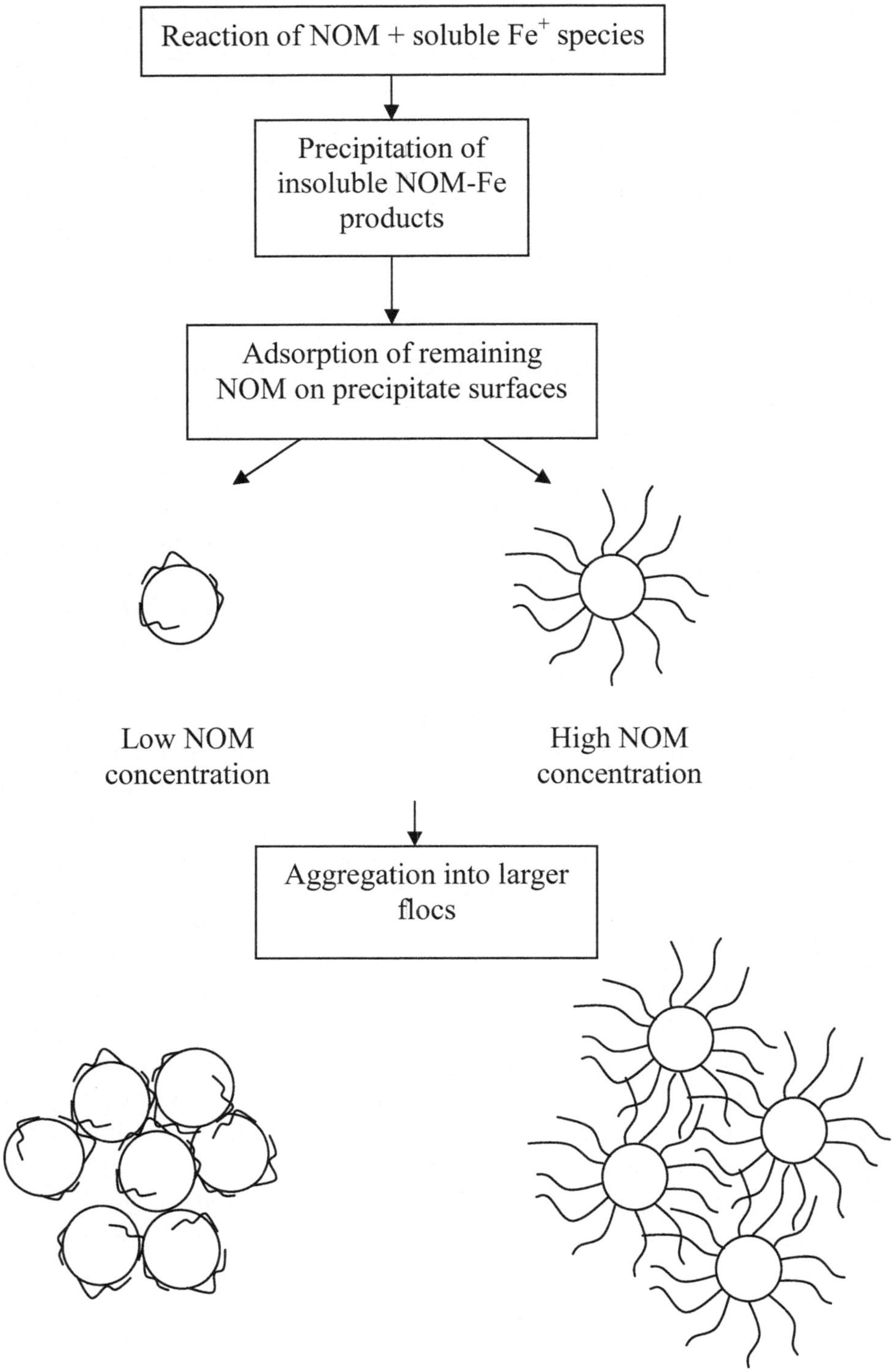

Figure 6.37 The proposed model for the formation of flocs at low and high Fe:DOC ratios.

An interesting point of note was that there was no clear relationship between zeta potential and the floc properties and it therefore seemed that the increased separation distance given by the previously described steric impact of adsorbed NOM seemed to dominate the resulting floc structure.

The impact of NOM on floc settling appeared to have two effects. Firstly there was the described change in fractal dimension and secondly there was a change in settling rate. The range of settling rates for the flocs under investigation show comparatively lower settling rates to turbidity dominated flocs but similar values to activated sludge flocs. The settling range of the flocs in this study and activated sludge flocs is 100-2000 μm s^{-1} (Li and Ganczarczyk, 1986; Wu and Lee, 1998) whilst similar sized kaolin flocs have a range up to 10,000 μm s^{-1} (Wu *et al.*, 2002). This is obviously as a result of turbidity particles in the floc giving increased floc density (kaolin dry solid density = 2600 kg m^{-3}), whilst both activated sludge flocs and NOM flocs have a high organic content which is of a lower density (humic acid dry solid density = 1750 kg m^{-3} (Hossain and Bache, 1991). The density of the Fe-organic material in the flocs is difficult to accurately establish, particularly the organic component which consists of a heterogeneous mix of molecules. However, an indication of the reduction in density with increasing organic fraction for the Fe-NOM flocs can be found from floc effective density. The effective density of flocs can be estimated from:

$$v = \frac{gd^2 \rho_e}{18 \mu \rho_w} \tag{6.21}$$

where v is the terminal settling velocity, g is gravitational acceleration, d is the particle diameter, μ is the water viscosity, ρ_e is the floc effective density and ρ_w is the density of the water. The assumption that Stokes law can apply to irregularly shaped and porous flocs is open to some debate, but the application to flocs is widespread (Bache et al., 1991; Gregory, 1997; Tang et al., 2002) and provides a useful relative comparison. The effective density in the absence of NOM (0 ratio) was ~1.8 kg m^{-3}, whilst this decreased to ~1.0 kg m^{-3} as the carbon fraction increased to a ratio of 3.8. Complex analysis of the dry solids in alum-humic flocs confirms this density reduction in other metal-organic floc systems where an approximation of 2000 kg m^{-3} has been cited as the density of the solid in alum-humic flocs (Hossain and Bache, 1991), with the density of aluminium hydroxide being around 2500 kg m^{-3} (Bache, 2004).

The floc breakage behaviour was more difficult to interpret given the spread and cross over of the data. However, when the breakage profiles were separated into those flocs that showed significant differences in fractal dimension from one another some more clear trends were seen. Smaller flocs generally do not break to the same extent as larger flocs given that larger flocs become more affected by the microscale eddies that are attributed to floc breakage. At small sizes, flocs are more likely to become entrained within eddies rather than be broken by them (Parker *et al.*, 1972). This explains why large flocs break more extensively and shows how the breakage with increasing shear gives a confusing indication of how strong flocs are. However, the underlying broken floc size of the high fractal flocs were not as small as the low fractal and high organic content flocs. Given the previous definition of floc strength being related to increased floc size for a given shear, it therefore appears that the lower fractal flocs have reduced strength, confirming that increased carbon in the floc reduces floc strength. The exception to this being at high shear of 150 and 200 rpm when the broken floc sizes tend to converge on a common median value of between 200-300 μm as the high applied energy is greater than the internal bonding of even the strongest flocs. This is likely to be a convergence on a similar sized microfloc particle.

The operational significance of the results suggests that the ratio of Fe:DOC is very important for the resulting floc structure. In the past, the main focus of attention has been getting the coagulant dose correct for NOM removal. However, the present study has identified that adding coagulant at a low dose can still give rise to good NOM removals, but give comparatively poor floc structure in terms of reduced floc size, settling rate, fractal dimension and strength. The broken floc size is of particular importance during solid-liquid separation because if small increases in shear during WTW unit processes give rise to floc breakage then downstream systems will be challenged by smaller particles. Reduced floc sizes give rise to lower sedimentation rates (Aguilar *et al.*, 2003) and reduces particle transport mechanisms during removal in filtration (Cleasby, 1999). It is the low fractal and low Fe:DOC ratios that give rise to this problem. For the Fe based coagulant used in this study, it appears that coagulant doses at and above a 1:1 ratio gives rise to improved floc structure.

Fractions

Raw water characterisation and coagulation

Long term monitoring of the raw water revealed the concentration of dissolved organics to vary between 6 and 14 mg.L^{-1} with the lower values typically measured in the late spring and summer months and the higher values in the autumn and winter periods. Characterisation of the waters at three separate periods revealed that the raw water was predominately hydrophobic in nature such that the hydrophobic fractions (HAF and FAF) made up 61 %, 79 % and 72 % of the measured material in sampled collected during June 2000, November 2000 and April 2002 respectively (Table 6.8). The observed variations in fractional make up were principally seen in the FAF fraction which varied between 42 and 61 % over the same sample periods. Similar results have been reported by Goslan (2003) over a 7 month period where the absolute concentration of each of the fractions remained relatively stable with the exception of the FAF fraction which varied considerably and matched the variations measured in the total DOC. Removal data on the individual fractions based on coagulation jar testing revealed that both the hydrophobic fractions were easily removed whereas the hydrophilic fractions (HPIA and HPINA) were less effectively treated. This was most apparent with the HPINA fraction with removal efficiencies between 14.5 % and 32.5 % across the sample period (Table 6.8).

Table 6.8
Comparison of raw water make up and removal

Fraction	June 2000		November 2000		April 2002	
	% by mass	Removal (%)	% by mass	Removal (%)	% by mass	Removal (%)
HAF	19	91.6	18	97.7	20	99.3
FAF	42	89.3	61	88.9	52	92.5
HPIA	8	95.6	8	71.7	12	69.7
HPINA	31	17.2	13	14.5	16	32.5
Total DOC (mg.L^{-1})	8.1	71.2	10.2	80	7.5	

Analysis of the grade efficiency profiles of the organic removals by size exclusion chromatography revealed that the main residual in all the fractions was in the sub 3000 Da size range indicating that the coagulation process is far more effective at removing the larger MW compounds. Currently work is being conducted to target the residual organic material by utilising adsorbent materials that are more effective over that specific size range but less effective at higher MW ranges. Initial results indicate that combinations of coagulants and magnetic ion exchange resins expand the overall range of MW sizes that can be removed and offer some potential in the future for a targeted solution to the problem.

Size and Strength

Comparison of the physical characteristics of the flocs formed by coagulating the individual fractions has indicated some clear differences between the types of flocs formed with each fraction. Whilst the floc growth profiles of the different fractions all followed a similar pattern with a rapid change for the first 4 minutes followed by a steady state phase (Figure 6.38) the growth rate and final floc size for the FAF flocs was significantly smaller than the flocs formed with the other fractions (statistically significant at a 95% confidence level). To illustrate, final floc sizes of 818 μm, 932 μm. 949 μm and 1020 μm were recorded with respect to the 50%ile size for the FAF, HAF, HPIA and HPNIA fractions respectively (Figure 6.38).

Greater observed differences were recorded with respect to the settling rate experiments where mean settling rates of similar sized flocs varied by more than 400% when comparing flocs formed from the different fractions. To illustrate, the settling rate of a 400 μm floc varied between 0.7 $m.h^{-1}$ (FAF) to 2.9 $m.h^{-1}$ (HAF) (Figure 6.38). This compared to mean settling rates at the same equivalent size of 1.3 $m.s^{-1}$ for raw water and 1.8 and 2.5 $m.s^{-1}$ for the HPIA and HPINA fractions respectively.

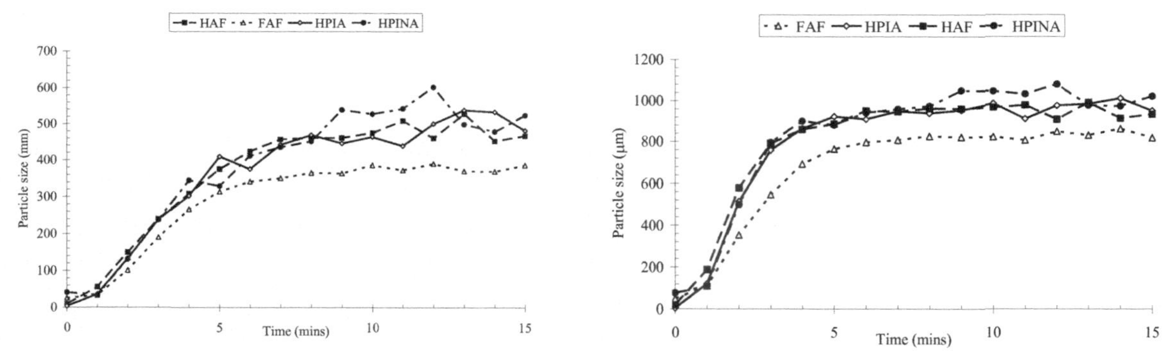

(a) 10%ile (b) 50%ile

Figure 6.38 Floc growth profiles for the different fractions dosed at the optimum concentration with an iron based coagulant.

Settling behaviour

The gradient of the log-log plots of settling rate vs size (Figure 6.39) can be interpreted as a structural parameter (settling rate fractal dimension - d_f). The slopes indicated that the most compact flocs were formed by coagulating the FAF (d_f = 2.24) and the most open flocs were formed by coagulating the HAF (d_f = 1.78). Comparison with flocs formed from the raw water

($d_f = 2.1$) revealed a structured that was characterised as being less compact that those formed by the FAF and more compact than those formed with any of the other fractions.

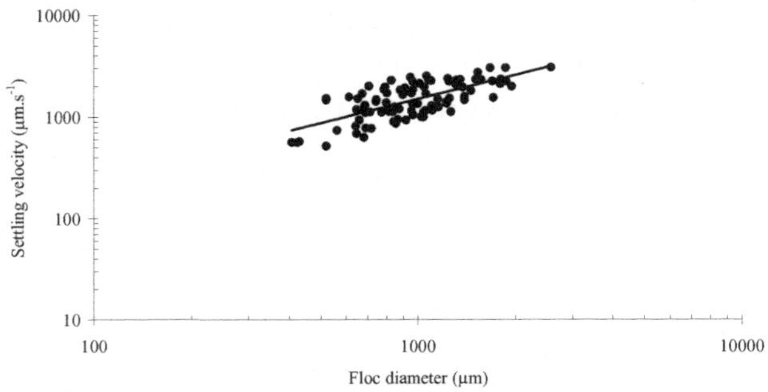

(a) HAF

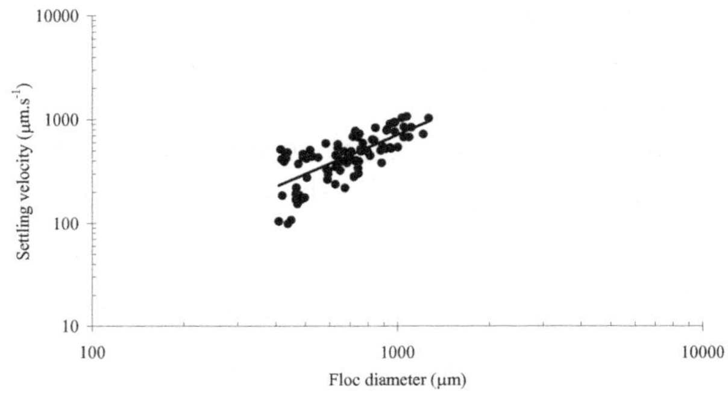

(b) FAF

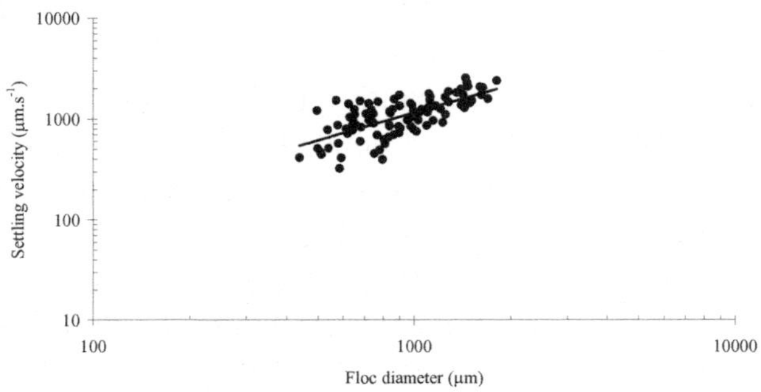

(c) HPIA

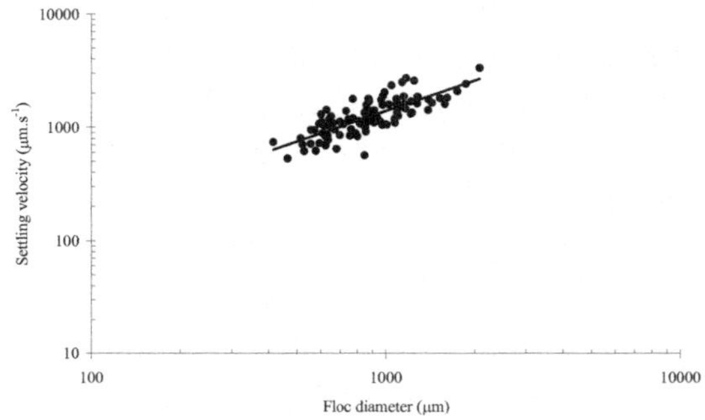

(d) HPINA

Figure 6.39 Settling rate profiles for the individual fractions generated from the April 2002 water coagulated with iron at the optimum dose

Microscopy

Further resolution of the flocs generated from the individual fractions revealed a difference in structural composition between the hydrophobic and hydrophilic fractions. The hydrophobic fractions were characterised by solid internal structures with a relatively smooth edge structure. This compared to the hydrophilic fractions which were characterised by more open internal structures with extensive edge morphology. The HPIA fractions revealed the greatest degree of internal holes and the most extensive amount of edge dendrites (Figure 6.40). The HPINA fraction showed a unique feature in that white spots were observed in all the images taken of the flocs formed with that fraction. XRD analysis of the spots revealed them to be concentrated areas of calcium although there appeared to be no specific pattern to the location of the calcium spots in relation to the structural characteristics of the flocs. Spot analysis within individual flocs with a 1 µm diameter spot zone has enabled the position of highly concentrated areas of iron to be identified in the flocs compared to highly concentrated areas of organics. However at the current time no significant difference has been identified with respect to the relative position of concentrated iron spots within the flocs and the structural characteristics of the fraction from which the flocs were formed.

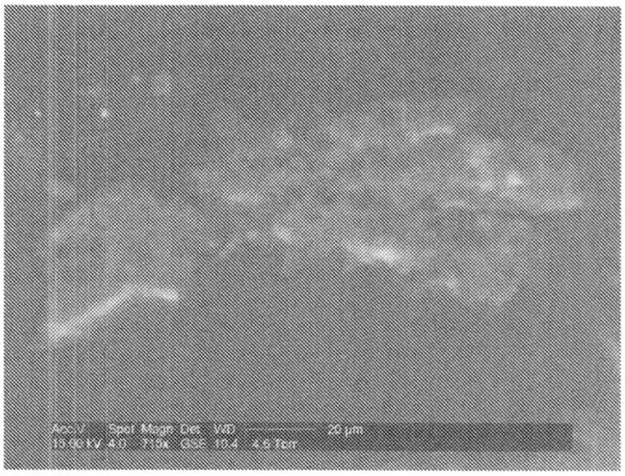

(a) HAF

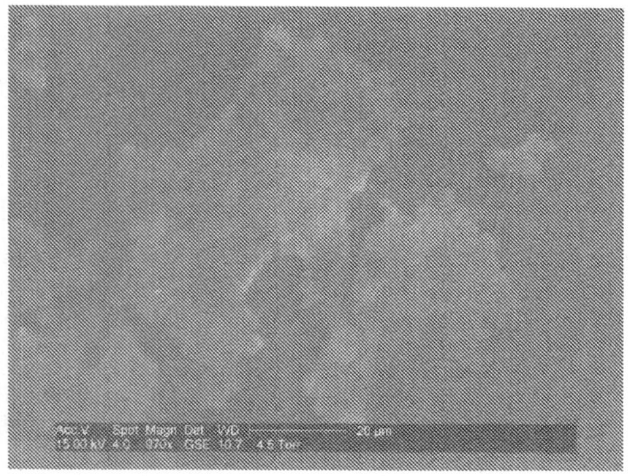

(b) FAF

(c) HPIA

(d) HPINA

Figure 6.40 Example ESEM images from the individual fractions generated from the April 2002 water coagulated with iron at the optimum dose

Rheological measurements

The effects of the fractions of WAN_t on the shear yield stress behaviour of AKP-30 alumina are presented in Figure 6.41. The fractions are designated as hydrophobic (H) and hydrophilic (HPIA and HPNIA) (see Chapter 3). WAN_{HPNIA} was high in ionic strength and desalted by gentle centrifugation and exchange of the supernatant for Milli-Q water, as outlined in the experimental section. Shear yield stress results for WAN_{HPNIA} are shown, both before and after desalting. All fractions of WANt shift the τ_{ymax} to lower pH. The size of the shifts for WAN fractions follow the order: HPNIA > HPIA > H. The shift caused by WANt lies between WAN_H and WAN_{HPIA} but closer to the latter. These results are consistent with the shifts in iep obtained from electrokinetic data.

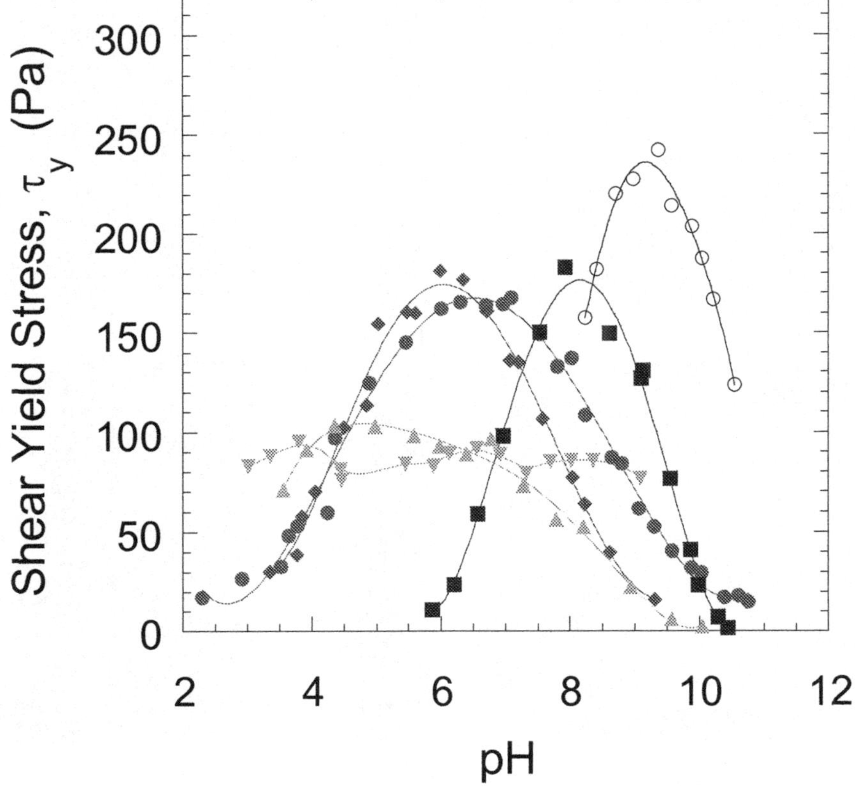

Figure 6.41 Shear yield stress of AKP-30 with Wanneroo NOM (WANt) and fractions as a function of pH. The alumina volume fraction was 0.250 and the electrolyte concentration was 0.010 M NaCl (except for the WAN_{HPNIA} suspensions which had no added electrolyte). The concentration of NOM fraction added was 0.050 % w/w TOC in each case. ○, no additive; ●, WAN_t; ■, WAN_H; ◆, WAN_{HPIA}; ▼, WAN_{HPNIA}; ▲, WAN_{HPNIA*}.

In addition, all fractions reduce the magnitude of τ_{ymax}. The extent of the reduction for WAN fractions follows the order: WAN_{HPNIA} >> WAN_{HPIA} = WAN_H and indicates that the thickness of the steric barrier follows the same order. WAN_t is marginally more effective in reducing the magnitude of τ_{ymax} than WAN_{HPIA} and WAN_H, possibly due to greater contributions from the hydrophilic fraction. It also produces a slightly broader curve which corresponds with the lower slope in the zeta potential versus pH curves for WAN_t.

The data for WAN_{HPNIA} prior to desalting displays a very flat shear yield stress response. Removal of the excess salt results in an extremely broad non-symmetrical peak. The broadness in the peak corresponds to the flat slope shown in the electrokinetic results. The maximum shear yield stress after desalting is not significantly different from the shear yield stress obtained prior to desalting. At high ionic strength the double layer is reduced substantially and any electrostatic interactions are shielded. In this situation, van der Waals forces dominate the interaction across most of the pH range. The fact that the maximum shear yield stress is only marginally affected by the change in ionic strength supports the premise that the dominant force operating in this region is van der Waals and that there is only slight electrostatic contribution to the interaction at that point. Above pH 8 there are substantial reductions in the shear yield stress after desalting due to removal of ionic screening and an increase in the electrostatic repulsion.

The effects of the fractions of HV_t on the shear yield stress behaviour of AKP-30 alumina are presented in Figure 6.42. All fractions of HV_t shift the τ_{ymax} to lower pH. The size of the shifts for HV fractions follow the order: HPNIA > HPIA > H. The shift caused by HV_t is greater than that caused by HV_{HPIA}. This is possibly due to the marginally higher dose of HV_t. It also indicates that this NOM is more highly charged, probably due to the large proportion of hydrophilic fraction present in HV_t.

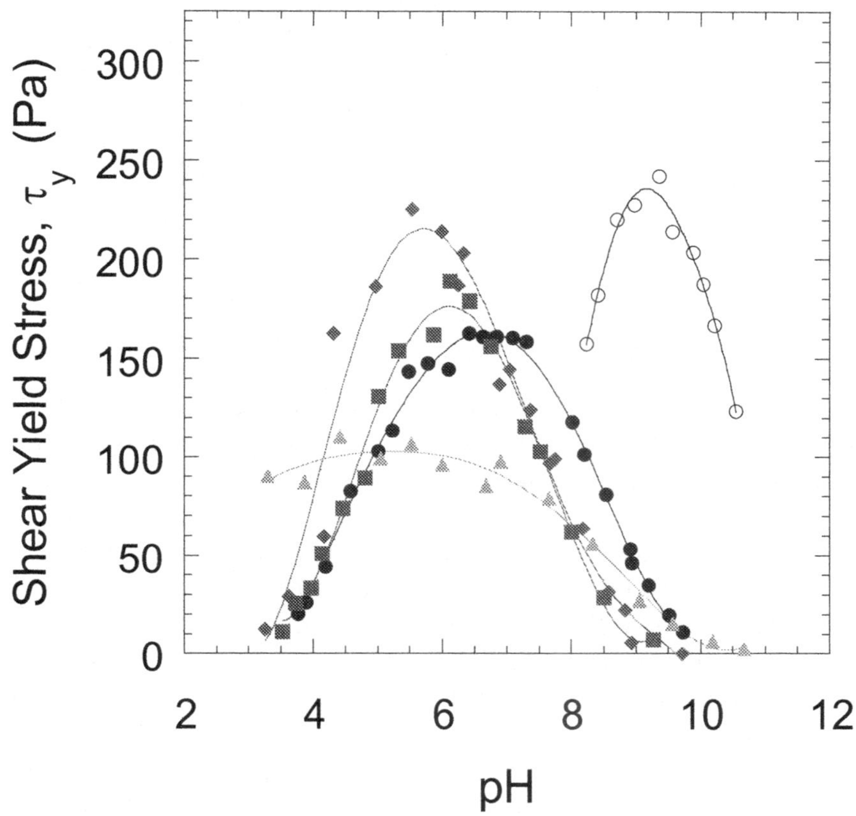

Figure 6.42 Shear yield stress of AKP-30 in the presence of Hope Valley NOM (HV_t) and fractions as a function of pH. The alumina volume fraction was 0.250 and the electrolyte concentration was 0.010 M NaCl (except for the HV_{HPNIA} suspension which had no added electrolyte). The concentration of NOM fraction added was 0.050 % w/w TOC in each case, except for HV_t, $[HV_t]$ = 0.056 % w/w TOC. ○, no additive; ●, HV_t; ■, HV_H; ◆, HV_{HPIA}; ▲, HV_{HPNIA}.

Further rheological testing was completed with other acid fractions from Wannerroo in which $Fe(OH)_3$-coated alumina was used as the substrate (Figure 6.43). These results show that the unfractionated Wanneroo (WAN_t) and both the hydrophobic (WAN_H) and hydrophilic (WAN_{HPNIA}) fractions all lower the iep and reduce the maximum shear yield stress. The hydrophobic fraction, WAN_H, shows less reduction of the maximum shear yield stress than either WAN_t or WAN_{HPNIA}, each of which appear to cause reductions in the maximum yield stress of approximately the same magnitude.

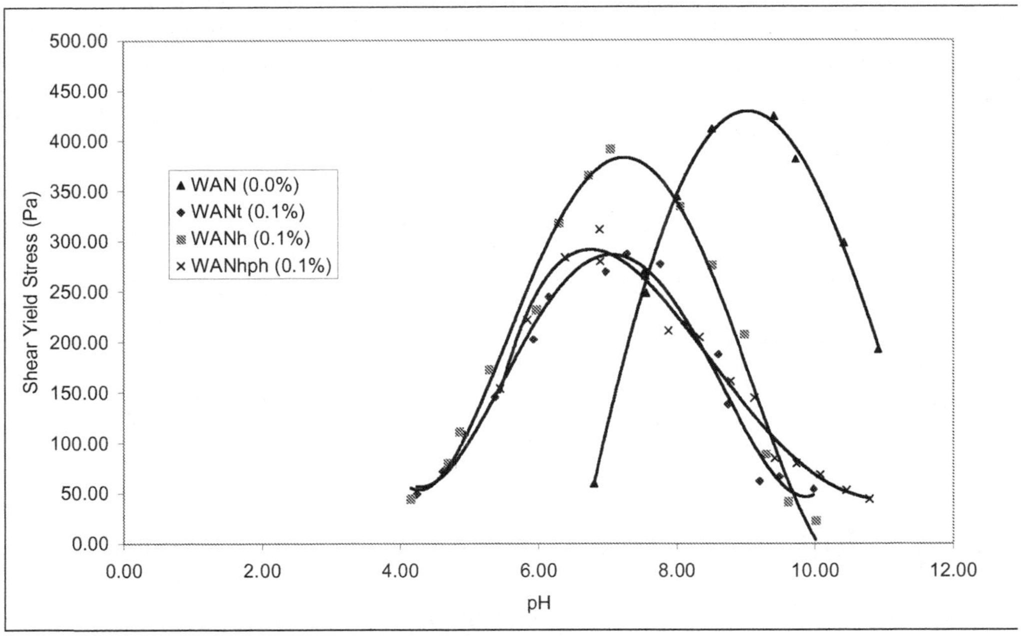

Figure 6.43 Shear yield stress of ferric hydroxide coated AKP-30 in the presence of Wanneroo NOM (WAN_t) and fractions as a function of pH. The solids volume fraction was 0.250 and the electrolyte concentration was 0.010 M NaCl.

Figure 6.44 shows data for some of the analogous test series completed on the Hope Valley NOM fractions using $Fe(OH)_3$-coated alumina. These results reveal that both the un-fractionated Hope Valley NOM (HV_t) and the Hope Valley hydrophilic fraction (HV_{HPNIA}) substantially lower the maximum shear yield stress and shift the iep to lower pH values. In contrast to the results obtained using the alumina substrate, however, the hydrophilic fraction is not substantially different from the un-fractionated material. The hydrophilic fraction in this case caused a slightly greater reduction in shear yield stress than HV_t and a marginally narrower peak overall. Whilst the trends in these results with regard to the different fractions are similar to those obtained for the Wanneroo NOM samples, as shown in Figure 6.3, there are significant variations which may simply reflect the differences in composition between these two water samples

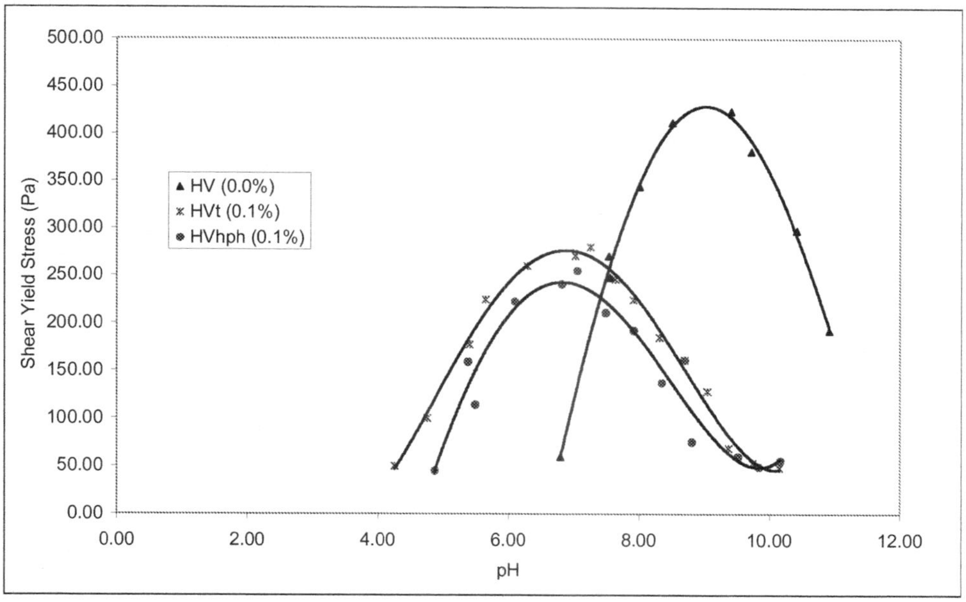

Figure 6.44 Shear yield stress behaviour of AKP-30 alumina in the presence of 7.6 w% ferric hydroxide, HVt and HVhph as a function of pH.

The effect of UK NOM samples

Figures 6.45 and 6.46 present the results and the yield stress data for the $Fe(OH)_3$–coated silica. The addition of UK NOM caused a shift in the suspension iep to lower pH values. Of the fractions tested, the UK_{HPIA} promoted the largest shift of the iep down to approximately pH 5.2 (from the base value of approximately pH 6.2 for the plain 2.5 w% $Fe(OH)_3$–coated silica), while the UK_{FAF} showed the least effect shifting the iep to approximately 5.6, with the iep for the UK_{HAF} dosed sample sitting somewhere in-between.

The shear yield stress results of Figure 6.46 do not show very good overlap or agreement with these electrokinetic results in that the UK_{HAF} appears to show a maximum in its yield stress curve at a significantly lower pH than either the UK_{FAF} or UK_{HPIA}. The latter two fractions display fairly similar curves. The UK_{HAF} also appears to possess a greater overall maximum than either of the other two fractions which possibly suggests that the dispersive strength of this fraction is somewhat less while the other two fractions promote greater repulsion/dispersion within the system and, subsequently, their lower yield stress at the same volume fraction is somewhat lower. The addition of the Wannaroo humic acid fraction with the same substrate and volume fraction reveal that this NOM material has a more severe effect on both the iep, shifting it further to the left than the HPIA, and the maximum yield stress, lowering it significantly.

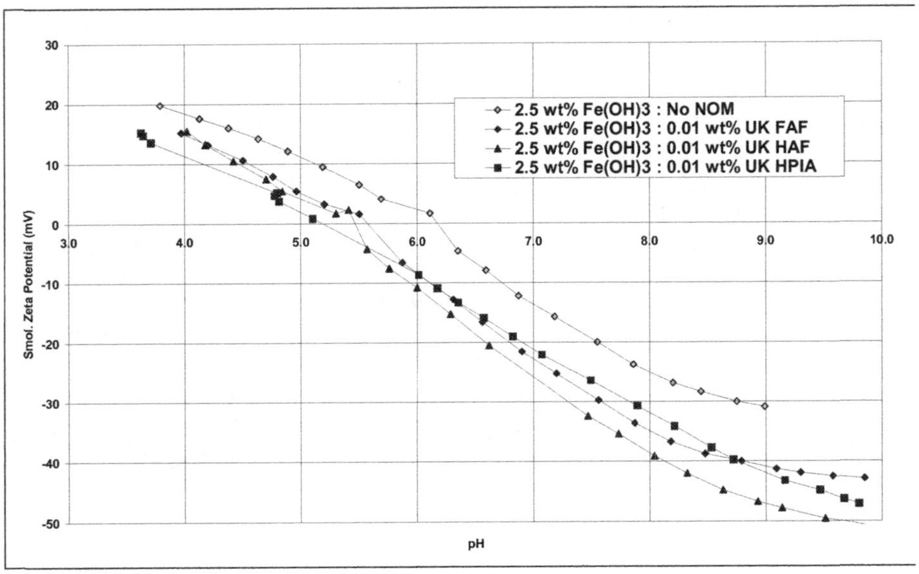

Figure 6.45 Comparison of the zeta potential behaviour of 2.5 w% $Fe(OH)_3$-coated silica as a function of pH with different UK NOM fractions. The total solids volume fraction is 0.02 in all cases.

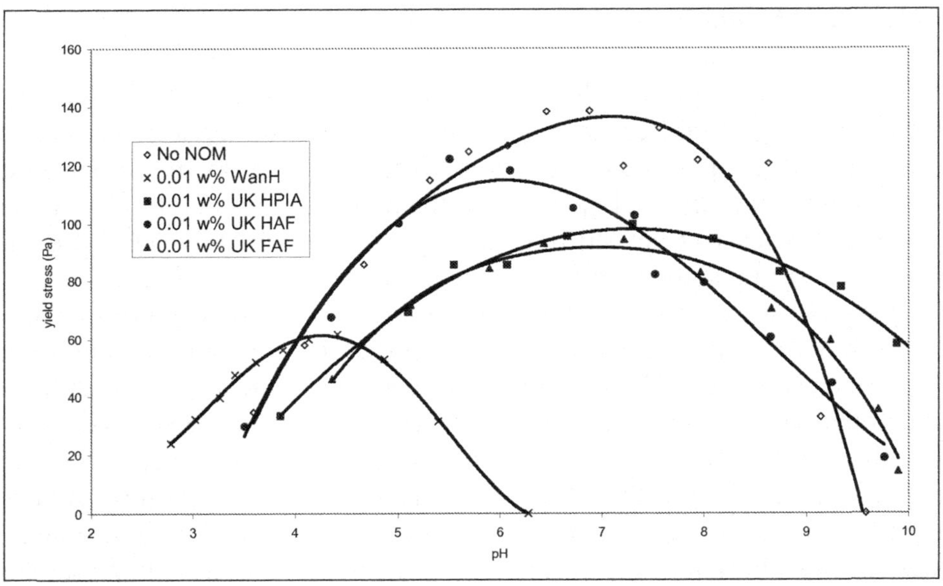

Figure 6.46 Comparing of the shear yield stress behaviour of 2.5 % $Fe(OH)_3$-coated silica as a function of pH with different UK NOM fractions. The volume fraction of solids in each case was 0.380.

The reasons for our selection of silica as the substrate and for the decision to use 2.5 w % $Fe(OH)_3$ have been outlined (section on methods). Unfortunately the charge density of this adsorbent is too low to permit the use of electroacoustic measurements. This is believed to be due to a combination of particle size and charge effects but in practice the surface potentials obtained are very close to zero and adversely affected by experimental error considerations. Hence no further electrokinetic measurements were made. To facilitate study of the effect of any additive on the rheological behaviour of this solid, it is important that the absolute magnitude of the shear yield stress maximum is sufficient that changes are outside the experimental error in the measurements. For this reason it was decided to use as high a value of the total solids volume as practically possible, which was determined to be 38% w/w.

The shear yield curve for this system (in the absence of any NOM) is shown in Figure 6.47, together with data for the same system at lower solids volumes. Whilst the pH values for the optimum shear yield stress remain similar for the three experiments there is a significant variation in the magnitude of the peak value which justifies our decision to use only the highest total solids volumes.

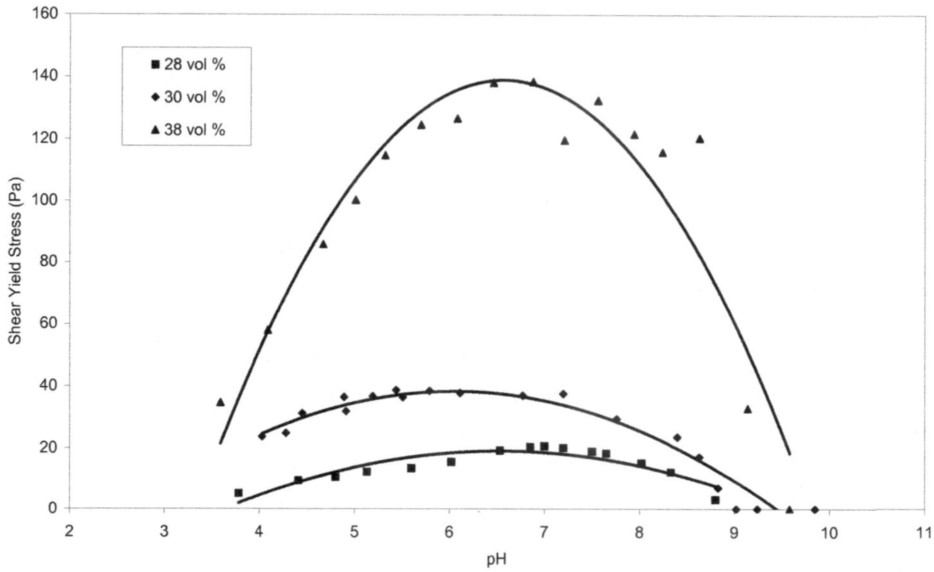

Figure 6.47 Shear Yield Stress behaviour of $Fe(OH)_3$ coated silica as a function of pH and total solids volumes in the absence of added NOM.

Figure 6.48 provides a summary of the effect that adding small amounts of each of the three UK NOM fractions has on the shear yield curves. With each fraction there was a significant decrease in the magnitude of the peak height indicating a reduction in the particle-particle interactions and there was a decrease in the IEP suggesting the uptake of anionic species at the surface.

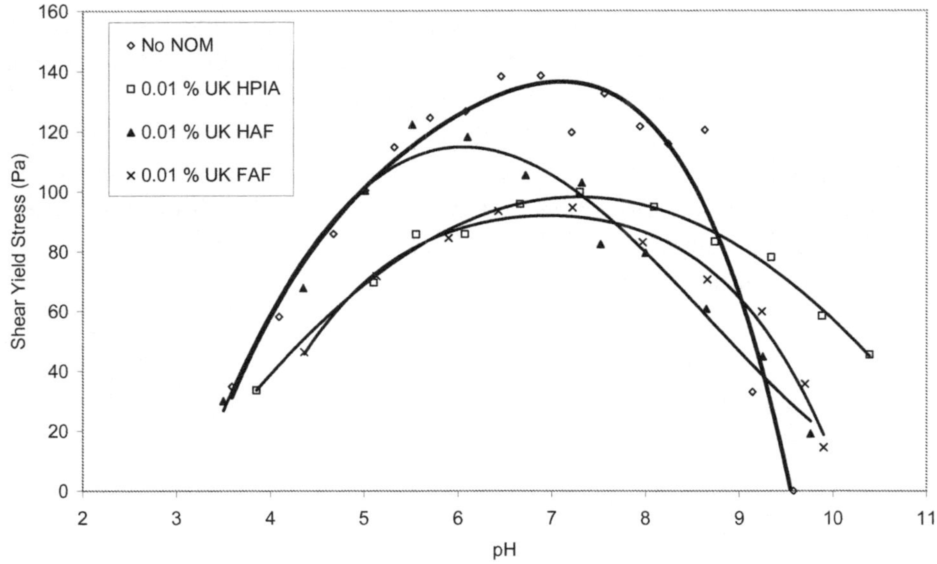

Figure 6.48 Shear yield stress of 2.5% $Fe(OH)_3$ coated silica (at 38% total solids volume) in the presence of 0.01 w% different UK NOM fractions.

In subsequent experiments with the same UK_{FAF} sample as above, the protocol was varied to allow for the organics to be contacted with adsorbent at higher pH values, ie the contact pH was approximately 10 rather than 7, and the results are shown in Figure 6.49. Whilst the peak heights are similar to the previous data there is evidence to suggest that the pH of contact is an important variable in the system, and is worthy of further examination.

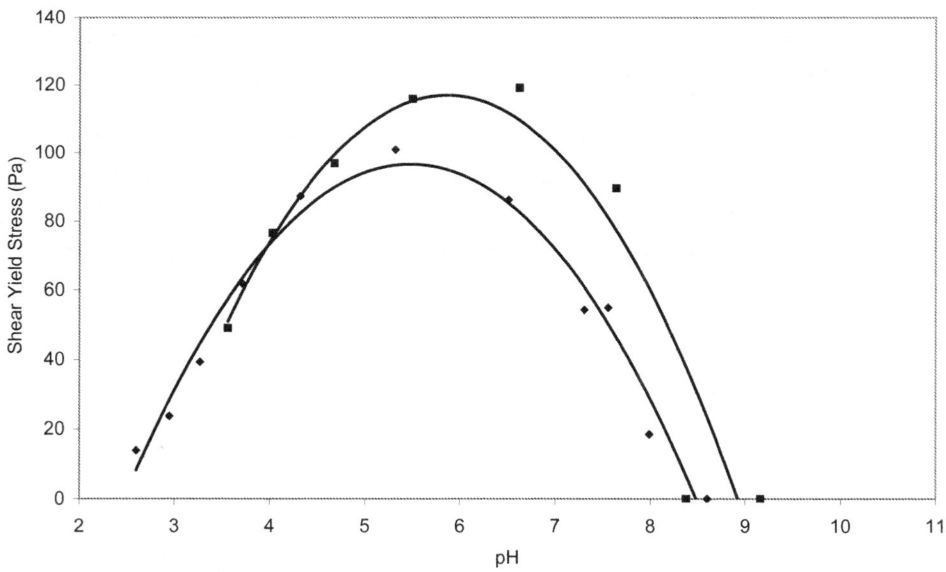

Figure 6.49 Shear yield stress of 2.5% $Fe(OH)_3$ coated silica (at 38% total solids volume) in the presence of 0.01 w% UK FAF.

Kinetic Aspects

The observed hystersis suggests possible kinetic effects which were further investigated in terms of the pH of contact, the duration of the interval between adding the NOM and beginning the rheology measurements (contact time) and the fraction of the NOM samples. The source of the NOM used was a Hope Valley MIEX regeneration effluent (Figure 6.50), similar to the Wanneroo sample but it should be remembered that the Hope Valley water is a surface water rather than a groundwater like Wanneroo. In the following experiments the rheometer runs were commenced as quickly as possible after the NOM was added but in practice the centrifugation stage required to eliminate the associated electrolytes takes a minimum of nearly 2 hours and it is only after the conductivity has been lowered to an acceptable value that the pH is raised from 7 to 10 and the rheological measurements are begun. The data shown are from duplicate runs and provide an indication of the extent of the experimental error.

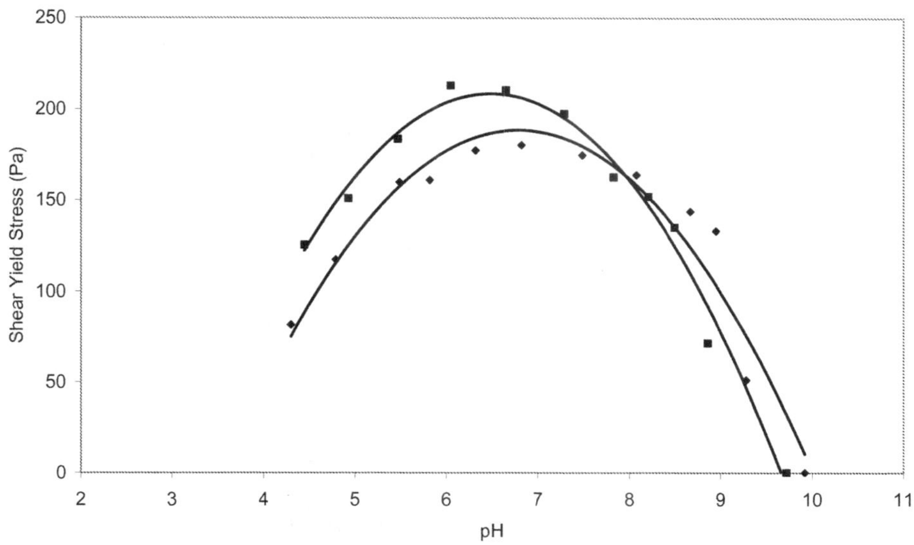

Figure 6.50 Shear yield stress of 2.5% $Fe(OH)_3$ coated silica (at 38% total solids volume) in the presence of 0.01 w% Hope Valley humic fraction.

In Figure 6.51 the curves illustrate the effect of delaying the onset of the rheological measurements by varying amounts of time. Even a slight delay (2 hours) in beginning the rheological run has a noticeable effect on the shear yield stress and when the sample is left at pH 10 for a number of days there is a significant change in both the magnitude of the peak height and the pH at which this occurred. The objective of the next series of experiments was to determine whether the kinetic effect seen after storage at pH 10 was also evident at other pH values. As shown in Figure 6.52 at pH 7 there was little difference between the data for the two experiments, one carried out immediately after centrifugation (which is identical for the same curve at pH 10, there being no additional time at pH 7) and the other left overnight at pH 7. For the final experiment, the pH of the slurry after centrifugation was decreased to pH 4 and then immediately raised to pH 10 to start the rheology measurements. The effect of this minimal contact time at low pH was sufficient to dramatically lower the magnitude of the yield stress recorded which in turn precluded the possibility that further time in this pH regime would have an observable effect. In summary there was no evidence of any kinetic effect at either of the two lower pH values.

The results essentially show that the interaction between NOM and the selected adsorbent is quite rapid at low and neutral pH values. There is no experimental evidence for any kinetic effect; whatever happens at the interface happens almost immediately. In fact any contact of NOM with the surface at pH 4 or 7 seems to result in a loaded adsorbent which does not undergo any subsequent change as the pH is raised. If the configuration of NOM on the surface later changes as a result of solution changes, it is only very slowly. In contrast there is an obvious kinetic effect at pH 10 which persists over several days and results in a significant decrease in both the iep and the peak height.

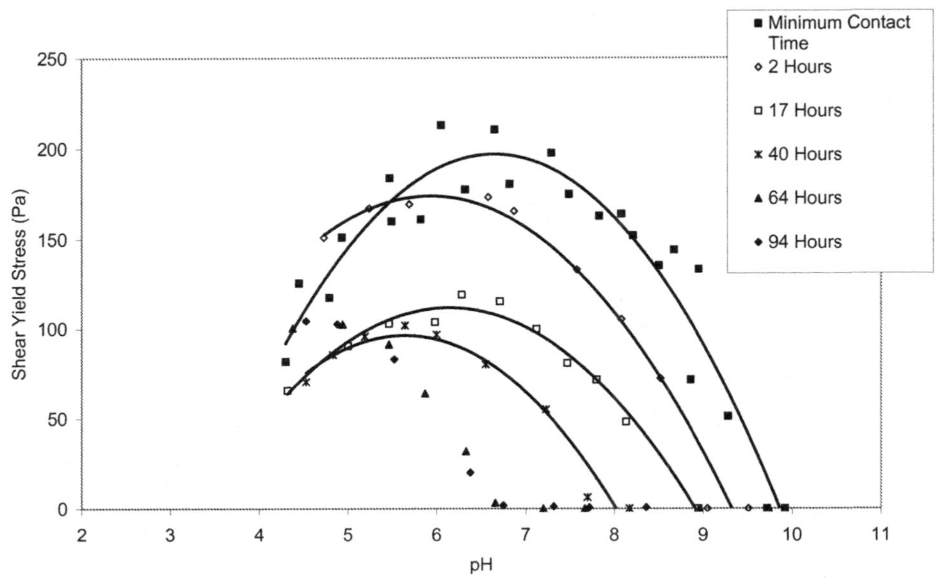

Figure 6.51 Shear yield stress of 2.5% $Fe(OH)_3$ coated silica (at 38% total solids volume) in the presence of 0.01 w% Hope Valley humic acid fraction in which the contact time has been varied.

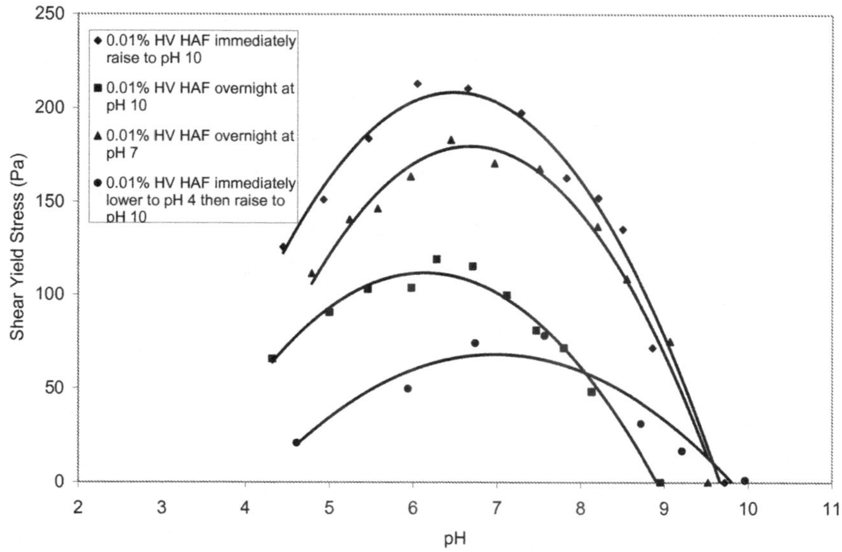

Figure 6.52 Shear yield stress of 2.5% $Fe(OH)_3$ coated silica (at 38% total solids volume) in the presence of 0.01 w% Hope Valley humic acid fraction in which both the contact pH and the contact time have been varied.

Dynamic yield stress measurements

An example of data from a dynamic yield stress measurement using the ARES rheometer is shown in Figure 6.53. The data for alumina, when compared to the data from vane yield stress experiments was found to be in agreement. An example of a data set as a function of pH for alumina is shown in Figure 6.54.

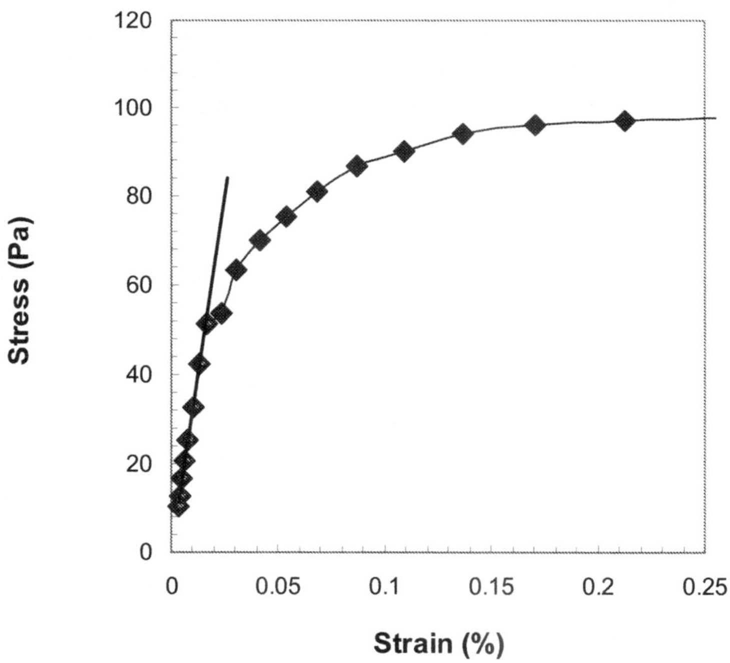

Figure 6.53 Example of the measurement of the shear yield stress of a coagulated alumina suspension as measured on the ARES constant-strain rheometer.

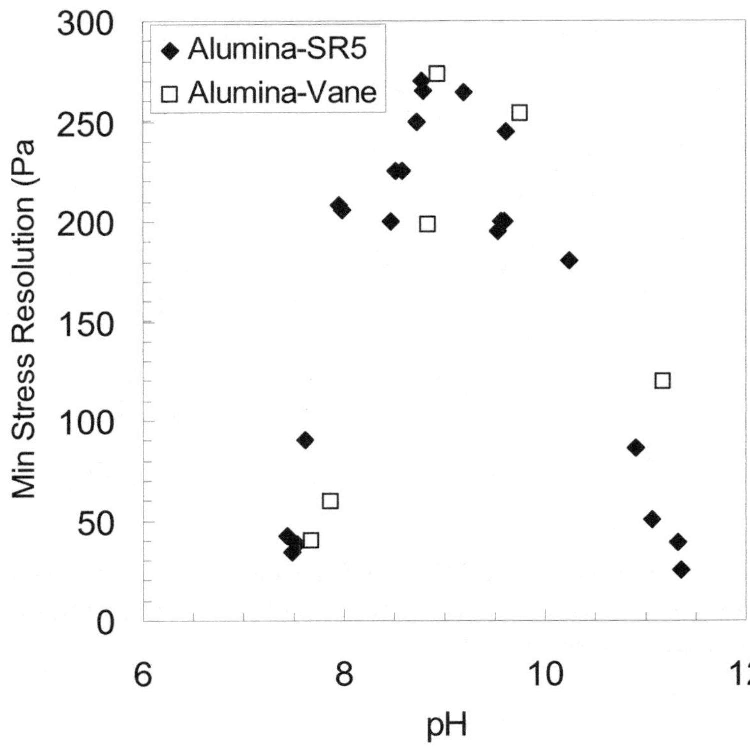

Figure 6.54 Comparison of shear yield stress measurement using the vane and constant strain rheometers.

The results for the addition of a range of NOM fractions at a constant concentration (150 mg l^{-1}) is shown in Figure 6.55. All fractions are observed to decrease the yield stress and cause a shift in the point of the maximum yield stress to a slightly lower pH. The largest shift was caused by the FAF component and there are indications that the molecule is desorbing at pHs above 9.

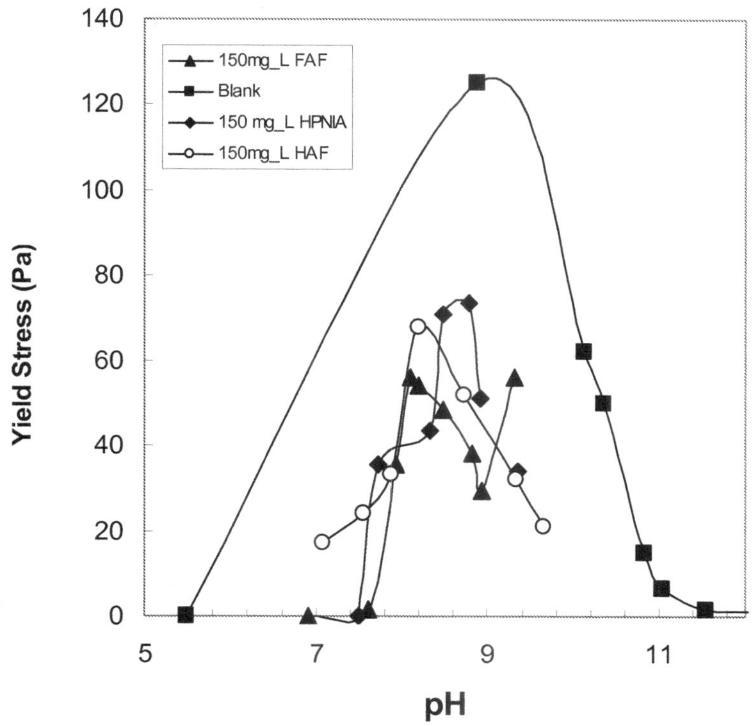

Figure 6.55 Shear yield stress behaviour of alumina suspensions in the presence of UK NOM samples.

There is only a slight shift in the point of maximum yield stress for the HPINA fraction and the HAF component is somewhere between the HPINA and the FAF. The differences between the fractions is not dramatic although a systematic difference is observed. The data indicates that electrostatic interactions are dominant in the adsorption process. This observation is based on the fact that strongly adsorbed inner sphere complexes produce significant adsorption and subsequent changes in the yield stress resulting in the complete displacement of the yield stress curve to lower pH's and complete retention of the parabolic shape. Such behaviour is not observed here. The iso-electric point for the alumina used here is approximately pH 9.2.

The next phase of the work looked at the nature of the failure. It has been shown over the past year that rheometry can be used to determine the mode of adsorption of a molecular species, be it flat on the surface or in loops and chains (only possible for larger molecules) from the type of yielding failure, be it ductile or brittle. As intimated earlier, experiments to determine the nature of the failure assume that the reference point is brittle failure. In this context, alumina has already been shown to yield in this manner at or about the iso-electric point of the solid. Using this as a reference, the yielding behaviour was examined on the SR-5 constant stress rheometer for the additives of interest. An example for a coagulated alumina suspension (left graph) and a flocculated alumina suspension (right graph) is shown in Figure 6.56. The failure in the coagulated case is brittle and in the presence of a high MW adsorbed species, ductile.

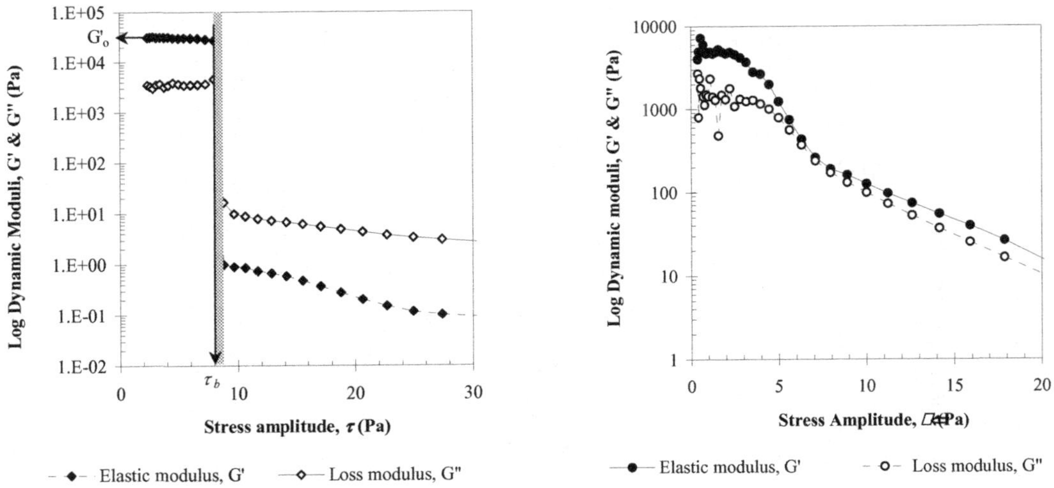

Figure 6.56 Elastic and loss modulus for the failure of coagulated (left) and flocculated (right) alumina suspensions.

The data for alumina in the presence of the UK NOM fractions is shown in Figure 6.57. A single pH was chosen (the point of maximum yield stress) and the additive addition rate was 150 mg l^{-1} in each case. The data shows that the linear viscoelastic regime (horizontal data prior to stress failure) is easy to observe for all samples except the HPINA. In the HPINA case the failure was more ductile although nothing like the case for a high MW polymeric additive. The data is consistent therefore with the concept that the HAF and FAF samples adsorb in a relatively flat configuration on the surface with electrostatics as the main driving force for adsorption. The data for HPINA is consistent with the adsorption being not as 'flat' as for the other components. The difference is subtle but consistent in both the lower shift in the point of maximum yield stress for the HPINA and the nature of the failure. There is nothing in the data to suggest that the fractions are behaving in an anomalous fashion.

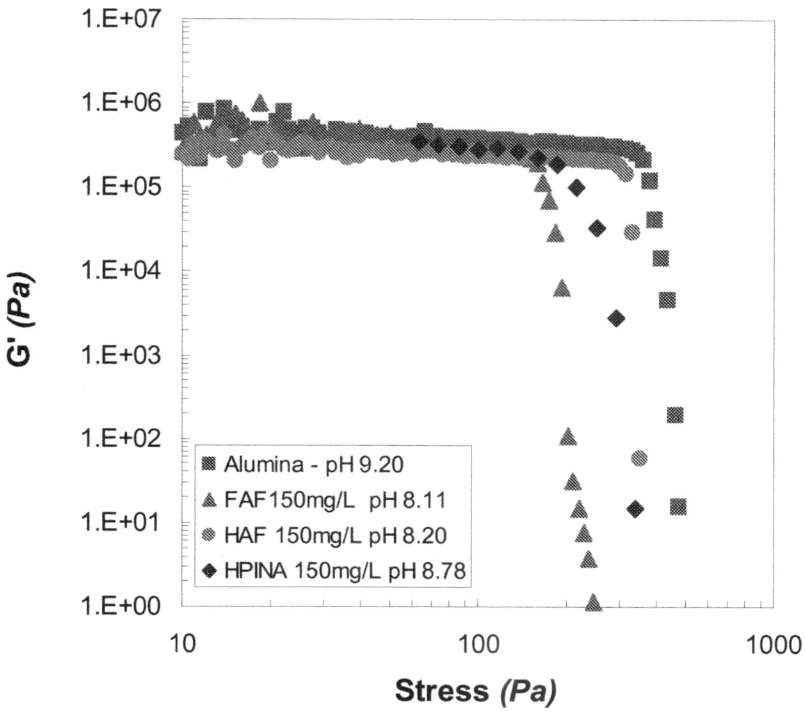

Figure 6.57 Elastic modulus as a function of stress for alumina in the presence of UK NOM fractions.

Adhesion force measurements

Adhesive force measurements were conducted on both dried and wet samples of the fractions. Samples were adsorbed onto either mica or the non specific ion exchange resins used during the fractionation process. In both cases the adhesive force between a silicon nitride tip and the dried material followed the sequence HPIA>>FAF>HAF. To illustrate, in the case of the ion exchange resins the average adhesive force between the tip and the sample was 156, 50 and 40 nN for the HPIA, FAF and HAF respectively (Figure 6.58). Comparison to the base substrate revealed a statistical difference to the adsorbed layers indicating the measured adhesive forces were due to the deposited material.

Equivalent measurement conducted on wet samples contained in low alkalinity water revealed the same basic tend but with which much lower actual forces measured (Figure 6.59). The adhesive force of the tip to the hydrophilic material was twice the force measured for the HAF and FAF fractions. The results of the two studies indicate that greater bonding strengths are generated with the hydrophilic components of the water.

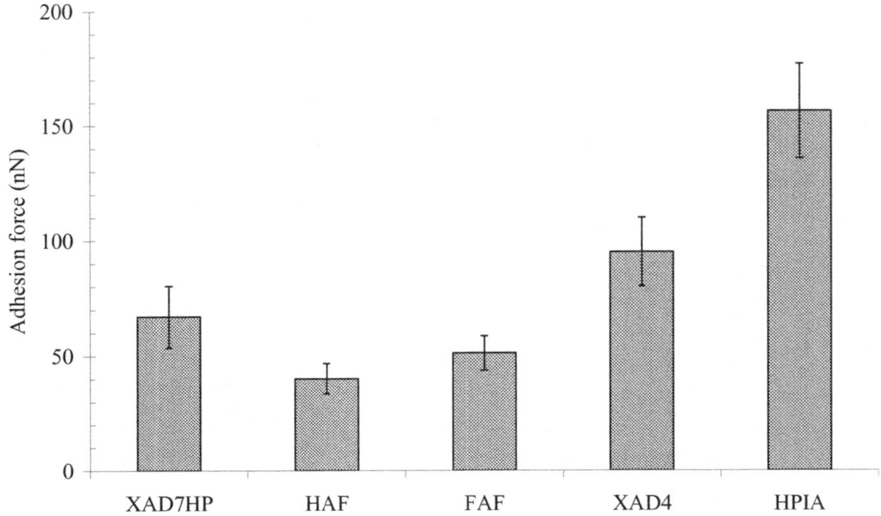

Figure 6.58 Adhesive force measurements of the dried fractions adsorbed onto the non specific ion exchange resins used during the fraction process measured in air

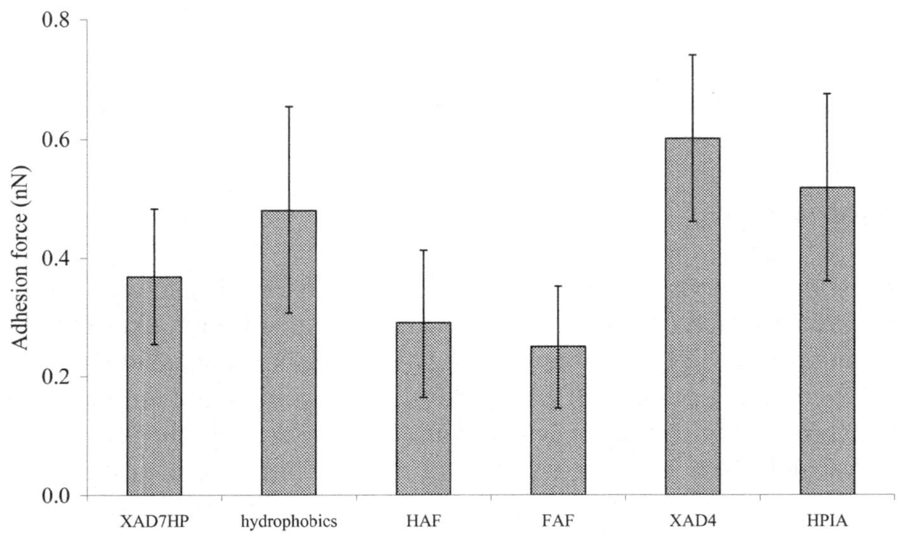

Figure 6.59 Adhesive force measurements of the wet fractions adsorbed onto the non specific ion exchange resins used during the fraction process measured in water of an equivalent alkalinity to the raw water.

Discussion

The reduction in τ_{max} for all NOM concentrations and fraction types indicates that the adsorption of NOM onto a substrate surface produces a steric barrier. The result of the steric barrier is to increase the separation between the contacting surfaces reducing the shear required to yield (i.e. weakening the strength of the floc). In most cases the observed changes in both pH and magnitude of τ_{max} occur in the sequence HPNIA>HPIA>HAF+FAF. Consequently the hydrophilic fractions are seen to have the greatest impact on the response of the surfaces. An explanation of the observed results is provided by a combination of the rheological and atomic force microscopy work. The dynamic rheological tests indicate that the hydrophilic fractions tend to orientate outwards as opposed to the hydrophobic fractions which orientate along the surface of the substrate. This is corroborated by the atomic force microscope measurements which reveal a stepped release curve during the hydrophilic force measurements characteristic of molecules orientated perpendicular to the surface of the substrate. Further, as the AFM data demonstrates that the hydrophilic fraction generates the largest adhesive bonding force, orientation of the molecules would seem to be more important than bonding strength.

The shift in both magnitude of τ_{max} and the pH at which it occurs is different for each of the source waters and substrates investigated. Similarly, differences in MW distribution and charge density are observed in the fractions collected at different times and from different locations (Chapter 4). The implication of such findings is that the distribution of organic molecules contained within each fraction varies from one sample to the next. Ultimately, this reduces the confidence with which the data can be interpreted to more global considerations of the importance of NOM on floc properties.

Comparison between the all the different diagnostic tests reveals an unclear picture of the significance of each fraction on the resultant floc properties. In particular the hydrophobic fractions (HAF + FAF) produce the extremes in each of the floc structural properties observed (size, compaction and settling rate). Such a view reflects the importance of the interactions between the different fractions during coagulation and suggests that small concentration of one fraction within the overall raw water may be responsible for the observed behaviour. Ultimately, the difficulties discussed suggest that continued work on floc diagnostics of fractionated water is unwise. More appropriate emphasis should be placed on assessing how marginal changes in make up in the water change overall properties.

CHAPTER 7
CONCLUSIONS AND FUTURE RESEARCH

CONCLUSIONS

Characterisation and Treatment

- Raw water characteristics vary considerably as a function of both location and time of sampling. The variation is observed in terms of UV adsorbance at 254 nm, DOC, hydrophobic/hydrophilic fraction composition and charge density. Consequently, stoichiometric dosing of coagulant based on flow changes does not provide a suitable control strategy for maintaining performance of the coagulation stage of the treatment process.
- The hydrophobic content of the waters, which is the main source of coagulant demand in the water, is easily removed by coagulation with inorganic salts. Measurement of the hydrophobic content and its charge density provides a relatively easy method to determine the likely charge and coagulant demand of any water. The hydrophobic content can also be conviently approximated through UV_{254} monitoring.
- The hydrophilic content and in particular the polar NOM is poorly removed by coagulation. Measurement of the polar NOM concentration of a raw water provides a good indicator of the achievable organic residual by coagulation.
- More detailed fractionation to further split the water appears to be of limited practical purpose due to the variation observed within individual fractions.
- Ferric based coagulants were able to routinely remove 0.5 $mg.L^{-1}$ DOC more than alum for all the waters tested.
- Comparison of moorland and snow melt waters indicated that during the period of elevated organics the characteristics of the changes were considerably different. In the case of the snow melt waters the changes in organic concentration occurred principally due to high MW hydrophobic fraction whereas in the moorland case the increase was observed across both high and low MW species and hydrophobic and hydrophilic material. The consequence of this is that in the case of elevated organic loading due to snow melt the achievable residual should not change much but the coagulant dose should increase considerably. Whereas a high achievable residual is likely to occur when the organic concentration increase on moorland sites.
- Good coagulation performance was shown to follow zeta potential measurements such that as long as the zeta potential was low the residual concentration was minimized. The exact range of suitable zeta potentials appears, in part, site specific but usually falls within the range $-10<\zeta<+3$ mV. The relationship between zeta potential and residual concentration held at all the sites studied and during all the sample periods.
- It is unimportant what combination of dose and pH is used as long as the zeta potential is within the required range. This provides a more robust method for assessing the suitability of a dose strategy for any given works.

- The exact value of the residual organic concentration achievable by coagulation is not dependent on the zeta potential but on the amount of polar NOM. The zeta potential merely informs the operator whether the plant is operating within its achievable optimum.
- Similar ranges of suitable zeta potential values exist for sedimentation, floatation and filtration. Whilst the exact range is system and site specific as a rule of thumb values between -5 and -10 mV appear to be widely suitable.
- The ion exchange resin MIEX® was shown to be effective in reducing UV_{254}, DOC and THMs for a range of raw waters. HPSEC data showed that the resin was particulary effective in removing mid to low MW organics and when used in combination with a low dose of coagulant proved to be an effective process for treating elevated organic content waters.

Floc diagnostics

- NOM dominated flocs grow to initially larger sizes, are more open structures and settle much more slowly than comparative turbidity dominated flocs. However, the NOM flocs also appear to have a higher resistance to exposure to increased shear levels.
- The overall ranking of the physical properties in relation to the choice of coagulant suggested the following sequence: MIEX® + Ferric > pDADMAC > Ferric > Ferric + pDADMAC > alum.
- pDADMAC addition as a floc aid did not significantly improve the physical properties of flocs when compared to floc coagulated with ferric alone. This was particularly evident in terms of the strength and regrowth potential of the systems.
- pDADMAC as a primary coagulant removes significant less DOC than the other options but did produce large flocs.
- All of the systems (apart from the combination of MIEX® and coagulant) had a very poor regrowth potential after exposure to high shear levels suggesting that avoidance of high shear zone is very important during the treatment of NOM dominated waters.
- MIEX® pretreatment appears to remove the mid to low molecular weight material prior to coagulation and this has a significant impact on the physical properties of the flocs. The MIEX® pretreated system generated the largest, most compact fastest settling flocs of all the system tested and visual observation suggests the flocs closely resembled the appearance of the turbidity dominated flocs. Perhaps the most important improvement was in terms of the flocs ability for regrowth after an exposure to high shear.
- Major changes in the physical properties of the flocs were observed as the mass ratio of NOM to coagulant (as M^{3+}) changed from 0 to 1:3.8 Fe:DOC. Coagulant doses at and slightly above a ratio of 1:1 appear to give rise to the most improved floc structures in terms of size, strength and settling rate. The difference appears to be due to the relative influence of polymer bridging on static repulsion which occurs as the relative concentration of NOM increases within the floc.
- Investigations into linking NOM character to floc properties revealed that molecular orientation of the adsorbed material is more important than the

adhesive bonding strength of the adsorbed layer. Consequently the hydrophilic, low MW material has the greatest influence in reducing floc properties as they tend to orientate perpendicular to the surface of the floc as opposed to hydrophobic material which orientates across the surface of the floc. The observed results are consistent with the improvements observed for the MIEX pretreated flocs as that process preferentially removes mid to low MW materials which tend to be the hydrophilic component within the floc.

- The variability of the material obtained during the fractionation process limits the usefulness of extended research in the properties of the individual fractions. This is especially evident in terms of the physical properties of the flocs formed by coagulation

FUTURE RESEARCH

1. The major limitation of the work was the absence of a truly difficult to treat water throughout the period of the project. Consequently the work should be repeated on a "first flush" water once it has been identified as difficult to treat by the operators.
2. The importance of NOM fractionation has been one of the central themes of this report but the technique remains a relatively slow and cumbersome activity. The need for a rapid and simple method of fractionation to determine the polar hydrophilic concentration would provide a straightforward method to determine the achievable organics residual by coagulation. This should be developed with a works operator in mind and be a single disposable cartridge.
3. The use of enhanced monitoring by fractionation, HPSEC and charge density measurement should be tested for its potential to rank available source waters in terms of treatability and coagulant demand. This could act as a decision support tool for supporting source water selection.
4. Alternative polymers should be tested to determine whether the results with respect to pDADMAC transfer to other polymers.
5. The link between zeta potential and treatment should be extended to include the impact on physical properties to determine the relative importance of organic concentration and net charge on the physical properties of the floc
6. The impact of residual particles after floc breakage should be studied to link together the floc diagnostic work directly to down stream treatment operation.
7. Treatment options for polar hydrophilic organic material should be studied such as oxidation and advanced oxidation processes, adsorbents and membranes. These treatment processes should be considered both as centralized and decentralized options for controlling polar NOM.

CHAPTER 8
RECOMMENDATIONS FOR UTILITIES

The current project has extended previous characterisation studies to include regular monitoring of raw waters in terms of both the distribution of hydrophobic and hydrophilic material in each sample and its charge density. The findings of this project show that the DOC, polarity and charge density of the waters vary considerably as both a function of location (source) and time of year or season. The knowledge gained has to be tempered with an understanding that:

(1) Two waters of the same DOC concentration can have very different coagulant demands and achievable residuals.

(2) For a single raw water the coagulant demand and achievable residual DOC can change throughout the year and this will not be observed through monitoring of bulk parameter such as colour and DOC.

(3) Stoichiometric dosing in proportion to flow does not provide a suitable means of coagulation control during periods of changing raw water quality characteristics (absolute or fractions).

(4) Improved dose control can be achieved by linking dose to the hydrophobic content of the water. This can be achieved through relatively simple monitoring of the UV_{254} which has been shown to give a reasonable correlation with hydrophobic material.

(5) HPSEC is a rapid and useful tool in optimising coagulation and related processes in relation to MW profile. It can provide insights into treatability as in general low MW material is difficult to remove by coagulation.

(6) Measurement of the hydrophobic/hydrophilic balance (by fractionation) and the charge density of raw waters provides a useful and reliable decision support tool for understanding coagulation of elevated organic waters. The hydrophobic proportion indicates the amount of material that can be removed by coagulation and allows the prediction of coagulant demand (though a charge density measurement). The hydrophilic proportion (specifically the polar fraction (HPNIA)) indicates the amount of material left in the water after coagulation. This residual organic concentration is the most reactive readily forming THM's (and HAA's) in the presence of disinfectants.

(7) A simple and rapid fractionation test should be developed using a combination of XAD4 and XAD8 resins. Measurement of raw water outlet (post filtration) DOC should be sufficient to give the fractionation information required.

Comparison of snow melt and moorland waters during periods of elevated organic concentrations reveals that in the case of the snow melt waters the increased DOC occurs principally in the hydrophobic, large MW bands. This indicates that the elevated organic levels should not significantly alter the achievable organic residual but should increase the coagulant demand considerably. In contrast, in the case of the moorland waters the increase in DOC occurs across all MWs and in both hydrophobic and hydrophilic components. Consequently, the impact is that the coagulant demand does not increase as significantly but the achievable organic residual is higher such that alternative treatment processes may become necessary. Throughout the course of the project neither the snow melt nor moorland source waters experienced elevated organic levels that where difficult to treat and as such the project was unable to conduct the research described during periods of operational difficulty. This remains an area of work that still needs to be performed in order to confirm the conclusion and recommendations outlined in this report.

The coagulation studies at both bench and pilot scale clearly indicated that the use of zeta potential is an extremely useful operational indicator in relation to the coagulation of coloured waters. The overall findings of importance are that zeta potential provides a universal indicator of the appropriateness of the treatment strategy irrespective of the chemical used or the system investigated. Of specific interest is:

(1) An operational range rather than a single specific value existed for each system within which the residual concentration was minimised and independent of the specific value for the zeta potential

(2) The range changed slightly depending on the specific system under investigation but always remained within the limits of -10 to +3 mV.

(3) The exact dose and pH was unimportant as long as the value of zeta potential was within the required range for maximum organics removal

(4) For the water tested here ferric coagulants removed around 0.5 $mg.L^{-1}$ more DOC than aluminium coagulants. PDADMAC gave no benefit over inorganic coagulants.

(5) Zeta potential operational ranges exist for flotation, sedimentation and filtration processes in combination with coagulation for organics removal. The exact range for each process is slightly different but all indicate that operating the coagulation/flocculation process to achieve a zeta potential of around -10mV should provide a useful guide value for optimal treatment of the downstream processes in terms of charge effects.

Consequently, zeta potential provides a straightforward means of control of the coagulation chemistry which requires semi regular rather than continuous monitoring.

The literature review identified ion exchange and coagulation as the two most suitable treatment option for elevated organic content waters. Here we have looked at the MIEX® process in comparison and in combination with inorganic coagulants. The combination of MIEX® with a reduced dose of ferric coagulant produced a very robust treatment process which performed well over a number of seasons and raw waters. If the residual organics after coagulation are high and likely to cause a THM problem then MIEX® as a pretreatment prior to coagulation should be considered as an alternative option.

(1) HPSEC in combination with DOC measurements should be used to optimize MIEX® and coagulant doses. The results presented here show that MIEX® selectively removes mid to low molecular weight organics and the coagulant the high molecular weight and highly charged organics. This leads to lower residual THMs even when equivalent DOCs removals are achieved

(2) MIEX® pilot plants trials should be used to support the findings of jar tests. There are a number of ongoing AwwaRF funded pilot plant trials on MIEX® which should be refereed to for more detail.

The current investigation has measured the physical properties of the flocculated materials in terms of size, strength, settling rate and shape factor for a range of treatment options and in terms of key operational concerns. The overall findings of importance are that NOM dominated flocs are very different in character to turbidity dominated flocs and that controlling the properties of the flocs can be achieved through selection of the appropriate coagulant/pre treatment or changing the ratio of DOC: coagulant. Specifically:

(1) NOM dominated flocs are much bigger, more open and settle much more slowly than turbidity dominated flocs although they appear to have a higher resistance to shear.
(2) Ferric flocs are bigger, stronger and settle quicker than Alum flocs
(3) pDADMAC does not significantly improve the physical properties of NOM dominated flocs compared to coagulant only systems.
(4) MIEX pretreatment generates the biggest, strongest and the most rapidly settling flocs of all the systems studied.
(5) Regrowth after exposure to high shear levels was poor for all systems studied except for the MIEX pretreatment system.
(6) As the ratio of NOM to coagulant increase the size, strength and settling rate of the flocs decreases.

Investigations linking the character of NOM to the physical properties of the flocs have revealed that molecular orientation rather than adhesive bonding strength is the more important parameter. Consequently, it is the hydrophilic, low MW, material that reduces the physical properties of the flocs. This supports the observation that the use of MIEX pretreatment produces the best flocs properties as it is known to remove mid to low MW components. The variability observed in the character of individual fractions has prohibited the confidence with which more specific conclusions can be drawn from the work on the factions.

The overall findings of the work suggest that improving the physical properties of flocs generated when treating organic rich waters can be achieved by:

(1) Using iron based coagulants rather than aluminium based coagulants or pDADMAC
(2) Reducing the organic to coagulant ratio to around 1:1
(3) Pre-treating the water to remove low to mid molecular weight organics using an ion exchange process

This project has shown how hydrophobic and charged organic matter can be monitored and removed. The hydrophilic (polar) organic matter is not significantly removed by either coagulation or MIEX® and required further investigation. If the hydrophilic content of your coagulated water is either too high and/or reactive then you are likely to have a THM issue. A previous AwwaRF study (*Polar NOM: Characterization, DBPs, Treatment #451*) has looked at this material in more detail. This project has found that monitoring tools such as HPSEC, hydrophobic/hydrophilic fractionation and zeta potential can be used to assess the impact individual sources/raw waters have on the treatability of the water and likely impact it will have on floc breakage and therefore could be used to optimize organics removal.

The results from the project suggest a number of checklists that can be utilised by operators when specific problems are encountered:

(1) If the treatment works fails to achieve the required organic residual concentration (Figure 8.1):

a. Check the post coagulation zeta potential to ensure the combination of dose and pH are in a suitable range. The target zeta potential is site specific but as a rule of thumb the zeta potential should not exceed -10 mV

b. If the zeta potential is in the appropriate range then fractionate the water to determine the polar hydrophilic concentration. If this level exceeds the residual organic target concentration then additional treatment processes are required in order of:

   I. ion exchange

   II. oxidation process

(2) If the physical properties of the flocs do not meet operational requirements (Figure 8.2):

a. Check the organic:coagulant mass ratio and adjust to a ratio around 1:1

b. Change the coagulant in order of preference:

   I. Ferric based coagulants

   II. Aluminium based coagulants

   III. pDADMAC

c. Pre treat the water with an ion exchange resin

(3) Where possible selection of source waters should be chosen by:

a. Deselecting waters with a high polar hydrophilic fraction if residual organic level is too high

b. Deselecting waters with a high hydrophobic fraction / charge density to reduce coagulant demand.

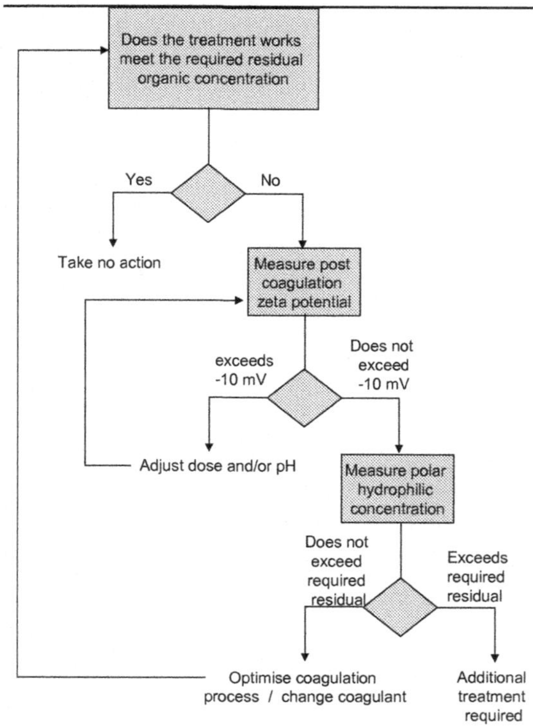

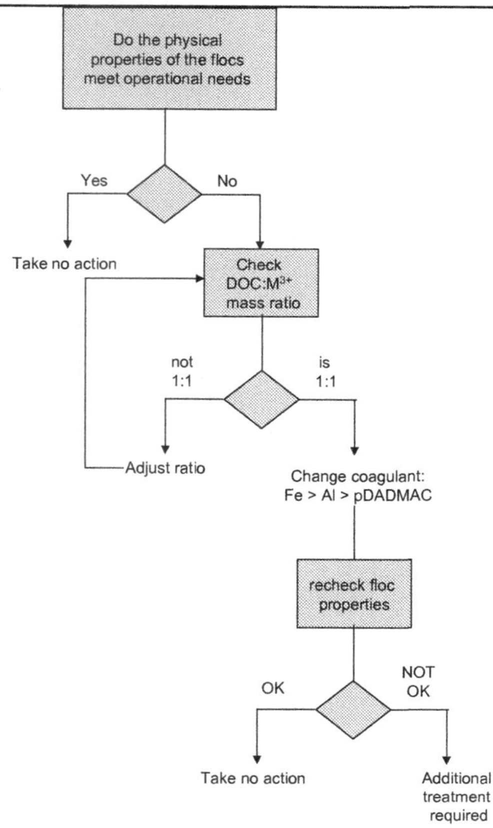

Figure 8.1 Operational check list for residual organic concentration

Figure 8.2 Operational check list for floc properties

APPENDIX
HPSEC COLUMN CALIBRATION

INTRODUCTION

With HPSEC it is known that large molecules are eluted from the column first and smaller molecules later (i.e. they have a greater retention time). However, it would be useful to know the apparent molecular weight (AMW) of each of the peaks. The sizes of dissolved organic substances are referred to as 'apparent molecular weights' since separations are calibrated with compounds of known molecular weights, not size. The AMW of each peak can be estimated using ultrafiltration (UF) membranes. The diffusive and advective transport of organics through UF membranes is influenced by a variety of factors (Logan and Jiang 1990). The most important of these factors is the quantification of the membrane rejection. The membrane rejection is the flux reduction through the membrane from concentration polarisation. A method presented by Logan and Jiang (1990) is used here to determine an experimental permeation coefficient that can be used to correct size distributions of dissolved organics by estimating the membrane rejection.

Before using UF fractions of Albert water, an attempt was made to calibrate the HPSEC column using Polystyrene Sulphonate (PSS) standards. A calibration curve was produced but it was not possible to infer any AMWs from the curve. It is believed that under or overestimation of the molecular weight (MW) occurred due to chemical interaction within the column. Some components can pass through the column more rapidly than expected due to ion exclusion or complex formation resulting in an overestimation of the component molecular weight (Knuutinen *et al.* 1988). Other components can be delayed by adsorption or electrostatic interaction with column packing leading to an underestimation of molecular weight (Miles and Brezonik 1983).

The use of UF for estimating molecular weight fractions can result in erroneous data. Broad nominal molecular weight cut-offs and solute interactions with membrane surfaces make the analysis of UF data for these solutes very difficult (Aiken 1984).

Experimental conditions

A sample of raw water taken in April 2002 from Albert Reservoir was used for the calibration. Ultrafiltrations were performed under nitrogen pressure (30 – 60 psi) in a stirred cell reactor (model 8200, Millipore, Massachusetts, USA). Membranes with MWCO values of 0.5, 1, 3, 5, 10 and 30 kDa were used (YM1 - YM30, Millipore, Massachusetts, USA, YC05, Australia). The YM membranes were made of regenerated cellulose and the YC membrane, cellulose acetate. The YM membranes were flushed with NaOH (0.1M), NaCl (0.5M) and deionised water to remove the wetting agents before use. The YC05 membrane was flushed with NaCl (1.0M) and deionised water.

110 mL samples of the raw water were filtered through each membrane at room temperature and five 20 mL samples of permeate collected (approximately 90% of the original sample). Each sample was analysed for DOC and UV absorbance at 254 nm as well as HPSEC analysis.

Molecular size distribution

It can be seen that the permeate concentrations of each sample increased in proportion to the volume of sample filtered (Figure 1). By using the data in Figure 1 and Equation 1 (Logan and Jiang 1990), the permeation coefficients were calculated (Table 1).

$$\ln C_p = \ln(pC_{r0}) + (p-1)\ln F \tag{1}$$

where C_p is the concentration of the permeate, p is the permeate coefficient, C_{r0} is the initial concentration of the sample (where $p > C_p/C_{r0}$) and F is the fraction of the original sample filtered. By plotting ln C_p against ln F, ln pC_{r0} is the y intercept and (p – 1) is the gradient (Figure 2).

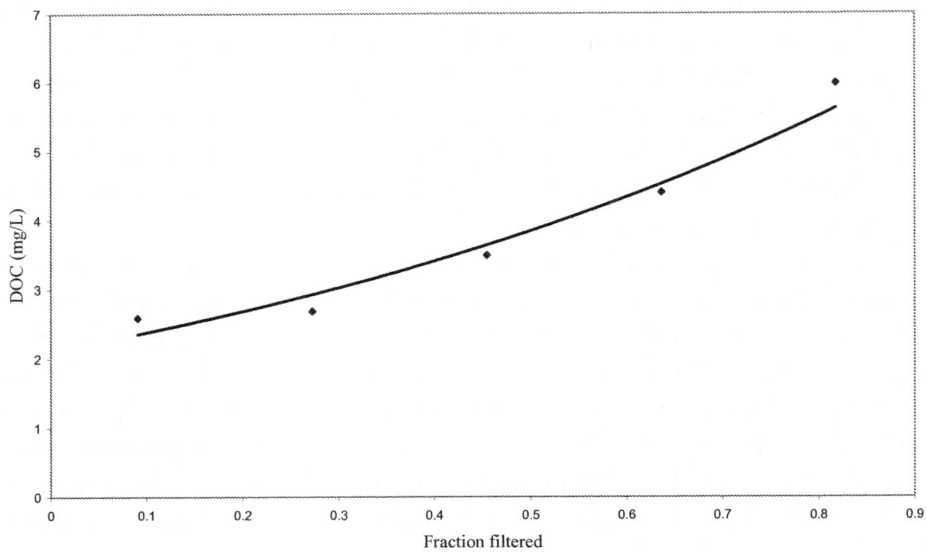

Figure 1 Permeate concentration as a function of fraction filtered

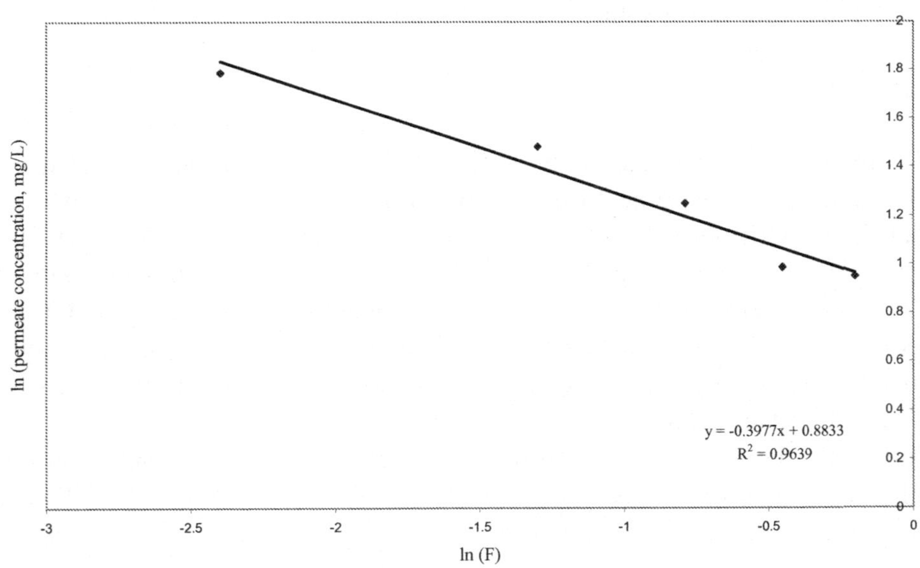

Figure 2 ln transformed data used to determine permeate coefficient

An average permeation coefficient, p, of 0.50 was calculated for all the membranes. This is similar to the expected value of p = 0.60 calculated from the manufacturer's rejection coefficient of 0.4.

Table 1
Permeate coefficients for each membrane for different natural water samples (adapted from Logan and Jiang 1990)

Membrane cut-off (kDa)	p reported (1)	p reported (2)	p calculated (3)
30	0.97	0.99	0.60
10	0.73	0.36	0.58
5	0.56	0.28	0.66
3	Nr	Nr	0.71
1	0.54	0.22	0.27
0.5	0.80	0.20	0.18

Nr – not reported
(1) – Orange County Water District (California groundwater)
(2) – Biscayne Aquifer (Florida groundwater)
(3) – Albert Raw Water April 2002

It is demonstrated that the average concentration of material measured after a batch filtration is a function of the volume collected as well as the permeation coefficient (equation 2). This is illustrated in Figure 2.

$$C_f = C_{r0}\frac{(1-F^p)}{(1-F)} \tag{2}$$

where C_f is the final concentration of the sample.

The extent that membrane rejection contributes to underestimation of material less than the membrane cut-off can be estimated using Equation 2. For example, when 90% of the sample has passed through the membrane and p = 0.60 (for the 30 kDa membrane), the collected filtrate concentration will be 61% of the true concentration of the material having a smaller size than the membrane cut-off. In general errors in C_{r0} are reduced when $p > 0.3$. A low p indicates that much of the material is about the same size as the membrane cut-off or that the material is strongly rejected by the membrane due to charge repulsion or some other effect. If $p < 0.2$ or > 0.9, the size distribution should not be adjusted. The unadjusted (white bars) and adjusted (black bars) molecular size distribution is shown in Figure 3. This adjustment is based on the UV absorbance of each sample, not the DOC concentration.

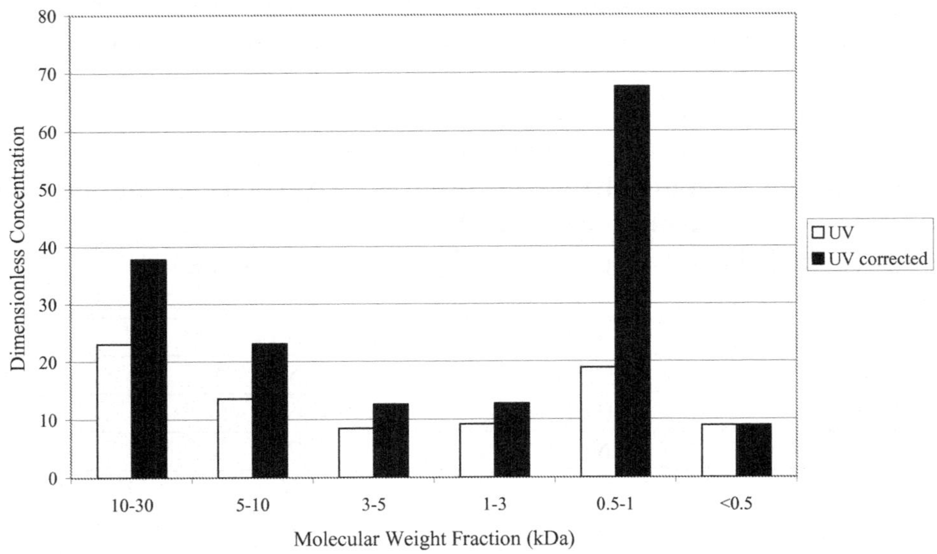

Figure 3 Molecular size distribution of Albert raw water (April 2002)

The UV absorbance measured is observed to fall as the MW falls. This has been reported in the literature. The exception to the trend is molecules with the size fraction 0.5 – 1.0 kDa which had a low p value compared with the higher MWCO membranes that were made of the same material. As stated above, a low p value indicates that the material is about the same size as the membrane cut-off or that the material is strongly rejected by the membrane. It seems unlikely that the material is strongly rejected by the membrane given that the membranes were all made from the same material with the exception of the 0.5 kDa MWCO membrane. It is assumed here that adjustment is not required for the 1 kDa MWCO membrane.

Column calibration

Each sample collected after 90% of the sample had been ultrafiltered was run on the HPSEC. Each chromatogram exhibited five distinct peaks. The retention times for these are shown in Table 2 alongside the assigned molecular weight range. These peaks corresponded to peaks observed in the raw water sample. As the membrane pore size decreased, the height of the peaks decreased. The remainder of NOM measured as the reduction of the heights of the peaks was plotted against the log MWCO value of the membranes as given by the manufacturers (Figure 4).

Table 2
Assigned molecular weight range for each peak

Peak	Retention Time (minutes)	AMW (kDa)
1	8.3 – 8.6	> 5
2	8.7 – 9.0	3.5 –5
3	9.1 – 9.7	2 – 3.5
4	9.7 – 10.4	1 – 2
5	10.5 – 10.8	0.5 – 1

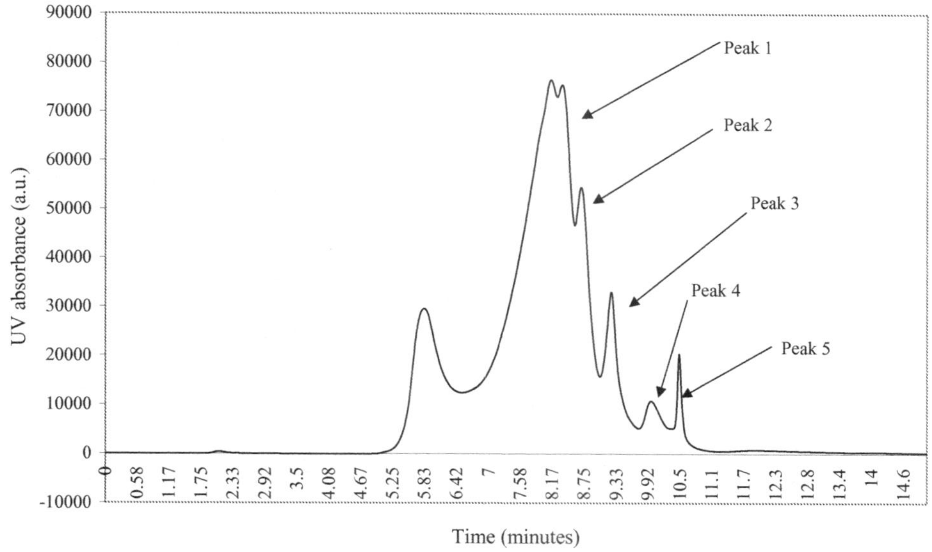

Figure 4 HPSEC chromatogram of Albert raw water with peaks assigned

The retention of the molecules depends on the pore size distribution of the membranes as well as the characteristics of the molecules, particularly the molecular volumes. Therefore, the retention of the molecules is not sharp. The 30 kDa membrane retains molecules from all size ranges as can be seen in Figure 5. The membrane cut-off values are usually defined as the mass of a molecule whose retention is 90% on that membrane. When this retention value of 90% is applied to the peaks in an HPSEC chromatogram, the following AMW ranges can be assigned to each peak (Table 2, Figure 4). A range for each peak is given as each peak does not represent one specific molecular weight.

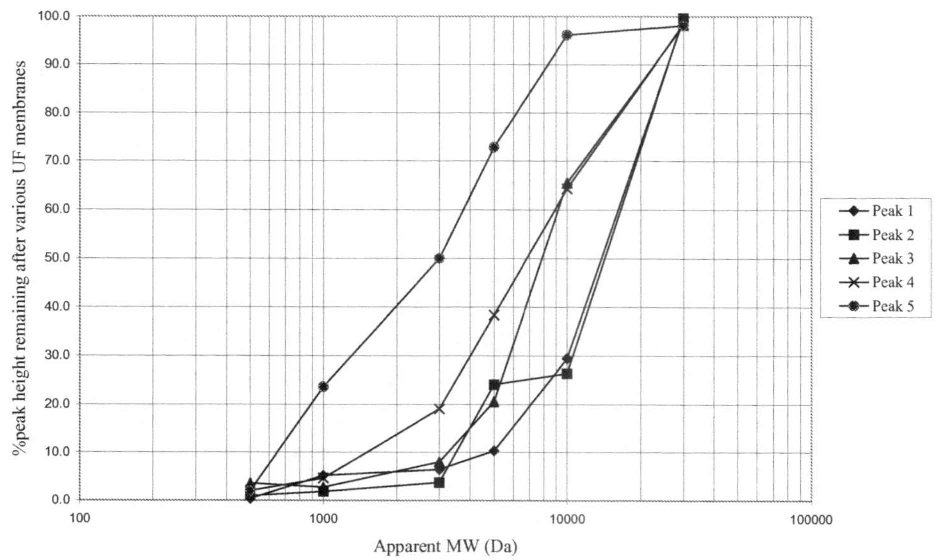

Figure 5 Remainder of NOM after various UF membranes measured as the reduction of peak heights in HPSEC traces

The shape of the raw water chromatogram is repeatable but the retention times for each peak can vary depending on the lengths of tubing used between the injector, the guard column and the column.

Sensitivity Analysis

It is possible that the TOC measurements made here could vary by ± 10%. The permeation coefficients were re-calculated from the TOC data with variance of ± 10%. The results are shown in Table 3.

Table 3
Variance of permeation coefficients

Membrane cut-off (kDa)	P	p + 10%	p – 10%
30	0.60	0.71	0.53
10	0.58	0.69	0.51
5	0.66	0.83	0.65
3	0.71	0.81	0.62
1	0.27	0.23	0.05
0.5	0.18	0.27	0.08

Figure 5 was redrawn using the varied permeation coefficients and the new AMWs were assigned to each peak. The results are reported as the AMW where the line for each peak cross the 10% line of the '%peak height remaining' and are presented in Table 4. The peak height values were not adjusted when the permeate coefficient values were less than 0.2.

Table 4
AMW values

Peak	Original values	+ 10%	- 10%
1	5000	3900	5000
2	3600	3400	3600
3	3250	3100	3400
4	1500	1500	1700
5	650	675	650

By varying the TOC values by ± 10%, the ranges given for peaks 2 to 4 (Table 4) are not altered. A variance of +10% for the TOC gives a significantly lower value for peak 1 that does not fit into the range. By carrying out the sensitivity analysis, the importance of assigning a range to each peak rather than an exact AMW value is illustrated. Again, it should be emphasised that the values for AMW calculated here should be taken as a rough guide.

Disadvantages of using UV detection

It should be noted that the analysis was carried out using an HPLC with UV detection at 254 nm. Such a system will not detect carbon single bonds. This may lead to an overestimation of the large molecule fractions. It has been pointed out that MWs measured with HPSEC with UV detection generated higher values than those measured with other methods such as ultracentrifugation and field flow fractionation. This was ascribed to higher MW fractions having a higher molar absorptivity (ε). Conversely, lower MW fractions with a lower ε appeared to be lower in concentration. A UV detector on an HPSEC quantifies the response intensity based on the ε. As a result, the MW determined with a UV detector at 254 nm is primarily the MW of only the high ε components such as humic and fulvic acids leading to inherent inaccuracy and over- or underestimation of MWs. Her *et al.* (2002a) concluded that MW estimation with a UV detector is problematic for a hetero-mixture of NOM and that a UV detector cannot be accurately used for quantitative measurements of NOM. With this in mind, the determination of AMWs made here should be used as a rough guide and not taken as exact values.

HPSEC calibration conclusions

UF cannot provide absolute values of molecular weight but can provide an 'index' of AMW that can be used for relative comparisons. The data obtained using the UF calibrated HPSEC should be interpreted by the observation of trends and not absolute values.

REFERENCES

Adachi, Y. and Tanaka, Y. (1997). Settling velocity of an aluminium-kaolinite floc. *Water Research* 31 (3), 449-454.

Afcharian, A., Levi, Y., Kiene, L. and Scribe, P. (1997). Fractionation of dissolved organic matter from surface waters using macroporous resins. *Water Research* 31 (12), 2989-2996.

Aguilar, M.I., Saez, J., Llorens, M., Soler, A. and Ortuno, J.F. (2003). Microscopic observation of particle reduction in slaughterhouse wastewater by coagulation-flocculation using ferric sulfate as coagulant and different coagulant aids. *Water Research* 37 (9), 2233-2241.

Aiken, G., Leenheer, J. and Woodside, G. (2004). Removal of dissolved organic matter in a subsurface recharge system. In: *Proceedings of the Natural Organic Matter Research: Innovations and Applications for Drinking Water Treatment.* Adelaide, Australia March 2-5, 2004.

Aiken, G.R. (1985). Isolation and concentration techniques for aquatic humic substances. In: *Humic Substances in Soil, Sediment and Water – Geochemistry, Isolation and Characterization.* Aiken, G.R., McKnight, D.M., Wershaw, R.L., and MacCarthy, P. (Eds.), John Wiley and Sons Limited, New York, USA. pp. 363-384.

Aiken, G.R. (1988). A Critical evaluation for the use of macroporous resins for the isolation of aquatic humic substances. In*: Humic Substances and their Role in the Environment.* Frimmel, F.H. and Christman, R.F. (Eds.), John Wiley and Sons Limited, New York, USA. pp. 15-28.

Aiken, G.R., McKnight, D.M., Thorn, K.A. and Thurman, E.M. (1992). Isolation of hydrophilic organic acids from water using nonionic macroporous resins. *Organic Geochemistry* 18 (4), 567-573.

Aiken, G.R., Thurman, E.M., Malcolm, R.L., and Walton, H.F. (1979). Comparison of XAD macroporous resins for the concentration of fulvic acid from aqueous solution. *Analytical Chemistry* 51 (11), 1799-1803.

Allen, T. (1997). Particle size measurement. In: *Powder Sampling and Particle Size Measurement*. Chapman and Hall, London.

Allgeier, S.C. and Summers, R.S. (1995). Evaluating NF for DBP control with the RBSMT. *Journal of the American Water Works Association* 87 (3), 87-99.

Amal, R., Raper, J.A. and Waite, T.D. (1992). The effect of fulvic acid adsorption on the aggregation kinetics and structure of haematite particles. *Journal of Colloid and Interface Science* 151 (1), 244-257.

Amirtharajah, A. and O'Melia, C.R. (1990). Coagulation processes: Destabilisation, mixing, and flocculation. In: *Water Quality and Treatment: A Handbook of Community Supplies.* McGraw-Hill Inc. New York, USA. pp.269-365.

Amy, G. L. and Chadik, P. A. (1982) PAC and polymer aided coagulation for THM control. In: *Proceedings of the National Conference on Environmental Engineering* 1982.

Amy, G.L. and Chadik, P.A. (1983). Cationic polyelectrolytes as primary coagulants for removing trihalomethane precursors, *Journal of the American Water Works Association* 75, 527-531.

Amy, G.L., Collins, M.R., Kuo, C.J., and King, P.H. (1987). Comparing gel permeation chromatography and ultrafiltration for the molecular weight characterisation of aquatic organic matter. *Journal of the American Water Works Association* 79 (1), 43-49.

Anderson, C.T. and Maier, W.J. (1979). Trace organics removal by anion exchange resins. *Journal of the American Water Works Association* 71 (5), 278-283.

Anderson, L.G., Broddevall, B., Lindberg, S. and Phillips, J. (1974). A new colour removal process: a field report. *TAPPI* 57 (4), 102-105.

Anderson, R.E., Axelsson H. and Lawrence, R.Y. (1982). Purification of kraft pulp bleach plant effluents using condensate resins. *Reactive Polymers* 1, 67-71.

Au, K.-K., Yang, S., and O'Melia, C.R. (1998). Adsorption of weak polyeletrolytes on metal oxide surfaces: a hybrid SC/SF approach. *Environmental Science and Technology* 32 (19), 2900-2908.

Avnea, M.J. and Koopal, L.K. (1999). Kinetics of humic acid adsorption at solid-water interfaces. *Environmental Science and technology* 32 (16), 2739-2744.

Bache D.H., Hossain, M.D., Al-Ani, S. H. and Jackson, P. J. (1991). Optimum coagulation conditions for a coloured water in terms of floc size, density and strength. *Water Supply* 9 (1), 93-102.

Bache, D.H. (2004). Floc rupture and turbulence: a framework for analysis. *Chemical Engineering Science* 59, 2521-2534.

Bache, D.H. and Al-Ani, S.H. (1989). Development of a system for evaluating floc strength. *Water Science and Technology* 21, 529-537.

Bache, D.H. and Papavasilopoulos, E. N. (2003). Dewatering of alumino-humic sludge: impacts of hydroxide. *Water Research* 37, 3289-3298.

Bache, D.H. and Rasool, E.R. (2001). Characteristics of alumino-humic flocs in relation to DAF performance. *Water Science and Technology* 43 (8), 203-208.

Bache, D.H., Johnson, C., McGilligan, J.F. and Rasool, E. (1997). A conceptual view of floc structure in the sweep floc domain. *Water Science and Technology* 36 (4), 49-56.

Bache, D.H., Rasool, E., Moffatt, D. and McGilligan, F.J. (1999). On the strength and character of alumino-humic flocs. *Water Science and Technology* 40 (9), 81-88.

Baham, J. and Sposito, G. (1994). Adsorption of dissolved organic-carbon extracted from sewage-sludge on montmorillonite and kaolinite in the presence of metal-ions. *Journal of Environmental Quality* 23 (1), 147-153.

Banks, J. and Wilson, D. (2002) Use of UV_{254} to predict the relationship between NOM and THMs on upland waters. In: *Proceedings of the International Conference on Natural Organic Matter Characterisation and Treatment*, Cranfield University, Cranfield, England, 15th May.

Beckett, R. (1990). The surface chemistry of humic substances in aquatic systems. In: *Surface and Colloid Chemistry in Natural Waters and Water Treatment,* R. Beckett (Ed.), Plenum, New York, pp. 3-20.

Beckett, R. and Le, N.P. (1990). The role of organic matter and ionic composition in determining the surface charge of suspended particles in natural waters. *Colloids and Surfaces* 44, 35-49.

Bell-Ajy, K., Abbaszadegan, M., Ibrahim, E., Verges, D. and LeChevallier, M. (2000). Conventional and optimized coagulation for NOM removal. *Journal of the American Water Works Association* 92 (10), 44-58.

Bellouti, M., Alves, M.M., Novais, J.M. and Mota, M. (1997). Flocs vs granules: differentiation by fractal dimension. *Water Research* 31, 1227-1231.

Bennett, D.M., Bolto, B.A., Dixon, D.R., Eldridge, R.J., Le, N.P. and Rye, C.S. (2000). Determining the fate of flocculants by fluorescent tagging. In: *Chemical Water and Wastewater Treatment VI*, H. H. Hahn, E. Hoffmann and H. Odegaard (Eds.), Springer, Berlin. pp.159-168.

Ben-Poorat, S., Kennedy, D.C., and Byington, C.H. (1987). Evaluation of a quarternary resin for the isolation or concentration of organic substances from water. In: *Organic pollutants in Water: Sampling, Analysis and Toxicity Testing.* Suffet, I.H. and Malaiyandi, M. (Eds.), American Chemical Society, Washington DC, USA. pp. 535-556.

Bernard, S., Chazal, P.H. and Mazet, M. (1997). Removal of organic compounds by adsorption on pyrolusite (β-MnO_2). *Water Research* 31 (5), 1216-1222.

Biggs, C.A. and Lant, P.A. (2000). Activated sludge flocculation: on-line determination of floc size and the effect of shear. *Water Research* 34 (9), 2542-2550.

Billica, J.A. and Gertig, K.R. (2000). Optimization of a coagulation process to treat high TOC, low alkalinity water and its impact on filtration performance. In: *Proceedings of the American Water Works Association annual conference*, Salt Lake City, UT. November 5-9, 2000.

Bjorklund, R.B., Karlsson, S., Boren, H., Allard, E. and Lundstrom, I. (2001). Photodesorption of fulvic acid from iron oxide surfaces into aqueous solution. *Applied Surface Science* 174, 166-175.

Black, D.L., McQuay, M.Q. and Bonin, M.P. (1996). Laser-based techniques for particle size measurement: a review of sizing methods and their industrial applications. *Progressive Energy and Combustion Science* 22, 267-306.

Blok, V.C., Slater, G.P., and Giblin, E.M. (1983). Comparison of sorption and extraction methods for recovery of trace organics from water. *Water Science and Technology* 15, 149-159.

Bloys Van Treslong, J. C. and A. J. Staverman, A. J. (1974). Polyethyleneimine II. Potentiometric titration behaviour in comparison with other weak polyelectrolytes. *Recueil des Travaux Chimiques des Pays-Bas* 93, 171-178.

Boening, P.H., Beckmann, D.D. and Snoeyink, V.L. (1980). Activated carbon versus resin absorption of humic substances. *Journal of the American Water Works Association* 72 (1), 54-59.

Boggs, S., Livermore D.G. and Seitz M.G. (1985). Humic macromolecules in natural waters, *Reviews in Macromolecular Chemistry and Physics* C25, 599-657.

Boller, M. and Blaser, S. (1998). Particles under stress. *Water Science and Technology* 37 (10), 9-29.

Bolto, B. A. (2003a). Removal of natural organic matter by cationic polymers. CSIRO Report.

Bolto, B.A. (1995). Soluble polymers in water purification, *Progress in Polymer Science* 20, 987-1041.

Bolto, B.A. (1996). Magnetic particle technology: desalination and water reuse applications. *Desalination* 106, 137-143.

Bolto, B.A. (2001). Cationic polymer and clay or metal oxide combinations for natural organic matter removal. *Water Research* 35 (11), 2669-2676.

Bolto, B.A. (2002). Removal of natural organic matter by ion exchange. CSIRO Report.

Bolto, B.A. (2003b). Removal of natural organic matter by adsorbents. CSIRO Report.

Bolto, B.A. and Levine, A.D. (2002). Reaction of polyelectrolytes with disinfectants to produce disinfection by-products. In: *Proceedings of the IWA Water Congress,* Melbourne, Australia. CD ROM. April 7-12, 2002.

Bolto, B.A., Abbt-Braun, G., Dixon, D.R., Eldridge, R.J., Frimmel, F., Hesse, S., King, S.J. and Toifl, M. (1999a). Experimental evaluation of cationic polyelectrolytes for removing natural organic matter from water. *Water Science and Technology* 40 (9), 71-79.

Bolto, B.A., Dixon, D.R., Eldridge, R.J. and King, S.J. (1998). The use of cationic polymers as primary coagulants in water treatment. In: *Chemical Water and Wastewater Treatment V.* H. H. Hahn, E. Hoffmann and H. Odegaard, (Eds.), Springer, Berlin, pp. 171-185

Bolto, B.A., Dixon, D.R., Eldridge, R.J. and King, S.J. (1999b) Unpublished results.

Bolto, B.A., Dixon, D.R., Eldridge, R.J. and King, S.J. (2001). Cationic polymer and clay or metal oxide combinations for NOM removal. *Water Research* 35, 2669-2676.

Bolto, B.A., Dixon, D.R., Eldridge, R.J., King, S.J. and Linge, K. (2002a). Removal of natural organic matter by ion exchange. *Water Research* 36, 5057-5065.

Bolto, B.A., Dixon, D.R., Eldridge, R.J., King, S.J. and Linge, K. (2002b). Removal of THM precursors by coagulation or adsorption processes. *Water Research* 36, 5066-5073.

Bolto, B.A., Eppinger, K.H., Ho, P.K.S., Jackson, M.B., Pilkington, N.H. and Siudak, R. (1978). An ion-exchange process with thermal regeneration XII. Desalting of sewage effluents. *Desalination,* 25, 45-59.

Bottero, J.Y., Tchoubar, D., Axelos, M.A.V., Quienne, P. and Fiessinger, F. (1990). Flocculation of silica colloids with hydroxy aluminium polycations. Relation between floc structure and aggregation mechanisms. *Langmuir* 6, 596-602.

Bourke, M. and Slunjski M. (1999). MIEX® DOC process launched in Western Australia. *Water* 26 (6), 17-20.

Bourke, M., Champlin, T., Perella, J., Topham, C. and Blattman, J. (2002) Use of a Magnetic Ion Exchange Resin to Improve DBP Precursor Removal and Reduce Coagulant Usage at Lee County's OLGA WTP. In: *Proceedings of the Florida Section American Water Works Association 2002 Annual Convention.* Palm Harbor, Fl, USA. Nov 11-13, 2002.

Bouyer, D., Line, A., Cockx, A. and Do-Quang, Z. (2001). Experimental analysis of floc size distribution and hydrodynamics in a jar-test. *Transactions of the Institution of Chemical Engineers.* 79 (A), 1017-1024.

Brakalov, L.B. (1987). A connection between the orthokinetic coagulation capture efficiency of aggregates and their maximum size. *Chemical Engineering Science* 42, 2373-2383.

Bratby, J. (1980). *Coagulation and Flocculation.* Uplands Press Ltd, Croydon, UK, pp 55-89.

Brattebø, H., Ødegaard, H. and Halle, O. (1987). Ion exchange for the removal of humic acids in water treatment. *Water Research* 21, 1045-1052.

Budd, G.C., Sinder, P.C., Edwards, J.C., Long, B.W., Meisch, M.E., Rexing, D.J., Owen, C.A., Cook, J.B. and Joslyn, B.L. (2003). Magnetized ion exchange (MIEX): Evaluation of process impacts in different sources. In: *Proceedings of the American Water Works Association Water Quality Technology Conference*, Philadelphia, PN. November 2-6.

Budd, G.V., Hess, A.F., Shorney-Darby, H., Neerman, J.J., Spencer, C.M., Bellamy, J.D. and Hargette, P.H. (2004). Coagulation applications for new treatment goals. *Journal of the American Water Works Association* 96 (2), 103-113.

Buleva, M. and Petkanchin, I. (1999). Interaction of humic substances with silica and alumina colloids: adsorption and stability – Electro-optical study. *Colloids and Surfaces A: Physicochemical and Engineering Aspects* 151 (1-2), 225-231.

Burgess, M.S. and Phipps, J.S. (2000). Flocculation of PCC Induced by Polymer/Microparticle Systems: Floc Characteristics. *Nordic Pulp and Paper Research Journal* 15 (5), 572-578.

Bursill, D.B., Hine, P.T. and Morran, J.Y. (1985). The effect on natural organics on water treatment processes. In: *Proceedings of the 11th Federal Convention of the. Australian Water and Wastewater Association.* Melbourne, Australia. pp. 197-204.

Buscall, R. (1993). The Hamaker coefficient for titanium dioxide (rutile) in liquid media. *Colloids and Surfaces A: Physicochemical and Engineering Aspects* 75, 269-272.

Bushell, G.C., Yan, Y.D., Woodfield, D., Raper, J. and Amal, R. (2002). On techniques for the measurement of the mass fractal dimension of aggregates. *Advances in Colloid and Interface Science* 95, 1-50.

Camp T. R. and Stein P. C. (1943) Velocity gradients and internal work in fluid motion. *Journal of the Boston Society of Civil Engineers* 30, 219–237.

Camp, T. (1968) Floc volume concentration. Journal of the American Water Works Association 60 (6), 656-673.

Çapar, G. and Yetiş, U. (2002). Removal of THM precursors: Ankara case study. *Water Research* 36 (5), 1379-1384.

Carlson, K.H. and Gregory, D. (2000). Optimizing water treatment with two stage Carroll, T., King, S., Gray, S.R., Bolto, B.A. and Booker, N.A. (2002). The fouling of microfiltration membranes by NOM after coagulation treatment. *Water Research* 34, 2861-2868.

Carlson, K.H., Elmund, G.K. and Gertig, K.R. (1994). Getting a jump on the information collection rule: Plant scale characterization of NOM and the relationship to DBP formation. In: *Proceedings of the American Water Works Association annual conference.* San Francisco, CA. November 6-10, 1994.

Carlson, K.H., Gregory, D. and MacMillan, S. (1996). The efficacy of pre-ozonation and enhanced coagulation for treating low alkalinity, high color water. In: *Proceedings of the American Water Works Association annual conference* Toronto, ON. June 26-27, 1996.

Carroll, T., King, S., Gray, S.R., Bolto, B.A. and Booker, N.A. (2000). The fouling of microfiltration membranes by NOM after coagulation treatment. *Water research* 34 (11), 2861-2868.

Carvalho M.S., Padmanabhan M. and Macosko C.W. (1994). Single point correction for parallel disk rheometry. *Journal of Rheology*, 38 (6), 1925-1936

Chadik, P. A. and Amy, G. L. (1983) Removing trihalomethane precursors from various natural waters by metal coagulants. *Journal of the American Water Works Association* 75 (10), 532-546.

Chae, S. (2002). Evaluation of drinking water treatment processes focusing on NOM removal and DBP formation. In: *Proceedings of the IWA Water Congress.* Melbourne, Australia. CD ROM. April 7-12, 2002.

Chaignon, V., Lartiges, B.S., El Samrani, A. and Mustin, C. (2002). Evolution of size distribution and transfer of mineral particles between flocs in activated sludges: an insight into floc exchange dynamics. *Water Research* 36, 676-684.

Chakraborti, R.J., Atkinson, J.F. and Van Benschoten, J.E. (2000). Characterisation of alum floc by image analysis. *Environmental Science and Technology* 34, 3969-3976.

Chakraborti, R.J., Gardner, K.H., Atkinson, J.F. and Van Benschoten, J.E. (2003). Changes in fractal dimension during aggregation. *Water Research* 37, 873-883.

Chang, Y., Li, C.-W. and Benjamin, M. M. (1997). Iron oxide-coated media for NOM sorption and particulate filtration. *Journal of the American Water Works Association* 89, 100-113.

Cheng, K.L. (1977). Separation of humic acid with XAD resins. *Mikrochimica Acta* II, 389-396

Cheng, R.C., Krasner, S.W., Green, J.F. and Wattier, K.L. (1995). Enhanced coagulation: A preliminary evaluation. *Journal of the American Water Works Association* 87 (2), 91-103.

Cheng, R.C., van Leeuwen, J., Fabris, R., King, S., Withers, N., Spark, K. and Drikas, M. (2000). Enhanced coagulation to meet DBP regulations for a California groundwater source: In: *Proceedings of the American Water Works Association Water Quality Technology Conference*. Salt Lake City, UT. 5-9 November.

Cheng, W.P. and Chi, F.H. (2002). A study of coagulation mechanisms of polyferric sulfate reacting with humic acid using a fluorescence-quenching method. *Water Research* 36 (18), 4583-4591.

Choo, K.H. and Kang, S.K. (2003). Removal of residual organic matter from secondary effluent by iron oxides adsorption. *Desalination* 154 (2), 139-146.

Chow, C. (1999) Personal communication, Australian Water Quality Centre.

Chow, C. W. K., Fabris, R. and Drikas, M. (2003). A new rapid fractionation analysis to characterize natural organic matter for the optimization of water treatment processes. *Water Research,* In press.

Chow, C., van Leeuwen, J., Fabris, R., King, S., Withers, N., Spark, K. and Drikas, M. (2000). Enhanced coagulation for removal of dissolved organic carbon – a fractionation approach. In: *Proceedings of the WaterTECH Conference, Australian Water and Wastewater Association*, Sydney, CD-ROM, April, 2000.

Chow, C.W.K., van Leeuwen, J.A., Drikas, M., Fabris, R., Spark, K.M. and Page, D.W. (1999). The impact of the character of natural organic matter in conventional treatment with alum. *Water Science and Technology* 40 (9), 97-104.

Christy, A.A. and Egeberg, P.K. (2000). Characterisation of natural organic matter from the Nordic Typing Project water samples by chemometric analysis of their near infrared spectral profiles. *Chemometrics and Intelligent Laboratory Systems* 50, 225-234.

Chu C.P., Lee D.J. and Peng X.F. (2004). Structure of conditioned sludge flocs. *Water Research*. 38 (8), 2125-2134.

Chu, C.P., Chang, B.V., Liao, G.S., Jean, D.S. and Lee, D.J. (2001). Observations on changes in ultrasonically treated waste-activated sludge. *Water Research* 35 (4), 1038-1046.

Churchman, G.J., Slade, P.G., Self, P.G., Keeling, J.L., Raven, M.D. and Kimber, R.W.L. (1996). Australian clays for the uptake of pollutants. In: *Proceedings of the Conference on Contaminants and the Soil Environment - Australasia-Pacific Region,* CSIRO, Adelaide.

Cleasby, J.L. and Logsdon G.S. (1999). Granular bed and precoat filtration. In: *Water Quality and Treatment: a Handbook of Community Supplies*. McGraw-Hill, New York.

Collins Green, R. (1997). Fluorescence polarization colloid charge titration development and application for feed forward coagulant control at water treatment facilities. MSc Thesis, University of Colorado, Boulder, CO.

Collins, M.R., Amy, G.L. and Steelink, C. (1986). Molecular weight distribution, carboxylic acidity and humic substances content of aquatic organic matter: implications for removal during water treatment. *Environmental Science and Technology* 20 (10), 1028-1032.

Cornelissen, A., Burnett, M.G., McCall, R.D. and Goddard, D.T. (1997). The structure of hydrous flocs by batch and continuous flow water treatment systems and obtained by optical, electron and atomic force microscopy. *Water Science and Technology* 34 (4), 41-48.

Cornwell, D.A. and Bishop, M.M. (1983). Determining velocity gradients in laboratory and full scale systems. *Journal of the American Water Works Association* 75 (9), 470-475.

Cousin, C.P. and Ganczarczyk, J. (1998). Effects of salinity on physical characteristics of activated sludge flocs. *Water Quality Research Journal of Canada* 33 (4), 565-587.

Cox, C. R. (1964). Operation and control of water treatment processes. *WHO Monograph Series No. 49,* WHO, Geneva.

Croué, J.-P., Martin, B., Deguin, A., and Legube, B. (1993a). Isolation and characterisation of dissolved hydrophobic and hydrophilic organic substances of a reservoir water. In: *Natural Organic Matter In Drinking Matter: Origin, Charaterisation and Removal, American Water Works Association Workshop.* Chamonix, France. September 19-22, 1993.

Croué, J.-P., Lefebvre, E., Martin, B. and Legube, B. (1993b). Removal of dissolved hydrophobic and hydrophilic organic substances during coagulation/ flocculation of surface waters. *Environmental Science and Technology* 27 (11), 143-152.

Croué, J.-P., Violleau, D., Bodaire, C. and Legube, B. (1999). Removal of hydrophobic and hydrophilic constituents by anion exchange resin. *Water Science and Technology* 40 (9), 207-214.

Crozes, G., White, P. and Marshall, M. (1995). Enhanced coagulation: Its effect on NOM removal and chemical costs. *Journal of the American Water Works Association* 87 (1), 78-89.

da Motta, M., Pons, M.-N., Roche, N., and Vivier, H. (2001). Characterisation of activated sludge by automated image analysis. *Biochemical Engineering Journal* 9 (3), 165-173.

Degrémont (1991). Main uses of activated carbon. In: *Water Treatment Handbook, Vol. 1*, Lavoisier Publishing, Paris, p.224.

Delphos, P.J., Harr, K. and Bianco, T.L.. (2001). Selective DBP precursor removal with an innovative ion exchange resin. In: *Proceedings of the American Water Works Association Water Quality Technology Conference.* Nashville, TN. November 11-15.

Dempsey, B.A., Ganho, R.M., and O'Melia, C.R. (1984). The coagulation of humic substances by means of aluminium salts. *Journal of the American Water Works Association* 76 (4), 141-150.

Deng, Y., Xiao, H. and Pelton, R. (1996). Temperature-sensitive flocculants based on poly(*N*-isopropylacrylamide-co-diallyldimethylammonium chloride). *Journal of Colloid and Interface Science* 179, 188-193.

Dennett, K.E., Amirtharajah, A., Moran, T.F. and Gould, J.P. (1996). Coagulation: its effect on organic matter. *Journal of the American Water Works Association* 88 (4), 129-142.

Dennett, K.E., Amirtharajah, A., Studstill, A., Moran, T.F. and Gould, J.P. (1995). *Humic substance removal and minimization of trihalomethanes by ferric chloride coagulation.* American Water Works Association Research Foundation Report, Denver, CO.

Dharmarajah, A.H. and Cleasby, J.L. (1986). Predicting the expansion behaviour of filter media. *Journal of the American Water Works Association* 78, 66-76.

Diamond, R.M. (1963). The aqueous solution behaviour of large univalent ions - A new type of ion pairing. *Journal of. Physical Chemistry* 67, 2513-2517.

Dixon, D.R. and Kolarik, L.O. (1990). The role of surface and colloid chemistry in the Sirofloc process. *Surface and Colloid Chemistry in Natural Waters and Water Treatment,* Beckett, R. (Ed.), Plenum, New York. pp. 103-118

Dolfing, J. (1997). Microbiological aspects of granular methanogenic sludge. PhD Thesis. Agricultural University Wageningen, Netherlands.

Drikas, M., Chow, C.W.K. and Cook, D. (2003). The impact of recalcitrant organic character on disinfection stability, trihalomethane formation and bacterial regrowth: An evaluation of magnetic ion exchange resin (MIEX®) and alum coagulation. *Journal of Water Supply: Research and Technology - AQUA* 52 (7), 475-487.

Drikas, M., Morran, J.Y., Cook, D. and Bursill, D.B. (2003b). Operating the MIEX® process with microfiltration or coagulation. In: *Proceedings of the American Water Works Association Water Quality Technology Conference*, 2-6 November. Philadelphia, PN, USA.

Droppo, I.G., Flannigan, D.T., Leppard, G.G. and Liss, S.N. (1996). Microbial floc stabilisation and preparation for structural analysis by correlative microscopy. *Water Science and Technology* 34 (5-6), 155-162.

Duan, J. and Gregory, J. (2003). Coagulation by hydrolysing metal salts. *Advances in Colloid and Interface Science* 100-102, 475-502.

Ducker, W. A., Senden, T. J. and Pashley, R. M. (1991) Direct measurement of colloidal forces using an atomic force microscope. *Nature* 353, 239-241.

Ducoste, J.J. and Clark, M.M. (1998). The influence of tank size and impeller geometry on turbulent flocculation: I Experimental. *Environmental Engineering Science* 15 (3), 215-224.

Duguet, J.-P. and Mallevialle, J. (1997). Influence of NOM on water treatment. In: *Proceedings of the 21st IWA Congress.* Madrid. September 20-26, 1997. p. SS13-8

Dulin,B. E. and Knocke, W. R. (1989). The impact of incorporated organic matter on the dewatering characteristics of aluminium hydroxide sludges. *Journal of the American Water Works Association* 81 (5), 74-79.

Edwards, M. (1997). Predicting DOC removal during enhanced coagulation. *Journal of the American Water Works Association* 89 (5), 78-89.

Edwards, M., Benjamin, M.M. and Ryan, J.N. (1996). Role of organic acidity in sorption of natural organic matter (NOM) to oxide surfaces. *Colloids and surfaces A: Physicochemical and engineering aspects* 107, 297-307.

Edzwald, J.K. (1993). Coagulation in drinking water treatment: particles, organics and coagulants. *Water Science and Technology* 27 (11), 21-35.

Edzwald, J.K. and Tobiason, J.E. (1999). Enhanced Coagulation and a broader view. *Water Science and Technology* 40 (9), 63-70.

Eikebrokk, B. and Saltnes, T. (2002). NOM removal from drinking water by chitosan coagulation and filtration through lightweight expanded clay aggregate filters. *Journal of Water Supply: Research and Technology - AQUA 51*, 323-332.

Eikkibrook, B. (1999). Coagulation-direct filtration of soft, low alkalinity humic waters. *Water Science and Technology* 40 (9), 55-62.

Elliott, H.A. and Huang, C.P. (1981) Adsorption characteristics of some Cu (II) complexes on aluminosilicates. *Water Research* 15 (7), 849-855.

Essamiani, K. and de Traversay, C. (2002). Optimisation of the flocculation process using computational fluid dynamics. In: *Chemical Water and Wastewater Treatment VII: Proceedings of the 10th Gothenburg Symposium*. IWA, Gothenburg, June 17-19, 2002.

Evans, S. and Maalman, T.F.J. (1979). Removal of organic acids from Rhine River water with weak base resins. *Environmental Science and Technology* 13, 741-743.

Farrow, J. and Warren, L. (1989). A new technique for characterising flocculated suspensions. *Dewatering Technology and Practice* 61-64.

Farrow, J. and Warren, L. (1993). Measurement of the size of aggregates in suspension. In: *Coagulation and Flocculation – Theory and Applications*. Marcel Dekker, New York. pp. 391-426

Faust, A.D. and Aly, O.M. (1983). *Chemistry of Water Treatment,* Butterworths, Boston, pp.326-328.

Fearing, D., Goslan, E., Banks, J., Wilson, D., Hillis, P., Campbell, A. and Parsons, S.A. (2004a). Staged coagulation for the treatment of refractory organics. *Environmental Engineering* 130, 975-982.

Fearing, D.A., Guyetand, S., Eroles, C.M., Jefferson, B., Banks, J., Wilson, D., Hillis, P., Campbell, A.T. and Parsons, S.A. (2004b). Combinations of Ferric and MIEX® for the treatment of humic rich waters. *Water Research*, 38 (10), *2551-2558.*

Fettig, J. (1999). Removal of humic substances by adsorption/ion exchange. *Water Science and Technology* 40 (9), 173-182.

Filius, J.D., Meeussen, C.L., Lumsdon, D.G., Hiemstra, T. and Riemsdijk, W.H. (2003). Modelling the binding of fulvic acid by goethite: The speciation of adsorbed FA molecules. *Geochimica et Cosmochimika Acta* 67, 1463-1474.

Fitzpatrick, S.B., Fradin, E. and Gregory, J. (2003). Temperature effects on flocculation using different coagulants. In: *Proceedings of the Nano and Micro Particles in Water and Wastewater Treatment Conference*. IWA, Zurich, September 2003.

Franchi, A. and O'Melia, C.R. (2003). Effects of natural organic matter and solution chemistry on the deposition and reentrainment of colloids in porous media. *Environmental Science and Technology* 37, 1122-1129.

Francois, R.J. (1987). Strength of aluminium hydroxide flocs. *Water Research* 21, 1023-1030.

Frimmel, F.H. and Abbt-Braun, G. (1999). Basic characterization of reference NOM from Central Europe - Similarities and differences. *Environment International* 25 (2/3), 191-207.

Giabbai, M.F., Chian, E.S.K., Reuter, J.H., Ringhand, H.P., and Kopfler, F.C. (1987). Evaluation of an integrated adsorption method for the isolation and concentration of trace organic substances from water. In: *Organic pollutants in water: Sampling, analysis and toxicity Testing.* Suffet, I.H. and Malaiyandi, M. (Eds.), American Chemical Society, Washington DC, USA. pp. 467-481

Gibbons J. and Laha S. (1999). Water purification systems: a comparative analysis based on the occurrence of disinfection by-products. *Environmental Pollution* 106 (3), 425-428.

Gibbs, R. J. (1983). Effect of natural organic matter on the coagulation of particles. *Environmental Science and Technology* 17, 237-240.

Glaser, H.T. and Edzwald, J.K. (1979). Coagulation and direct filtration of humic substances with polyethyleneimine, *Environmental Science and Technology* 13, 299-305.

Gorczyca, B. and Ganczarczyk, J. (1999). Structure and porosity of alum coagulation flocs. *Water Quality Research Journal of Canada* 34, 653-666.

Goslan E H, Banks J, Wilson D, Hillis P, Campbell A T and Parsons S A. (2004). A model for predicting dissolved organic carbon distribution in a reservoir water using fluorescence spectroscopy. *Water Research* 38, 783-791.

Goslan, E.H. (2003). Natural organic matter character and reactivity: assessing seasonal variation in a moorland water. PhD thesis, Cranfield University, UK.

Goslan, E.H., Fearing, D.A., Banks, J., Wilson, D., Hillis, P., Campbell, A.T. and Parsons, S.A. (2002). Seasonal variations in the disinfection by-product precursor profile of a reservoir water. *Journal of Water Supply: Research and Technology - AQUA* 51, 475-482.

Gottlieb M. (1996) The Reversible Removal of Naturally Occuring Organics Using Resins Regenerated with Sodium Chloride. *Ultrapure Water* 14 (8), 53-58.

Gottlieb, M.C. (1995). Improved organic trap resin. *Industrial Water Treatment,* 61 (3), 41-52.

Gregor, J.E., Nokes, C.J., and Fenton, E. (1997). Optimising natural organic matter removal from low turbidity waters by controlled pH adjustment of aluminium coagulation. *Water Research* 31 (12), 2949-2958.

Gregory, D. and Carlson, K. (2003). *The impact of chemical sequencing on filtration performance.* American Water Works Association Research Foundation Report, Denver, CO.

Gregory, D. and Carlson, K. (2004). Relationship of pH and floc formation kinetics to granular media filtration performance. *Environmental Science and Technology* 37 (7), 1398-1403.

Gregory, J. (1985). The action of polymeric flocculants. In: *Flocculation, Sedimentation and Consolidation*, B. M. Moudigil and P. Somasundaran, (Eds.), Engineering Foundation, Washington. p.125.

Gregory, J. (1985). Turbidity fluctuations in flowing suspensions. *Journal of Colloid and Interface Science* 105, 676-684.

Gregory, J. (1988). Polymer adsorption and flocculation in sheared suspensions. *Journal of Colloid and Interface Science* 31, 231-253.

Gregory, J. (1989). Fundamentals of flocculation. *Critical Reviews in Environmental Control* 19, 185-230.

Gregory, J. (1997). The density of particle aggregates. *Water Science and Technology* 36 (4), 1-13.

Gregory, J. (1998). The role of floc density in solid-liquid separation. *Filtration and Separation* 35, 367-371.

Gregory, J. (2003). Monitoring floc formation and breakage. In: *Proceedings of the Nano and Micro Particles in Water and Wastewater Treatment Conference.* IWA, Zurich, September 2003.

Gregory, J. and Chung, H. (1995) *Journal of Water Supply: Research and Technology – Aqua* Continuous monitoring of floc properties in stirred suspensions 44 (3), 125-131.

Gregory, J. and Dupont, V. (2001). Properties of flocs produced by water treatment coagulants. *Water Science and Technology* 44 (10), 231-236.

Gregory, J. and Nelson, D.W. (1986). Monitoring of aggregates in flowing suspension. *Colloids and Surfaces* 18, 175-188.

Gregory, R. and Dillon, G. (2002). Minimising sludge production at water-treatment plants. *Journal of the Chartered Institute of Water and Environmental Management* 16 (3), 174-179.

Grim, R.E. (1968). *Clay Mineralogy,* 2nd Ed, McGraw-Hill, New York, pp. 189, 308.

Gu, B., Schmitt, J., Chen, Z., Liang, L. and MacCarthy, J. (1995). Adsorption and desorption of different organic matter fractions on iron oxide. *Environmental science and Technology* 59 (2), 219-229.

Guan, J, Waite, D., Amal, R. (1998). Rapid structure characterization of bacterial aggregates. *Environmental Science and Technology* 32 (23), 3735-3742.

Gustafson, R.L. and Lirio, J.A. (1968). Adsorption of organic ions by anion-exchange resins. *Industrial and Engineering Chemistry Product Research and Development* 7, 116-120.

Hainthaler, M., Ho, L., Newcombe, G. and Kaeding, U. (2003). A spectroscopic and molecular weight investigation of the effect of six water treatment processes on Myponga Reservoir NOM. In: *Proceedings of the American Water Works Association Water Quality Technology Conference*, 2-6 November. Philadelphia, PN, USA.

Han, S., Kim, S., Lim, H., Choi, W., Park, H., Yon, J. and Hyeon, T. (2003). New nanoporous carbon materials with high adsorption capacity and rapid adsorption kinetics for removing humic acids. *Microporous and Mesoporous Materials* 58, 131-135.

Healy, T.W., Herring, A.P. and Fuerstenau, D.W. (1966). The effect of crystal structure on the surface properties of a series of manganese dioxides. *Journal of Colloid and Interface Science* 21, 435-444.

Heijman, S.G.J., van Paassen, A.M., van der Meer, W.G.J. and Hopman, R. (1999). Absorptive removal of natural organic matter during drinking water treatment. *Water Science and Technology* 40 (9), 183-190.

Hennis, S., Brown, G., Vasconcellos, S.R. and Gucciardi, J.M. (2001). Polymer technology boosts wastewater treatment performance. *Chemical Processing* 64, 32-36.

Hesse, S., Kleiser, G. and Frimmel, F.H. (1999). Characterisation of refractory organic matter (ROS) in water treatment. *Water Science and Technology* 40 (9), 1-7.

Hodgkin, J.H. and Eibl, R. (1988). Gold extraction with polydiallylamine resins. *Reactive Polymers* 9, 285-291.

Höll, W. H. (1995). Personal communication.

Hongve, D., Baann, J., Becher, G. and Beckmann, O.-A. (1999). Experiences from operation and regeneration of an anionic exchanger for NOM removal. *Water Science and Technology* 40 (9), 215-221.

Hossain, M.D. & Bache, D.H. (1991). Composition of Alum Flocs Derived from a Coloured, Low Turbidity Water. *Journal of Water Supply: Research and Technology - AQUA* 40 (5), 298.

Hsia, T.H., Lo, S.L., Lin, C.F., and Lee, D.Y. (1993). Chemical and spectroscopic evidence for the specific adsorption of chromate on hydrous iron oxide. *Chemosphere*, 26 (10), 1897-1904.

Huang, C. and Shiu, H. (1996). Interaction between alum and organics in coagulation. *Colloids and surfaces A: Physicochemical and engineering aspects* 113 (1-2), 155-163.

Huang, H. (1994). Fractal properties of flocs formed by fluid shear and differential settling. *Physical Fluids* 6 (10), 3229-3234.

Huang, H.-J. and Yeh, H.-H. (1993). Organic fractionation for water treatment process evaluation. In: *Proceedings of the Water Quality and Technology Conference, Pt I.* American Water Works Association, Denver.

Huang, P., Pan, J.R. and Huang, S. (1999). Collision efficiencies of algae and kaolin in depth filters: the effect of surface properties of particles. *Water Research* 33 (5), 1278-1286.

Huck, P.M., Coffey, B.M., Emelko, M.B., Maurizio, D.D., Slawson, R.B., Anderson, W.B., van den Eover, J., Douglas, I.P. and O'Melia, C.R. (2002). Effects of filter operation on *Cryptosporidium* removal. *Journal of the American Water Works Association* 94, 97-111.

Hundt, T.R. and O'Melia, C.R. (1988). Settling aluminium-fulvic acid interactions: mechanisms and applications. *Journal of the American Water Works Association* 80 (4), 176-186.

Hunter, R.J. (1981). *Zeta potential in colloid science.* Academic Press Inc., London.

Hurst, A.M., Edwards, M.J., Chipps, M., Jefferson, B. and Parsons, S.A. (2004). The impact of rainstorm events on coagulation and clarifier performance in potable water treatment. *Science of the Total Environment* 321 (1-2), 219-230.

Hutchison, P.R. and Healy, T.W. (1990). Coagulation and flocculation – destabilizing practices? (with particular reference to metal ion coagulants). In: *Surface and Colloid Chemistry in Natural Waters and Water Treatment,* R. Beckett, (Ed.), Plenum, New York. p.119.

Hwang, C.J., Krasner, S.W., Sclimenti, M.J., Amy, G.L., Dickenson, E., Bruchet, A., Prompsy, C., Filippi, G., Croué, J.-P., Violleau, D. and Leenheer, J.A. (2001). Polar NOM: *Characterisation, DPBs, treatment.* American Water Works Association Research Foundation Report for Project #451. American Water Works Association Research Foundation, Colorado.

Imai, A., Matsushige, K. and Nagai, T. (2003). Trihalomethane formation potential of dissolved organic matter in a shallow eutrophic lake. *Water Research* 37 (17), 4284-4294.

Irving, J., Dale, J.A. and Heller, T.L. (1997). The properties and performance of Type III strong base anion exchange resin Purolite A-555. In: *Proceedings of the IWA 58th Annual Meeting,* Pittsburg, pp. 16-26.

Isais, N.P. and Al Rammah, A.S. (1999). Pretreatment of sewage effluent for reverse osmosis for use as boiler feed water. *Desalination & Water Reuse* 9 (1), 37.

Jackson, G.E. (1980). Granular media filtration in water and wastewater treatment: Part II. *CRC Critical Reviews in Environmental Control* 11, 1-36.

Jardine, P.M., Weber, N.L. and McCarthy. J.F. (1982). Mechanism of dissolved organic carbon adsorption on soil. *Soil Science Society of America Journal* 53, 1378-1385.

Jefferson, B. and Parsons, S.A. (In Press). *Agglomeration of Clay Particles*. Thomas Telford, London.

Jefferson, B., Jarvis, P., Sharp, E., Wilson, S. and Parsons, S.A. (2004). Flocs through the looking glass. *Water Science and Technology* 50(12), 47-54.

Jensen, C.H., and Diamond, R.M. (1965). Anion exchange in concentrated solutions. *Journal of Physical Chemistry* 69, 3440-3445.

Jiang, J-Q. and Graham, N.J.D. (1998). Pre-polymerized inorganic coagulants and phosphorus removal by coagulation- A review. *Water SA* 24 (3), 237-244.

Johannsen, K., Assenmacher, M., Kleiser, M., Abbt-Braun, G., Sontheimer, H. and Frimmel, F.H. (1993). Influence of the molecular size on the adsorption of humic compounds. *Vom Wasser* 81, 185-196.

Johnson, C.P., Li, X. and Logan, B.E. (1996). Settling velocities of fractal aggregates. *Environmental Science and Technology* 30 (6), 1911-1918.

Johnson, P.N. and Amirtharajah, A. (1983). Ferric chloride and alum as single and dual Coagulants. *Journal of the American Water Works Association* 75 (5), 232-239.

Johnson, S.B., Scales, P.J., and Healy, T.W. (1999). The binding of monovalent electrolyte ions on α-alumina. I. Electroacoustic studies at high electrolyte concentrations. *Langmuir*, 15 (8), 2836-2843.

Jorand, F., Zartarian, F, Thomas, F, Block, J. C., Bottero, J. Y., Villemin, G., Urbain, V. and Manem, J. (1995). Chemical and structural (2D) linkage between bacteria within activated sludge flocs. *Water Research* 29 (7), 1639-1647.

Kainulainen, T., Tuhkanen, T., Vartiainen, T., Heinonen-Tanski, H. and Kalliokoski, P. (1994). The effect of different oxidation and filtration processes on the molecular size distribution of humic material. *Water Science and Technolology* 30 (9), 169-174.

Kaiser K., Guggenberger, M., Kaupenjohann, M. and Zech, W. (2002). Refractory organic substances in aggregated forest soils – retention verses translocation. In: *Refractory Organic Substances in the Environment*. Frimmel, F. H. Abbt-Braun, G., Heumann, K. G., Hock, B., Ludemann, H. D. and Spiteller, M., (Eds.), Wiley-VCH, Weinheim. p 411.

Kaiser, K. and Zech, W. (1998). Release of natural organic matter sorbed to soil minerals. In: *Proceedings of the 9th International Meeting International Humic Substances Society,* Adelaide, South Australia.

Kam, S-K. and Gregory, J. (1999). Charge determination of synthetic cationic polyelectrolytes by colloid titration. *Colloids and Surfaces A: Physicochemical and Engineering Aspects 159(1), 165-179.*

Kam, S-K. and Gregory, J. (2001). The interaction of humic substances with cationic polyelectrolytes. *Water Research* 35 (15), 3557-3566.

Kancharla, V., Ngo, H., Vigneswaran, S and Bolto, B. A. (1997) The use of polyelectrolyte in downflow filtration in a dual system of floating media and sand. In: *Proceedings of the AWWA 17th Federal Convention* 1997, Melbourne, Australia.

Kaneco, S., Itoh, K., Katsumata, H., Suzuki, T., Masuyama, K., Funasaka, K., Hatano, K. and Ohta, K. (2003) Removal of natural organic polyelectrolytes by adsorption onto tobermorite. *Environmental Science and Technology* 37 (7), 1448 -1451.

Kapur, P. C. (1999). Some Applications of the AMPC model of the shear yield stress of particulate fluids. *Colloids and Surfaces A: Physicochemical and Engineering Aspects* 146, 25-32.

Kapur, P.C., Scales, P.J., Boger, D.V., and Healy, T.W. (1997). Yield stress of suspensions loaded with size distributed particles. *American Institute of Chemical Engineers Journal* 43 (5), 1171-1179.

Kastl, G., Sathasivian, A., Fisher, I. and van Leeuwen, J. (2004). Modeling DOC removal by enhanced coagulation. *Journal of the American Water Works Association* 96 (2), 79-89.

Katsumata, H., Kaneco, S., Matsuno, R., Itoh, K., Masuyama, K., Suzuki, T., Funasaka, K. and Ohta, K. (2003). Removal of organic polyelectrolytes and their metal complexes by adsorption onto xonotlite. *Chemosphere* 52, 909-915.

Kavanaugh, M.C. (1978). Modified coagulation for improved removal of trihalomethane precursors. *Journal of the American Water Works Association* 70 (11), 613-620.

Kawamura, S. (1976). Considerations on improving flocculation. *Journal of the American Works Association* 68, 328-336.

Kaye, B.H. (1989). *A Random Walk Through Fractal Dimensions.* VCH Verlagsgesellschaft, Weinheim, Germany.

Keeney, M.E. (1987). Removal of humates from Bayer process liquors. *Australian Patent Application* 83078/87.

Kim, B.Y., Snoeyink, V.L. and Saunders, F.M. (1976). Adsorption of organic compounds by synthetic resins. *Journal of the Water Pollution Control Federation* 48 (1), 120-133.

Kim, E. K. and Walker, H. W. (2001) Effect of cationic polymer additives on the adsorption of humic acid onto iron oxide particles. *Colloids and Surfaces A: Physicochemical and Engineering Aspects* 194 (1), 123-131.

Kim, P.H.-S. and Symons, J. M. (1991). Using anion exchange resins to remove THM precursors. *Journal of the American Water Works Association* 83 (12), 61-68.

Kim, S.H., Moon, B.H. and Lee, H.I. (2001). Effects of pH and dosage on pollutant removal and floc structure during coagulation. *Microchemical Journal* 68, 197-203.

Kitis, M., Karanfil, T., Kilduff, J.E., and Wigton, A. (2001). The reactivity of natural organic matter to disinfection by-products formation and its relation to specific ultraviolet absorbance. *Water Science and Technology* 43 (2), 9-16.

Kitis, M., Karanfil, T., Wigton, A. and Kilduff, J.E. (2002). Probing reactivity of dissolved organic matter for disinfection by-product formation using XAD-8 resin adsorption and ultrafiltration fractionation. *Water Research* 36, 3834-3848.

Klimpel, R.C. and Hogg, R. (1991). Evaluation of floc structures. *Colloids and Surfaces* 55, 279-288.

Kobayashi, M. (2004). Breakup and strength of polystyrene latex flocs subjected to a converging cell. *Colloids and Surfaces A: Physicochemical and Engineering Aspects* 235, 73-78.

Kolarik, L.O. (1983). Colour and turbidity removal with reusable magnetite particles. IV. Alkali activated magnetite - a new solid, reusable coagulant-adsorbent. *WaterResearch* 17, 141-147.

Kölle, W. (1979). Humic acid removal with macroreticular ion exchange resins at Hanover. In: *Proceedings of the Symposium on Practical Application of Adsorption Techniques in Drinking Water Treatment.* US EPA, Washington.

Kölle, W. (1984). Humic acid removal by ion exchange. *Report EPA/570/9-84-005,* US EPA, Cincinnati.

Korshin, G.V., Benjamin, M.M. and Sletten, R.S. (1997). Adsorption of natural organic matter (NOM) on iron oxide: Effects on NOM composition and formation of organo-halide compounds during chlorination. *Water Research* 31 (7), 1643-1650.

Krasner, S.W. and Amy, G. (1995). Jar-Test Evaluations of Enhanced Coagulation. *Journal of the American Water Works Association* 87 (10), 93-107.

Krasner, S.W., Amy, G. and Zhu, H.W. (1994). Use of some simple NOM characterization techniques to evaluate enhanced coagulation treatability. In: *American Water Works Association Enhanced coagulation research workshop.* Denver, CO.

Krasner, S.W., Croué, J.-P., Buffle, J. and Perdue, E.M. (1996). Three Approaches for Characterizing NOM. *Journal of the American Water Works Association* 88 (6), 66-79.

Kretzschmar, R. and Sticher, H. (1997). Transport of humic coated oxide colloids in a sandy soil: Influence of Ca^{2+} and trace metals. *Environmental Science and Technology* 31 (12), 3497-3504.

Kunin, R. and Yarnell, P. (1997). Four Decades of Macroreticular (Macroporous) Resins. *Ultrapure Water* October, 58-64.

Kunte, H. and Pfeiffer, E.H. (1985). Isolation of Organic Material from Water. *Bulletin of Environmental Contamination and Toxicology* 34 650-655.

Lange, R., Tattersall, J., Botica, C. and Smith, P. (2001). Design and construction of the world's first large scale MIEX® water treatment plant. In: *Proceedings of the. 19th Federal Australian Water Association Convention,* Canberra, April 1-4, 2001.

Lartiges, B., Bottero, J.Y., Democrate, C. and Coupel, J. F. (1995) Optimizing flocculant demand by following floc size distribution. *Journal of Water Supply: Research and Technology – Aqua* 44 (5), 219-223.

Lawrence, J. (1989). *Analysis of Trace Organics in the Aquatic Environment,* Afghan B.K. and Chau, A.S.Y., (Eds.), CRC Press, Boca Raton. p.313.

Le Cloirec, P., Lelacheur, R.M., Johnson, J.D., and Christman, R.F. (1990). Resin concentration of by-products from chlorination of aquatic humic substances. *Water Research* 24 1151-1155.

Lee, C.H. and Liu, J.C. (2001). Sludge dewaterability and floc structure in dual polymer conditioning. *Advances in Environmental Research* 5, 129-136.

Lee, D.J., Chen, G.W., Liao, Y.C. and Hsieh, C.C. (1996). On the free-settling test for estimating activated sludge floc density. *Water Research* 30 (3), 541-550.

Lee, J.F., Liao, P.-M., Seng D.-H. and Wen, P.-T. (1998). Behaviour of organic polymers in drinking water purification. *Chemosphere* 37 (6), 1045-1061.

Lee, S. Y., Cho, J., Shin, H., Son, B. and Chae, S. (2003). Investigation of NOM size, structure and functionality (SSF): impact on water treatment process with respect to disinfection by-products formation. *Journal of Water Supply: Research and Technology - AQUA* 52, 555-564.

Lee, S.H., Shin, W.S., Shin, M.C., Choi, S.J. and Park, L.S. (2001). Improvement of water treatment performance by using polyamine flocculants. *Environmental Technol*ogy 22, 653-659.

Leenheer, J.A. (1981). Comprehensive approach to preparative isolation and fractionation of dissolved organic carbon from natural waters and wastewaters. *Environmenta Science and. Technology* 15 (5), 578-587.

Leenheer, J.A. (1985). Fractionation techniques for aquatic humic substances. In: *Humic Substances in Soil, Sediment and Water. Geochemistry, Isolation and Characterization.* Aiken, G.R., McKnight, D.M., Wershaw, R.L., and MacCarthy, P. (Eds.), John Wiley and Sons Limited, New York, USA.

Leenheer, J.A. and Huffman, E.W.D. (1976). Classification of organic solutes in water by using macroreticular resins. *Journal of Research U.S. Geological Survey* 4 (6), 737-751.

Leenheer, J.A., Wershaw, R.L. and Reddy, M.M. (1995a). Strong acid, carboxyl group structures in fulvic acid from the Suwannee River, Georgia. 1. Minor structures. *Environmental Science and Technology* 29 (2), 393-398.

Leenheer, J.A., Wershaw, R.L. and Reddy, M.M. (1995b). Strong acid, carboxyl group structures in fulvic acid from the Suwannee River, Georgia. 2. Major structures. *Environmental Science and Technology* 29 (2), 399-405

Leentvaar, J. and Rebhun, M. (1983). Strength of ferric hydroxide flocs. *Water Research* 17, 895-902.

Lefebvre, E., and Legube, B. (1993). Coagulation-flocculation par le chlorure ferrique de quelques acides organiques et phenols en solution aqueuse. *Water Research* 27, 433-447.

Leitschkis, I. and Rammensee, W. (1999). Influence of zeta potential on the filtration efficiency of aqueous suspension on precoat filters. *Colloids and Surfaces A: Physicochemical and engineering aspects* 159, 405-412.

Leong, Y.K., Boger, D.V., Scales, P.J., Healy, T.W., and Buscall, R. (1993a). Control of the rheology of concentrated aqueous colloidal systems by steric and hydrophobic forces. *Journal of the Chemical Society - Chemical Communications* 7, 639-641.

Leong, Y.K., Scales, P.J., Healy, T.W. and Boger, D.V. (1995). Interparticle forces arising from adsorbed polyelectrolytes in colloidal suspensions. *Colloids and Surfaces A: Physicochemical and Engineering Aspects* 95, 43-52.

Leong, Y.K., Scales, P.J., Healy, T.W., Boger, D.V., and Buscall, R. (1993b). Rheological evidence of adsorbate-mediated short-range steric forces in concentrated dispersions. *Journal of the Chemical Society - Faraday Transactions* 89 (14), 2473-2478.

Leu, R.-J. and Ghosh, M.M. (1988). Polyelectrolyte characteristics and flocculation. *Journal of the American Water Works Association* 80 (4), 159-167.

Levich, V. G. (1962) Physico-chemical hydrodynamics. Prentice Hall, Englewood Cliffs, US.

Li, D.H. and Ganczarczyk, J.J. (1986) Physical characteristics of activated sludge Flocs. *CRC Critical Reviews in Environmental Control* 17 (1), 53-87.

Li, F., Yuasa, A., Ebie, K., Azuma, Y., Hagishita, T. and Matsui, Y. (2002). Factors affecting the adsorption capacity of dissolved organic matter onto activated carbon: modified isotherm analysis. *Water Research* 36 (18), 4592-4604.

Li, Q., Snoeyink, V. L. and Campos, C. (2003b) Pore blockage effect of NOM on atrazine adsorption kinetics of PAC: the roles of NOM molecular weight and PAC pore size distribution. *Water Research* 37 (20), 4863-4872.

Li, S., Tuan, V. A., Noble, R. D. and Falconer, J. L. (2003a) MTBE adsorption on to all silica ß zeolite. *Environmental Science and Technology* 37 (17), 4007-10.

Lin, W.W., Sung, S.S., Lee, D.J., Chen, Y.P., Chen, D.S. and Lee, S.F. (2002). Coagulation of Humic-Kaolin-PACl aggregates. *Water Science and Technology* 47 (1), 145-152.

Lind, C. B. (1995) Experiments in TOC removal by polyaluminium hydroxychloride and enhanced coagulants. In: *Proceedings of the American Water Works Association Annual Conference.* Anaheim, CA, USA. 17-21 June.

Ljubas, D., Ruzinski, N. and Dobrovic, S. (1999). The influence of coagulant and adsorbent dosing sequence on natural organic matter removal from lake water. *Strojarstvo* 41(5-6), 191-200.

Logsdon, G.S., Chipps, M.J. and Rachwal, A.J. (2002). *Filter maintenance and operations guidance manual.* American Water Works Research Foundation Report. Denver, CO.

Luckham P.F. and Rossi S. (1999). The colloidal rheological properties of bentonite suspensions. *Advances in Colloid and Interface Science* 82, 43-92.

Macosko C.H.W. (1994) *Rheology: Principles, Measurements, and Applications*. John Wiley & Sons.

Madsen, M. and Schlundt, J. (1989). Low technology water purification by bentonite clay flocculations performed in Sudanese villages: Bacteriological examinations. *Water Research* 23, 873-879.

Mahiuddin, S., Suryanarayan, I., Dutta, N.N. and Borthakur, P.C. (1992). Adsorption studies of sodium humate on Indian iron ore fines. *Colloids & Surfaces* 64, 177-184.

Makharadze, G.A., Reviya, R.L., and Supatashvili, G.D. (2000). Separation of the components of fulvic acids from river water by high-performance liquid chromatography. *Russian Journal of Physical Chemistry* 74 (11), 1897-1901.

Malcolm, R. L. (1991) Factors to be considered in the isolation and characterization of aquatic humic substances. In: *Humic substances in the aquatic and terrestrial environment* (ed. B. Allard, H. Boren & A. Grimvall), pp. 9–36. Proceedings of the. International Symposium, August 21–23, 1989, Linkoping, Sweden.

Malcolm, R.L. (1985). Geochemistry of stream fulvic and humic substances. In:. *Humic substances in soil sediment and water. Geochemistry isolation and characterization.* Aiken, G.R., McKnight, D.M., Wershaw, R.L. and MacCarthy, P. (Eds.), Wiley, New York. pp.53-86.

Malcolm, R.L. and MacCarthy, P. (1992). Quantitative evaluation of XAD-8 and XAD-4 resins used in tandem for removing organic solutes from water. *Environmental International* 18, 597-607.

Maloney, R.J., Home, G.P. and Stockley, M. (1989). The SIROFLOC process: development, construction and operating experience of a 20,000 m^3/day potable water plant. *Institute of Chemical Engineering, Symposium Series* 13, 93-104.

Mamchenko, A.V. (1993). Sorption of humous compounds by ion-exchangers. *Khimya i Tekhnolgiya Vody* 15, 270-294.

Manning, A.J. and Dyer, K.R. (1999). A laboratory examination of floc characteristics with regard to turbulent shearing. *Marine Technology* 160, 147-170.

Marhaba, T.F. and Pipada, N.S. (2000). Coagulation: Effectiveness in removing dissolved organic matter fractions. *Environmental Engineering Science* 17 (2), 107-115.

Marhaba, T.F., Van, D. and Lippincott, R.L. (2000). Changes in NOM fractions through treatment: a comparison of ozonation and chlorination. *Ozone Science and Engineering* 22, 249-266.

Matilainen, A., Lindqvist, N., Korhonen, S. and Tuhkanen, T. (2002). Removal of NOM in the different stages of the water treatment process. *Environment International* 28 (6), 457-465.

McCarthy, J.F. and Zachara, J.M. (1989). Subsurface transport of contaminants. *Environmental Science and Technology* 23 (5), 496-502.

McCurdy, K., Carlson, K. and Gregory, G. (2004). Floc morphology and cyclic shearing recovery: comparison of alum and polyaluminium chloride coagulants. *Water Research* 38 (2), 486-494.

McGarvey, F.X. (1995). Operating experiences with a new organic trap resin: prepared Discussion. In: *Proceedings of the International Water Conference.* Engineering Society. W. Penn., Pittsburg, p. 11.

McKnight, D.M., Bencala, K.E., Zellweger, G.W., Aiken, G.R., Feder, G.L. and Thorn, K.A. (1992). Sorption of dissolved organic carbon by hydrous aluminium and iron oxides occurring at the confluence of Deer Creek with the Snake River, Summit County, Colorado. *Environmental Science and Technology* 26 (7), 1388-1396.

McLaughlin, H.S. (1995). Regenerate activated carbon using organic solvents. *Chemical Engineering Progress* 91 (7), 45-53.

McLean, C.D., Ong, A.K. and Solomon, D.H. (1976). Cyclopolymerisation VI. Preparation and properties of crosslinked polyamines by cyclopolymerisation. *Journal of Macromolecular Science-Chemstry*A10, 857-873.

McMeen, C.R. and Benjamin, M.M. (1997). NOM removal by slow sand filtration through iron oxide-coated olivine. *Journal of the American Water Works Association* 89, 57-71.

Medvedev, M .I., Koryukova, L. V., Berestetsky, V. I., Meleshevich, S.I . (1992) Sorptional removal of natural organic substances from regenerated coagulant. *Khimiya i tekhnologiya vody Kiev* 14 (10), 781-786.

Meltzer, T.H. and Jornitz, M.W. (2002). Membranes: Filter applications for organic removal. *Ultrapure Water* 19 (4), 39-43.

Meyers, P.S. (1995). Operating experiences with a new organic trap resin. In: *Proceedings of the International Water Conference.* Engineering Society. W. Penn., Pittsburg, pp. 1-10.

Mikkelsen, L.H. (2003). Applications and limitations of the colloid titration method for measuring activated sludge surface charges. *Water Research* 37 (10), 2458-2466.

Mikkelsen, L.H. and Keiding, K. (2002). The shear sensitivity of activated sludge: an evaluation of the possibility for a standardised floc strength test. *Water Research* 36 (12), 2931-2940.

Miyahara, K., Adachi, Y., Nakaishi, K. and Ohtsubo, M. (2002) Settling velocity of a sodium montmorillonite floc under high ionic strength. *Colloids and Surfaces A: Physicochemical Aspects* 196, 87-91.

Montgomery, J.M. (1985). *Water Treatment Principles and Design.* John Wiley and Sons. New York, USA.

Moore, B.C., Cannon, F.S., Westrick, J.A., Metz, D.H., Shrive, C.A., Demarco, J. and Hartman, D.J. (2001). Changes in GAC pore structure during full-scale water treatment at Cincinnati: a comparison between virgin and thermally reactivated GAC. *Carbon* 39 (6), 789-807.

Murcott, S. and Harleman, D.R.F. (1993). Jar tests on the natural polymer chitosan. In: *Proceedings of the Water Quality Technology Conerence,* Part I, American Water Works Association, Denver, pp. 417-444.

Murphy, E.M., Zachara, J.M. and Smith, S.C. (1990). Influence of mineral-bound humic substances on the sorption of hydrophobic organic compounds. *Environmental Science and Technology* 24 (10), 1507-1516.

Muzzarelli, R. A. A. (1977) *Chitin*. Pergamon Press, Oxford. pp 356.

Naumczyk, J., Szpyrkowicz, J.L. and Grandi, F.Z. (1989). Organics isolation from fresh and drinking waters by macroporous anion-exchange resins. *Water Research* 23, 1593-1597.

Newcombe, G. (1999). Charge vs. porosity – some influences on the adsorption of natural organic matter by activated carbon. *Water Science & Technology* 40 (9), 191-198.

Newcombe, G., Drikas, M. and Hayes R. (1997). Influence of characterised natural organic matter on activated carbon adsorption: II. Effect on pore volume distribution and adsorption of 2 methylisoborneol. *Water Research* 31 (5), 1065-1073.

Nguyen Quoc, D. and Bogor, D.V. (1983). Yield stress measurement for concentrated suspensions. *Journal of Rheology*, 27 (4), 321-349.

Nguyen Quoc, D. and Bogor, D.V. (1985). Direct yield stress measurement with vane method. *Journal of Rheology* 29 (3), 335-347.

Nguyen, M.-L., Baker, L.A. and Westerhoff, P. (2002). DOC and DBP precursors in western US watersheds and reservoirs. *Journal of the American Water Works Association* 94, 98-112.

Niskavaara, H., Reimann, C., Chekushin V. and Kashulina, G. (1997). Seasonal variability of total and easily leachable element contents in topsoils (0–5 cm) from eight catchments in the European Arctic (Finland, Norway and Russia) *Environmental Pollution* 96 (2), 261-274.

Nissinen, T.K., Miettinen, I.T., Martikainen, P.J. and Vartiainen, T. (2001). Molecular size distribution of natural organic matter in raw and drinking waters. *Chemosphere* 45, 865-873.

Njam, I.N., Snoeyink, V.L., Lykins, B.W. and Adams, J.Q. (1991). Using powdered activated carbon: a critical review. *Journal of the American Water Works Association* 83 (1), 65-76.

Nokes, C. J., Fenton, E. and Randall, C. J. (1999) Modelling the formation of brominated trihalomethanes in chlorinated drinking waters. *Water Research* 33 (17), 3557-3568.

Nozaic, D.J., Freese, S.D. and Thompson, P. (2001). Long term experience in the use of polymeric coagulants at Umgeni Water. *Water Science and Technology: Water Supply* 1(1), 43-50.

O'Melia, C.R., Becker, W.C. and Au, K.K. (1999). Removal of humic substances by coagulation. *Water Science and Technology* 40 (9), 47-54.

Ødegaard, H., Eikebrokk, B. and Storhaug, R. (1999). Processes for the removal of humic substances from water - an overview based on Norwegian experiences. *Water Science and Technology* 40 (9), 37-46.

Önkal-Engin, G., Wibulswas, R. and White, D.A. (2000). Humic acid uptake from aqueous media using hydrotalcites and modified montmorillonite. *Environmental Technology* 21, 167-175.

Othman, M.Z., Roddick, F.A. and Hobday, M.D. (2000). Evaluation of Victorian low rank coal-based adsorbents for the removal of organic compounds from aqueous systems. *Water Research* 34 (18), 4351-4358.

Owen, D.M., Amy, G.L. and Chowdhury, Z.K. (1993). *Characterization of natural organic matter and its relationship to treatability.* American Water Works Association Research Foundation Report, American Water Works Association, Denver, CO.

Owen, D.M., Amy, G.L., Chowdhury, Z.K., Paode, R., McCoy G. and Viscosil K. (1995). NOM characterisation and treatability. *Journal of the American Water Works Association* 87 (1), 46-63.

Owen, D.M., Chowdhury, Z.K., Summers. R.S., Hooper, S.M., Solarik, G. and Gray, K. (1998). *Removal of DBP Precursors by GAC Adsorption.* American Water Works Association Research Foundation Report, American Water Works Association, Denver, CO.

Parazak, D.P., Burkhardt, C.W., McCarthy, K.J. and Stehlin, M.P. (1988). Hydrophobic flocculation. *Journal of Colloid and Interface Science* 123, 59-72.

Parfitt, R.L., Fraser, A.R. and Farmer, V.C. (1977). Adsorption on hydrous oxides. III. Fulvic acid and humic acid on goethite, gibbsite and imogolite. *Journal of Soil Science* 28, 289-296.

Parker, D.S., Kaufman, W.J. and Jenkins, D. (1972). Floc breakup in turbulent flocculation processes. *Journal of the Sanitary Engineering Division: Proceedings of the American Society of Civil Engineers* SA 1, 79-99.

Parsons, S.A., Jefferson, B., Dixon, D. and Harbour, P. (2003). Treatment of elevated organic carbon content waters. *First Periodic Report of AWWARF Tailored Collaboration Project RFP TC 2974,* American Water Works Association, Denver. pp. 9-11

Parsons, S.A., Sharp, E., Bourgeois, W., Goslan, E. and Jefferson, B. (2004). Application of charge measurement to water treatment processes. In: *Proceedings of the IWA World Water Congress and Exhibition*, Marrakech, Morocco. September 19-24, 2004.

Pelekani C., Drikas M. and Bursill D. (2001) Application of Direct Filtration to MIEX® Treated River Murray Water. In: *Proceedings of the 19th Federal Australian Water Association Convention* 2001, Canberra, Australia.

Pelekani, C., Newcombe, G., Snoeyink, V.L., Hepplewhite, C., Assemi, S. and Beckett, R. (1999). Characterisation of natural organic matter using high performance size exclusion chromatography. *Environmental Science and. Technology* 33 (16), 2807-2813.

Petrovic, M., Kastelan-Macan, M. and Horvat, A.J.M. (1999). Interactive sorption of metal ions and humic acids onto mineral particles. *Water, Air & Soil Pollution* 111, 41-56.

Peuravouri, J., Ingham, P., Pihlaja, K. and Koivikko, R. (2001). Comparisons of sorption of aquatic humic matter by DAX-8 and XAD-8 resins from solid-state ^{13}C NMR spectroscopy's point of view. *Talanta* 55, 733-742.

Peuravuori, J. and Pihlaja, K. (1997) Isolation and characterization of natural organic matter from lake water: Comparison of isolation with solid adsorption and tangential membrane filtration. *Environment International* 23 (4), 441-451.

Pignon, H., Brasquet, C. and Le Cloirec, P. (2000) Coupling ultrafiltration and adsorption onto activated carbon cloth: application to the treatment of highly coloured wastewaters. *Water Science and Technology* 42 (5-6), 355–362.

Pignon, H., Faur-brsquet, C. and Lecloirec, P. (2002). Treatment of complex solutions by the coupling of ultrafiltration and adsorption onto activated carbon cloth. *Environmental Technology* 21 (11), 1261-1270.

Randtke, S.J. (1988). Organic contaminant removal by coagulation and related processes. *Journal of the American Water Works Association* 80 (5), 40-56.

Ransom, B., Shea, K.F., Burkett, P.J., Bennett, R.H., and Baerwald, R. (1998). Comparison of pelagic and nephloid layer marine snow: Implications for carbon cycling. *Marine Geology* 150, 39-50.

Ratnaweera, H., Hiller, N. and Bunse, U. (1999). Comparison of the coagulation behaviour of different Norweigen aquatic NOM sources. *Environment International* 25 (2-3), 347-355.

Rattanakawin, C. and Hogg, R. (2001). Aggregate size distributions in flocculation. *Colloids and Surfaces A: Physicochemical and Engineering Aspects* 177, 87-98.

Rebhun, M. and Lurie, M. (1993). Control of organic matter by coagulation and floc separation. *Water Science and Technology* 27 (11), 1-20.

Reckhow, D.A. and Singer, P.C. (1990). Chlorination by-products in drinking waters: From formation potentials to finished water concentrations. *Journal of the American Water Works Association* 82 (4), 173-180.

Reckhow, D.A., Singer, P.C., and Malcolm, R.L. (1990). Chlorination of humic materials: Byproduct formation and chemical interpretations. *Environmental Science and Technology* 24 1655-1664.

Rengasamy, P. and Churchman, G.J. (1999). Cation exchange capacity, exchangeable cations and sodicity. In: *Soil Analysis: an Interpretation Manual,* Peverill K. I., Sparrow L. A. and Reuter D. J. (Eds.), CSIRO, Melbourne. pp. 147-157.

Rook, J.J. and Evans, S. (1979). Removal of trihalomethane precursors from surface waters using weak base resins. *Journal of the American Water Works Association* 71, 520-524.

Rout, D., Verma, R. and Agarwal, S.K. (1999). Polyelectrolyte treatment – an approach for water quality improvement. *Water Science and Technology* 40 (2), 137-141.

Saleh, F.Y., Ong, W.C.A., and Chang, D.Y. (1989). Structural features of aquatic fulvic acids - analytical and preparative reversed-phase high-performance liquid-chromatography separation with photodiode array detection. *Analytical Chemistry* 61 (24), 2792-2800.

Sanks, R.L. (1973). Ion exchange colour and mineral removal from kraft bleach wastes. *Report EPA/R2/73-255,* US EPA, Cincinnnati, Ohio.

Sawyer, C.N., McCarty, P.L., and Parklin, G.F. (1994). *Chemistry for Environmental Engineering*. McGraw-Hill, Inc., New York, USA.

Scales, P.J., Johnson, S.B., Kapur, P.C., and Healy, T.W. (1998). The shear yield stress of partially flocculated suspensions. *American Institute of Chemical Engineers Journal* 44, 538-544.

Schlauch, R.M. (1981). Coagulation for gravity type clarification and thickening. In: *Polyelectrolytes for Water and Wastewater Treatment.* Schwoyer, W.K., (Ed.), CRC Press, Boca Raton, Florida, p. 91.

Schmita, K. H. and Wells, M. J. M. (2002) Preferential adsorption of fluorescing fulvic and humic acid components on activated carbon using flow field-flow fractionation analysis. *Journal of Environmental Monitoring* 4, 75-84.

Schulthess, C.P. and Huang, C.P. (1991). Humic and fulvic acid adsorption by silicon and aluminium oxide surfaces on clay minerals. *Soil Science Society of America Journal* 55, 34-42.

Scott, M.J., Jones, M.N., Woof, C. and Tipping, E. (1998). Concentrations and fluxes of dissolved organic carbon in drainage water from an upland peat system. *Environment International* 24 (5-6), 537-546.

Scott, M.J., Jones, M.N., Woof, C., Simon, B. and Tipping, E. (2001). The molecular properties of humic substances isolated from a UK upland peat system: A temporal investigation. *Environment International* 27 (6), 449-454.

Seida, Y. and Nakano, Y. (2000) Removal of humic substances by layered double hydroxide containing iron. *Water Research* 34, 1487-1494.

Seida, Y. and Nakano, Y. (2002). Removal of humic substances by layered double hydroxide containing iron. *Water Research* 34, 1487-1494.

Selomulya, C., Amal, R., Bushell, G., and Waite, T.D. (2001). Evidence of shear rate dependence on restructuring and break-up of latex aggregates. *Journal of Colloid and Interface Science* 236, 67-77.

Semmens, M.J., and Ayers, K. (1985). Removal by coagulation of trace organics from Mississippi River water. *Journal of the American Water Works Association* 77 (5), 79-84.

Sharp, E.L., Parsons, S.A. and Jefferson, B. (2004). The effects of changing NOM composition and characteristics on coagulation performance, optimization and control. *Water Science and Technology: Water Supply* 4(4), 95-102.

Siddiqui, M., Amy, G., Ryan, J., and Odem, W. (2000). Membranes for the control of natural organic matter from surface waters. *Water Research* 34 (13), 3355-3370.

Sinasbaugh III, R.L., Hoehn, R.C., Knocke, W.R. and Kinkins III, A.E. (1986). Removal of dissolved organic carbon by coagulation with iron sulfate. *Journal of the American Water Works Association* 78 (5), 74-82.

Singer, P.C. (1999). Humic substances as precursors for potentially harmful disinfection by-products. *Water, Science and Technology* 40 (9), 25-30.

Singer, P.C. and Bilyk, K. (2002). Enhanced coagulation using a magnetic ion exchange resin. *Water Research* 36 (16), 4009-4022.

Slunjski, M., Ngyen, H., Ballard, M., Eldridge, R., Morran, J., Drikas, M., O'Leary, B. and Smith, P. (2002). MIEX® - Good research commercialised. *Water* 29 (2), 42-51.

Smith, P., Botica, C., Lange, R. and Tattersall, J. (2001). Design and construction of the worlds first large scale MIEX® water treatment plant. In: *Proceedings of the 19th Federal AWA Convention*. Canberra, ACT, Australia. 1-4 April.

Specht, C.H., Kumke, M.U. and Frimmel, F.H. (2000). Characterization of NOM adsorption to clay minerals by size exclusion chromatography. *Water Research* 34 (16), 4063-4069.

Spicer, P.T. and Pratsinis, S.E. (1996). Shear-induced flocculation: the evolution of floc structure and the shape of the size distribution at steady state. *Water Research* 3 (5), 1049-1056.

Spicer, P.T., Pratsinis, S.E., Raper, J., Amal, R., Bushell, G., and Meesters, G. (1998). Effect of shear schedule on particle size, density and structure during flocculation in stirred tanks. *Powder Technology* 97, 26-34.

Sposito, G. (1992). Characterisation of particle surface charge. In: *Environmental Particles,* vol. 1, Buffle, J. and van Leewen, H. P. (Eds.), Lewis, Boca Raton, pp. 291-314.

Stumm, W. (1992). *Chemistry of the Solid-Water Interface,* Wiley, New York. p.114.

Stumm, W. and Morgan, J.J. (1981). Aquatic chemistry. An introduction emphasizing chemical equilibria in natural waters. (2nd Ed.) John Wiley and Sons Inc., New York.

Summers, R.S. and Roberts, P.V. (1998). Activated carbon adsorption of humic substances. 2: Size exclusion and electrostatic interactions. *Journal of Colloid and Interface Science* 122 (2), 382-397.

Świetlik, J., Dąbrowska, A., Raczyk-Stanisławiak, U. and Nawrocki, J. (2004). Reactivity of natural organic matter with chlorine dioxide and ozone. *Water Research* 38 (3), 547-558.

Symons, J.M., Fu, P. L-K. and Kim, P.H.-S. (1992). The use of anion exchange resins for the removal of natural organic matter from municipal water. *Proceedings of the International Water Conference.* Engineering Society. W. Penn., Pittsburg, pp. 92-123.

Symons, J.M., Fu, P.L-K. and Kim, P.H.-S. (1995). Using anion exchange resins to remove THM precursors. In: *Ion Exchange Technology.* Sengupta A. K., (Ed.), Technomic, Lancaster, Penn., pp. 149-192.

Tang, P., Greenwood, J. and Raper, J.A. (2002). A model to describe the settling behaviour of fractal objects. *Journal of Colloid and Interface Science* 247, 210-219.

Tang, S. (1999). Prediction of fractal properties of polystyrene aggregates. *Colloids and Surfaces A: Physicochemical and Engineering Aspects* 157, 185-192.

Thebault, P., Cases, J. M. and Feisinger, F. (1981) Mechanism underlying the removal of organic micropollutants during flocculation by and aluminium or iron salt. *Water Research* 15, 183-189.

Thomas, D. N. (1999) Flocculation modelling in wastewater treatment. PhD Thesis, Cranfield University, UK.

Thomas, D.N., Judd, S.J., and Fawcett, N. (1999). Flocculation modelling: A review. *Water Research* 33 (7), 1579-1592.

Thurman, E. M. (1985). *Organic Geochemistry of Natural Waters.* Martinus Nijhoff-Dr Junk Publishers, Dortrecht. p.3.

Thurman, E.M. and Malcolm, R.L. (1981). Preparative isolation of aquatic humic substances. *Environmental Science and Technology* 15 (4), 463-466.

Tiller, C.L. and O'Melia, C.R. (1993). Natural organic matter and colloidal stability: models and measurements. *Colloids and surfaces A: Physicochemical and engineering aspects* 73, 449-454.

Tilsley, G M (1979). Interaction of organic matter with anion resins. *Chemistry and Industry* 3, 142-149

Tipping, E. (1993). Modeling ion binding by humic acids. *Colloids and surfaces A: Physicochemical and engineering aspects* 73, 117-131.

Tipping, E. and Cooke, D. (1982). The effect of adsorbed humic substances on the surface charge of goethite (α-FeOOH) in freshwater. *Geochimica et Cosmochimika Acta* 46, 75-80.

Tombácz, E., Filipcsei, G., Szekeres, M. and Gingl, Z. (1999). Particle aggregation in complex aquatic systems. *Colloids & Surfaces A: Physicochemical & Engineering Aspects* 151, 233-244.

Tseng, T., Segal, B.D. and Edwards, M. (2000). Increasing alkalinity to reduce turbidity. *Journal of the American Water Works Association* 92 (6), 44-54.

Turchiuli, C. and Gargues, C. Influence of structural properties of alum and ferric flocs on sludge dewaterability. *Chemical Engineering Journal*, In press.

van Leeuwen, J., Chow, C., Fabris, R., Wither, N., Page, D. and Drikas, M. (2002). Application of a fractionation technique for the better understanding of the removal of NOM by alum coagulation. In: *Proceedings of the IWA Water Congress,* Melbourne, CD ROM, April 7-12, 2002.

Van Rossum, P., Webb, R. G. (1978) Isolation of organic water pollutants by XAD resins and carbon. *Journal of Chromatography* 150, 381-392.

Van Velsen, A.F.M., van der Vos, G., Boersma, R. and de Reuver, J.L. (1991). High gradient magnetic separation technique for wastewater treatment. *Water Science and Technology* 24, 195-203.

Vartiainen, T., Liimatainen, A. and Kauranen, P. (1987). The use of TSK Size exclusion columns in determination of the quality and quantity of humus in raw waters and drinking waters. *Science of the Total Environment* 62, 75-84.

Vermeer, A.W.P. and Koopal, L.K. (1998). Adsorption of humic acid to mineral particles. 2. Polydispersity effects with polyelectrolyte adsorption. *Langmuir* 14, 4210-4216.

Vilge-Ritter, A., Rose, J., Masion, A., Bottero, J.Y. and Laine, J.M. (1999). Chemistry and structure of aggregates formed with Fe-salts and natural organic matter. *Colloids and Surfaces A: Physicochemical and Engineering Aspects* 147, 297-308.

Volk, C., Bell, K., Ibrahim, E., Verges, D., Amy, G. and Lechevallier, M. (2000). Impact of enhanced and optimized coagulation on the removal of organic matter and its biodegradable fraction. *Water Research* 34 (12), 3247-3257.

Vuorio, E., Vahala, R., Rintala, J., and Laukkanen, R. (1998) The Evaluation of drinking water treatment performed with HPSEC. *Environment International* 24 (5/6), 617-623.

Wahlroos, G. (1991). The realization of activated-carbon filtration in a water purification process for highly humic water by prior two-stage Iron(III) salt coagulation. *Journal of Water Supply: Research and Technology - AQUA* 40 (4), 211-216.

Waite T.D., Amal, R., Ngo, H. and Vigneswaren, S. (1993). Effects of adsorbed organic-matter on size, structure and filterability of iron oxyhydroxide flocs, Water Science and Technology 27 (11), 133-142.

Waite, T. D. (1999). Measurement and implications of floc structure in water and wastewater treatment. *Colloids and Surfaces A: Physicochemical and Engineering Aspects* 151, 27-41.

Waite, T.D., Cleaver, J.K. and Beattie, J.K. (2001). Aggregation kinetics and fractal structure of γ-alumina assemblages. *Colloid and Interface Science* 241, 333-339.

Walker, H. W. and Bob, M. M. (2001). Stability of particle flocs upon addition of natural organic matter under quiescent conditions. *Water Research* 35 (4), 875-882.

Walker, H.W. and Kim, E.K. (2001). Influence of flocculant aids and NOM characteristics on the removal of DBP precursors during enhanced coagulation. In: *Proceedings of the American Water Works Association Water Quality Technology Conference*, Nashville, TN, USA. November 11-15.

Wang, L.-L., Chin, Y.-P. and Traina, S.J. (1997). Adsorption of poly(maleic acid) and an aquatic fulvic acid by goethite. *Geochimica et Cosmochimika Acta* 61, 5313-5324.

Wang, X.C., Jin, P.K. and Gregory, J. (2002). Structure of Al-humic flocs and their removal at slightly acidic and neutral pH. *Water Science and Technology: Water Supply* 2 (2), 99-106.

Warton, B., Zappia, L., Heitz, A., Joll, C., Masters, D., O'Leary, B., Franzmann, P. and Kagi, R. (2004). MIEX® treatment for NOM removal and improved drinking water quality. In: *Proceedings of NOM Research: Innovations and Applications for Drinking Water Treatment Conference*. Victor Harbor, SA, Australia. March 2-5.

Weber, W.J., Pirbazari, M., Long, J.B. and Barton, D.A. (1981). Potential mechanisms for the removal of humic acids from water by activated carbon. *Activated Carbon Adsorption of Organics from the Aqueous Phase*, vol. 1, Suffet, I. W. and McGuire, M. J. (Eds.), Ann Arbor Science, Michigan, pp. 317-336.

Weiss, D. A., Bolto, B. A., McNeill, R., MacPherson, A. S., Siudak, R., Swinton, E. A. and Willis, D. (1966) An ion-exchange process with thermal regeneration. V. Multistage operation. *Australian Journal of Chemistry* 19 (5), 791-796.

Wen, H.J. and Lee, D.J. (1998). Strength of cationic polymer-flocculated clay flocs. *Advances in Environmental Research* 2, 390-396.

White, M.C., Thompson, J.D., Harrington, G.W. and Singer, P.C. (1997). Evaluating criteria for enhanced coagulation compliance. *Journal of the American Water Works Association* 89 (5), 64-77.

Wiese, G.R., and Healy, T.W. (1975). Coagulation and electrokinetic behaviour of Ti O_2 and Al_2O_3 colloidal dispersions. *Journal of Colloid and Interface Science* 51 (3), 427-433.

Wilen, B-M., Jin, B. and Lant, P. (2003). Impacts of structural characteristics on activated sludge floc stability. *Water Research* 37 (15), 3632-3645.

Wilkinson, K.J., Negre, J.C. and Buffle, J. (1997). Coagulation of colloidal material in surface waters: the role of natural organic matter. *Journal of Contaminant Hydrology* 26, 229-243.

Willis, D. (1966). An ion-exchange process with thermal regeneration. II. Properties of weakly basic resins. *Australian Journal of Chemistry* 19, 561-587.

Wilson, D. (2004). Personal Communication.

Wong, S., Hanna, J.V., King, S.J., Carroll, T., Eldridge, R.J., Dixon, D.R., Bolto, B.A., Abbt-Braun, G., Frimmel, F.H. and Hesse, S. (2002). Fractionation of natural organic matter in drinking water and characterization by ^{13}C cross polarisation magic angle spinning NMR spectroscopy and size exclusion chromatography. *Environmental Science and Technology* 36 (16), 3497-3503.

Worrall, F. and Burt, T. (2004). Predicting the future DOC flux from upland peat catchments. *Journal of Hydrology* 300(1-4), 126-139.

Wu R. M., Lee, D. J., Waite, T. D., Guan, J. J. (2002). Multilevel structure of sludge flocs. *Journal of Colloid and Interface Science* 252 (2), 383-392.

Wu, C.C., Huang, C., Lee, D. J. (1997). Effects of polymer dosage on alum sludge dewatering characteristics and physical properties. *Colloids and Surfaces A: Physicochemical and Engineering Aspects* 122 (1-3), 89-96.

Wu, C.C., Wu, J.J., and Huang, R.Y. (2003). Floc strength and dewatering efficiency of alum sludge. *Advances in Environmental Research* 7, 617-621.

Yan, Y.D., Burns, J.L., Jameson, G.J., and Biggs, S. (2000). The structure and strength of depletion force induced particle aggregates. *Chemical Engineering Journal* 80, 23-30.

Yates, D.E. and Healy, T. W. (1975). Mechanism of anion adsorption at the ferric and chromic oxide/water interfaces. *Journal of Colloid and Interface Science* 52 (2), 222-228.

Yeh, H.-H., Huang W.-J. and Chen H.-W. (1994). The reaction between chlorine and various organic fractions extracted from source water. In: *Proceedings of the Water Technology Conference of the American Water Works Association*. San Francisco, pp. 357-370.

Yeung, A.K., Gibbs, A., and Pelton, R.P. (1997). Effect of shear on the strength of polymer-induced flocs. *Journal of Colloid and Interface Science* 196, 113-115.

Yeung, A.K.C. and Pelton, R. (1996). Micromechanics: a new approach to studying the strength and break-up of flocs. *Journal of Colloid and Interface Science* 184, 579-585.

Ying, Y.D., Lei, D. and Wania, F. (2004). Is rain or snow a more efficient scavenger of organic chemicals? *Atmospheric Environment* 38 (22), 3557-3571.

Yoshimura A.S. and Prud'Homme R.K. (1988). Wall slip effects on dynamic oscillatory measurements. *Journal of Rheology* 32 (6), 575-584.

Yukselen, M.A. and Gregory, J. (2004). The reversibility of floc breakage. *International Journal of Mineral Processing* 73, 251-259.

Zhang, Z., Sisk, M.L., Mashmoushy, H. and Thomas, C.R. (1999). Characterisation of the breaking force of latex particle aggregates by micromanipulation. *Particle and Particle Systems Characterisation* 16, 278-283.

Zhou, Q., Cabaniss, S.E. and Maurice, P.A. (2000). Considerations in the use of high-pressure size exclusion chromatography (HPSEC) for determining molecular weights of aquatic humic substances. *Water Research* 34 (14), 3505-3514.

ABBREVIATIONS

ACE	Acetate carbon equivalents
ALUM	Aluminium sulfate
BSS	British standard sieve
BV	Bed volumes
CCC	Colloid charge capacity
CCD	Close-coupled device
CD	Charge density
CFD	Computational fluid dynamics
CHAR	Charged hydrophilic compounds
COD	Chemical oxygen demand
CPAM	Cationic polyacrylamide
CPMA	Cationic polymethacrylate
Da	Daltons
DAF	Dissolved air flotation
DBP	Disinfection by-product
D/DBP	Disinfectants/Disinfection by-products
DOC	Dissolved organic carbon
FAF	Fulvic acid fraction
FFFF	Flow field-flow fractionation
FCWTF	Fort Collins Water Treatment Facility
FI	Flocculation index
GAC	Granular activated carbon
HAF	Humic acid fraction
HPA	Hydrophobic acids
HPI-A	Hydrophilic acid
HPI-NA	Hydrophilic non acid
HPLC	High Performance Liquid Chromatography
HPOA	Hydrophobic acid
HPON	Hydrophobic neutral
HPSEC	High performance size exclusion chromatography
HV	Hope valley
IC	Inorganic carbon
IEP	Iso electric point
MIEX®	Magnetic ion exchange resin
MW	Molecular weight
NEUT	Neutral hydrophilic compounds
NOM	Natural organic matter
NPOC	Non purgeable organic carbon
PAC	Powdered activated carbon
PACl	Poly aluminium chloride
PCU	Platinum color units
PDA	Photometric dispersion analyser
PDADMAC	Polydiallyldimethylammonium chloride

PEI	Polyethyleneimine
POM	Particulate organic matter
PSD	Particle size distribution
RO	Reverse osmosis
THM	Trihalomethane
THMFP	Trihalomethane formation potential
SEC	Size exclusion chromatography
SHA	Slightly hydrophobic acids
SUVA	Specific Ultra Violet Absorbance
TC	Total carbon
TOC	Total organic carbon
TPA	Transphilic acids
USEPA	United Stated Environmental Protection Agency
VHA	Very hydrophobic acids
WAN	Waneroo
WHA	Weakly hydrophobic acid
WTW	Water treatment works

IWA Publishing's authorised EU representative for General Product Safety Regulations is Diane D'Arras, 15 rue Duret, 75116 Paris, France, e-mail: safety@iwap.co.uk.

Printed and bound by CPI Group (UK) Ltd, Croydon, CR0 4YY
17/09/2026
02218804-0006